Springer Proceedings in Mathematics & Statistics

Volume 98

This book series features volumes composed of selected contributions from workshops and conferences in all areas of current research in mathematics and statistics, including operation research and optimization. In addition to an overall evaluation of the interest, scientific quality, and timeliness of each proposal at the hands of the publisher, individual contributions are all refereed to the high quality standards of leading journals in the field. Thus, this series provides the research community with well-edited, authoritative reports on developments in the most exciting areas of mathematical and statistical research today.

More information about this series at http://www.springer.com/series/10533

Pierre Cartier • A.D.R. Choudary
Michel Waldschmidt
Editors

Mathematics in the 21st Century

6th World Conference, Lahore, March 2013

Springer

Editors
Pierre Cartier
Institut des Hautes Études
Scientifiques (IHÉS)
Bures-sur-Yvette, France

A.D.R. Choudary
Abdus Salam School
of Mathematical Sciences
Lahore, Pakistan

Michel Waldschmidt
Université Pierre et Marie Curie (Paris VI)
Paris Cedex 05, France

ISSN 2194-1009 ISSN 2194-1017 (electronic)
ISBN 978-3-0348-0858-3 ISBN 978-3-0348-0859-0 (eBook)
DOI 10.1007/978-3-0348-0859-0
Springer Basel Heidelberg New York Dordrecht London

Library of Congress Control Number: 2014950900

Printed on acid-free paper

Springer is part of Springer Science+Business Media (www.springer.com)

Foreword: Mathematics for a New Century

The 6th World Conference on 21st Century Mathematics 2013 took place in Lahore (Pakistan) from March 6 to 9, 2013, in the Abdus Salam School of Mathematical Sciences (ASSMS). It was a successful event bringing together many scientists from all over the world and a large audience of local students and colleagues. Despite the recurring political unrest in Pakistan, it was a peaceful event, under the expert guidance of Dr. A. D. Raza Choudary.

I take this opportunity to comment on the developments of mathematical sciences in the past century. Everyone associates the second International Congress of Mathematicians (ICM), that took place in Paris, August 1900, with the famous address of David Hilbert (the 23 problems of Hilbert). The audience was mostly European (French, Germans, Italians, British) and the towering figures were Poincaré from France, and Hilbert from Germany. Both were universal scientists, and their work extends from mechanics to philosophy, through algebra, geometry, number theory, and analysis. They participated in strong debates about the nature of mathematical objects, Poincaré being basically pragmatist and Hilbert formalist—reflecting perhaps the different philosophical traditions of their respective countries. Both were aware of the importance of the creation, by Georg Cantor, of the theory of sets. Let us also mention that a young assistant to the British embassy in Paris, Bertrand Russell, followed with great interest the various lectures: this was the beginning of his career as a logician and a philosopher. At that time, set theory was plagued by the so-called paradoxes or inconsistencies: Russell invented the theory of types as an alternative to set theory—more about that later.

As I already mentioned, science was European at the time, even dominated by scientists with a German cultural background (whether Germans, Danes, Swedish, Hungarians, even Russians). There were very few scientists both in northern and Latin America; a few Japanese in the Meiji era used German, and a few Indians used English to communicate. A century later, mathematics is truly international and we remember ICM 2002 in Beijing, ICM 2010 in Hyderabad and ICM 2014 in Seoul. The number of world practitioners in mathematics increased in a century from maybe a thousand to hundred thousands. Mathematics in Vietnam, for instance, did not exist in 1970, and in 2012 the joint meeting of the mathematical societies of France and Vietnam gathered 500 Vietnamese and 100 French participants. Science in Latin America fared well in the last 50 years, after a modest start in Sao Paulo around 1950 (remember the visits of André Weil, Oscar Zariski, Jean Dieudonné and Alexander Grothendieck). This trend has been supported by various international organizations, among them the International Mathematical Union (IMU) recreated after the Second World War, the SISSA[1] in Trieste created by Abdus Salam, and the CIMPA[2] (created in Nice at the request of UNESCO for organizing summer schools all over the world). As a note of comfort, let us mention the quick and rather unexpected development of science in India (the Tata Institute of Fundamental Research, Mumbai, and now a number of organizations like NISER[3]) as well as in China (due to the rebirth of this big and ancient country). The challenges facing us next are in the Middle East (including Pakistan and its neighbours) and in Africa, especially tropical and East Africa. I take great comfort by observing many projects in this direction.

So, mathematics at the turn of the twenty-first century is truly international. Another welcome development is the increasing number of women studying mathematics. I have been surprised by the number of women attending my lectures in Pakistan, as well as in Kurdistan and in Algeria. I have been told of similar patterns in Iran. Even in a so-called advanced country like France, the progress has been slow: my mother did not have a personal checkbook before 1948; my mother-in-law, a widow, was not the legal guardian of her daughter, my wife; the sections for boys and for girls of the École Normale Supérieure merged in 1990 only, and in the most prestigious École Polytechnique, there are approximately 20 % of female students! It is also recently that a woman has joined the Fields medallists!

Another important development has been the gradual change of emphasis: what is really important in mathematics? As Hilbert stated it repeatedly: “No one should take us outside Cantor’s paradise”. One of the initial successes of Hilbert was his book on geometry, where he revised Euclids’ axiomatics for geometry, by taking into account the critical study (by Pasch, Peano, etc.) motivated by the advent of non-Euclidean geometry. The dream of Hilbert was a complete exposition of mathematics via the axiomatic method and the use of set theory. There were a few initial successes, like Hausdorff book on topology, followed by Banach and his normed vector spaces, and even more importantly, the “Modern Algebra” by van der Waerden (called simply “Algebra” in later editions). The philosophy of Hilbert is best described by the inscription on his tomb (in Göttingen):

> Wir müssen wissen,
> Wir werden wissen[4]

[1]Scuola Internazionale Superiore di Studi Avanzati.

[2]Centre International de Mathématiques Pures et Appliquées.

[3]National Institute of Science Education and Research.

[4]We must know, we shall know.

Hilbert was convinced that, using the axiomatic method, every mathematical problem could be solved. In his list of 23 problems, the sixth is called "Axiomatics of physics", and his own version of Einstein's general relativity is presented as an axiomatic theory. We are less ambitious now, especially after Gödel's discovery of the incompleteness of all formal axiomatic systems, and more dramatically after Cohen's proof of the undecidability of the *continuum hypothesis*.

These limitations did not hamper Bourbaki's enterprise, whose goal was the materialization of Hilbert's dream. In approximately 50 years (from 1934 to 1983), the group of 10–15 (mostly) French mathematicians, with varying membership, published an encyclopedic treatise with eight complete series (ranging from set theory to Lie groups) and the beginning of two more. The recent reprint consists of 30 volumes, totalling slightly less than 10,000 pages. The initial ambition was to cover *all* existing mathematics; too big! But after the foundations (from set theory to Lebesgue integration), Bourbaki published two very successful series on "Commutative Algebra" and "Lie Groups and Algebras". This was supposed to be the starting point for developments in algebraic topology, differential geometry and also algebraic number theory. Despite many unpublished drafts on these subjects, the momentum was exhausted after 50 years and led to the advent of a fourth generation of collaborators. To compensate for this shortage, a number of important books in the Bourbaki spirit were published under their own names by members of Bourbaki, or later by disciples.[5]

The famous British historian Eric Hobsbawm published a book entitled *The Short 20th Century (1914–1991)* describing in his own terms: "a century of ideologies". This fits quite well with the development of mathematics: the twentieth century was an epoch of formalism and axiomatics. There were great successes despite a number of limitations, and I will try now to describe the birth of a new era.

There are a number of challenges, each one requiring a sharp turn. The first is coming from the inside, with the birth and development of *category theory*. This was created by S. MacLane and S. Eilenberg around 1940, as a tool to be used in algebraic topology. It was enormously developed by Ch. Ehresmann and A. Grothendieck in the 1960s and it is now one of the most vigorous branch of mathematics. Theoretically, it can be described as some kind of exotic extension of group theory: a category is a set with a partially defined inner operation allowing many units. A *functor* is nothing but a homomorphism between two objects of this kind. There is an interesting combinatorial theory developed along these lines, but the most useful applications transcend this narrow domain. Among the so-called inconsistencies of set theory, the Russell paradox is paramount, and rests on the illegal assumption of a set of all sets. But, practitioners of category theory have no hesitation at mentioning the category of all sets, and worst, the category of all categories. Grothendieck invented a beautiful escape from this tangle, namely *universes*. A *universe* is a set U such that the collection of U sets (that is the

[5]Let us mention R. Godement's *Theory of sheaves* and A. Weil's *Basic number theory* among many others by Chevalley, Serre.

elements of U) obeys all the properties attributed to sets. The category of all U sets is now a legitimate object. All that is required is to assume the existence of such universes; but this is tantamount to assume the existence of so-called large cardinals. Here we penetrate into the muddy waters of set theory, the logical marsh. It could very well be that some property of categories is true for one universe, but not for all.[6] Very few people (Grothendieck himself in SGA4, as well as Demazure and Gabriel in their treatise on algebraic groups) put this burden on their shoulders, the price to be paid being heavy. In view of the big successes of category theory in its many applications, most practitioners of mathematics, especially topologists, are tempted to follow the advice of Charles de Gaulle: "L'intendance suivra".[7]

History of mathematics teaches us to be hopeful and brave. In the eighteenth century, mathematicians like Euler, Lagrange and many others developed calculus, differential geometry, and mechanics, using mathematics with shaky foundations. Berkeley had already pinpointed the inconsistencies of the notion of infinitesimal. For the cure, we had to wait until the middle of the nineteenth century with Cauchy, Weierstrass and Dedekind. Nevertheless, the concept of infinitesimal is still widely used by many physicists and engineers. One of the greatest innovations in the eighteenth century was the *calculus of variations*. There the foundations were even more shaky, and Hilbert mentioned this among his 23 problems. After two centuries of struggle, we have reliable foundations for ordinary calculus and calculus of variations, and a very large part of the discoveries of Euler and his followers has been salvaged. So we can be hopeful.

A possible cure for this disease could be offered by the *theory of types*, created by Bertrand Russell. To explain the difference between sets and types, I will use the *parable of the green cats*. In the fairy tale version, the *set* of green cats exists because the king was able to gather all of them in a big room. In the entomologist's version, the *type* of green cats is a box in a museum with the proper name, to accommodate all green cats to be caught. In more serious terms, a set is closed, defined by the collection of all its members; a type is open ended waiting for the creation or discovery of new members. Recently, Voevodsky in Princeton made a serious attempt by developing his theory of *homotopical types*. Each type is open ended as well as the collection of types. The main difficulty is to define equality of two types: it cannot be static, it has to be dynamic; that is, an equality statement is an evolving proof. This view is comforted by a recent discovery in logics—-that *proofs* and *programmes* are virtually the same objects.

This kind of problems is closely connected with the advent of *certified software*. This is a practical problem. In modern technology, many big systems have been fully automatized. Running a nuclear plant, controlling a spaceship, monitoring the flights in an airport or the trains in the railway system, requires huge programmes. I have been told that two million instructions are not uncommon. Who can write such a programme knowing that any serious bug is a threat to safety? For

[6] This is to be expected, after the incompleteness theorem of Gödel!

[7] The supply shall follow the fighters!

mathematicians, a similar challenge is the existence of *monstrous proofs*: the four-colour problem, Wiles' proof of Fermat's last theorem, the sphere packing problem, and the classification of finite simple groups. The printed version of Wiles' proof runs over 600 pages, but it is a "human" proof not using computers in a serious way. The classification of finite simple groups consists of more than 10,000 published pages. The four-colour problem and the sphere-packing conjecture use extensive computer calculations, both combinatorial and numerical.

Of course, the dream of a mathematician is to have a beautiful and easy-to-follow proof, which could be printed as one of the "Proofs from the Book".[8] Everyone hopes that the problems mentioned above will receive such a proof, but I am doubtful about such a possibility. Anyhow, refusing to accept that kind of proof would seriously hamper the development of mathematics, since obviously we will have more and more of such proofs in coming years. Maybe the duty of mathematicians will be to create some kind of astronomical clocks, imitating nature and mind operations, to be run and watched. A possible cure exists already with the existence of *proof assistants*, like[9] the French version COQ. The ambition is to have a certified encyclopedia of mathematics: it contains a core (or nucleus) of about 500 instructions written in *C*++, rather easy to check by human means, containing all the basic syntactical rules. Then it develops like onion rings—each level referring to its inner level. I have been told that mathematics at the level of second university year is already available in such systems. Sooner or later, we shall write our proofs in our standard half-formal way, but the referees will use such proof assistants to certify our paper. So Hilbert's dream of a mechanizable axiomatic system may come true. So far, the solution of the four-colour problem, as well as the 250-page-long proof of the theorem of Feit and Thompson ("every finite simple group has an even order") have been completely checked by such methods.

Coming back to types, they are already a standard tool in computer science: even in old-fashioned languages like FORTRAN, we would declare

```
real x
integer p,
```

etc. The recent systems are based on the typed λ-calculus, like LISP and the followers. It seems to me that a serious proposal for revision of the foundation of mathematics would be to replace *set theory* by *type theory*. For instance, declaring a type SET, a type CAT (category), a type CAT/CAT (category of all categories) is without flaw as long as you do not insist on gathering all sets in a single room, etc. The *topos* of Grothendieck can be viewed as unorthodox models of set theory. Their flexibility should allow them to help in this search for new foundations. Maybe what is at stake is to develop a purely syntactical mathematics, without any underlying *ontology*.

---

[8]To use Erdős' terminology: I refer you to Aigner's book (published by Springer) with the same title.

[9]I mention also HOLIGHT developed by T. Hales.

In the beginning of this century, we have no shortage of big problems awaiting solutions. I have mentioned already the search for new foundations of the mathematics building. The *Millenium Prizes* are here to remind us that, even after the solution of Poincaré's conjecture by Perelman, we know little about the Riemann hypothesis as well as about the solutions of Navier-Stokes equations. In combinatorics, as well as in number theory, many more or less reasonable conjectures await a proof. In recent decades, arithmetic algebraic geometry developed in a rather strange way, by piling one conjecture above another one, for instance, Hodge conjecture and the related standard conjectures of Grothendieck, Langlands functoriality conjecture. In some sense, it is a new style of mathematical research.

To go outside pure mathematics, the marriage between mathematics and physics initiated by Galileo 400 years ago was very successful in the hands of Newton, Euler, Gauss, Poincaré, etc. but there were talks of divorce in the 1950s and 1960s. At least in France, Bourbaki had much contempt and ignorance about mathematical physics, and very influential physicists like Yves Rocard pretended to use only the sliding rule and second order ordinary differential equations. These views have been challenged; physicists and engineers came to value sophisticated mathematics, and mathematicians, "contaminated" to some extent by the remnants of Gelfand's school, are not ashamed of looking for inspiration at the major problems raised by quantum physics.

This should be broadened, and biology, ecology, climate systems, and Darwin's evolution theory offer big mathematical challenges. Even if the concept of *interdisciplinarity* has been overused and abused, it is clear that we need a new generation of polyglot mathematicians.

I would like to conclude by trying to describe what could be major challenges in our working habits of mathematicians. The wildest dreams of philosophers like Teilhard de Chardin and Carl Jung are coming true. The first one invented the notion of noosphere, the second one that of collective unconscious. What they predicted was the existence of a new level of consciousness above the consciousness of individuals. The boldest representation was the one by Teilhard, who viewed evolution as increasing levels of consciousness, starting with amoebas and ending with a final stage called the omega point, one of the most reasonable descriptions of a God. Teilhard based his views on his experience as a paleontologist, as well as the coming of radiocommunications in the 1930s. The growing evidence of life in many places in the universe supports the idea that evolution is part of the general development of the universe, and may be an important component of cosmology.

In more down-to-earth terms, we are entering the age of *instant communication*. The most diverse information at the tip of our fingers through repositories of knowledge like Wikipedia or arXiv has changed the daily life of mathematicians. I remember fondly that, while vacationing in a very remote place in the Caribbean Islands, it took me no more than 10 min to locate a paper with the only information of the name of the authors—a Chinese Li among millions of other Lis—and that it dealt with a new proof of an identity by D. Zagier. Videoseminars between Tokyo,

Beijing, and IHÉS in France have been in operation for some years. Writing a joint paper between an American, an Indian, and a German is made easy using email and Skype. One could multiply the examples.

The advent of big systems like MAPLE, GAP, MATHEMATICA, PARI, user-friendly systems, made possible the birth of experimental mathematics, and the purest of the mathematicians are aware of these new tools. Let an old man look enthusiastically towards the life of his mathematical grandchildren, or great-grandchildren!

Bures-sur-Yvette, France
March 1, 2014

Pierre Cartier

Preface

The growing international visibility, and successes, of the world conferences on the theme "21st Century Mathematics" in 2004, 2005, 2007, 2009, and 2011 meant that the 6th edition of the World Conference on 21st Century Mathematics (WC 2013), held at the Abdus Salam School of Mathematical Sciences (ASSMS), Lahore, Pakistan, in March 2013, would be the largest in scale and scope yet. And it indeed proved to be so. With over 120 invited and guest speakers from all over the world and over 400 participants from across the region and beyond, it can safely be said that WC 2013 was one of the largest mathematical meetings in the 2013 calendar year.

The main goal of these conferences is to provide a platform for young mathematicians in Pakistan to get in touch with researchers of world renown and to encourage Pakistani institutions of higher education and research. To this end, the importance of WC 2013 can be gauged from the fact that previous editions of this conference led to a marked increase in the awareness and enthusiasm for mathematical education among universities in Pakistan. Consequently, a large number of scientific events were organized by universities and centres in Pakistan, and even in neighbouring countries.

The articles in this volume are carefully selected through the peer review process. These articles are strongly representative of all the major mathematical themes of the conference and are written precisely with the goal of spreading awareness of mainstream mathematics to the relatively younger audience. It is hoped that these efforts will inspire a generation of young mathematicians across the region.

It is a pleasure to thank all the authors for their contributions and their cooperation, the publisher for producing this volume, and the scientific committee of WC 2013 for their unstinted support. If there is one individual that I would like to especially thank, it is Pierre Cartier. His towering presence and mesmeric lectures made WC 2013 truly memorable.

Lahore, Pakistan
March 17, 2014

A.D.R. Choudary

Contents

Contributors

Martin Bača Department of Applied Mathematics and Informatics, Technical University, Košice, Slovakia

Yasir Bashir Abdus Salam School of Mathematical Sciences, GC University, Lahore, Pakistan

Edy Tri Baskoro Faculty of Mathematics and Natural Sciences, Combinatorial Mathematics Research Group, Institut Teknologi Bandung (ITB), Bandung, Indonesia

Massimiliano Ferrara Department SSGES, University "Mediterranea" of Reggio Calabria, Reggio Calabria, Italy

Constantin Fetecau Department of Mathematics, Technical University of Iasi, Iasi, Romania

Academy of Romanian Scientists, Bucuresti, Romania

Abdus Salam School of Mathematical Sciences, GC University, Lahore, Pakistan

Corina Fetecau Department of Theoretical Mechanics, Technical University of Iasi, Iasi, Romania

Kâzım İlhan İkeda Department of Mathematics, Yeditepe University, Ataşehir, Istanbul, Turkey

Hông Vân Lê Institute of Mathematics of ASCR, Praha 1, Czech Republic

Olav Arnfinn Laudal Matematisk institutt, University of Oslo, Blindern, Oslo, Norway

Renaud Leplaideur Laboratoire de Mathématiques de Bretagne Atlantique, UMR 6205 Université de Brest, Brest, France

Muhammad Faisal Nadeem Abdus Salam School of Mathematical Sciences, GC University, Lahore, Pakistan

Ragni Piene CMA/Department of mathematics, University of Oslo, Oslo, Norway

Ayesha Shabbir Abdus Salam School of Mathematical Sciences, GC University, Lahore, Pakistan

Hironori Shiga Graduate School of Science, Chiba University, Chiba, Japan

Abdus Salam School of Mathematical Sciences, Lahore, Pakistan

Petr Somberg Mathematical Institute, Charles University, Praha 8, Czech Republic

Anton Stefanescu Faculty of Mathematics and Computer Science, University of Bucharest, Bucharest, Romania

Ioan Tomescu Faculty of Mathematics and Computer Science, University of Bucharest, Bucharest, Romania

Abdus Salam School of Mathematical Sciences, GC University, Lahore, Pakistan

Jiři Vanžura Institute of Mathematics of ASCR, Brno, Czech Republic

Dumitru Vieru Abdus Salam School of Mathematical Sciences, GC University, Lahore, Pakistan

Department of Theoretical Mechanics, Technical University of Iasi, Iasi, Romania

Michel Waldschmidt Université Pierre et Marie Curie-Paris 6, Institut de Mathématiques de Jussieu IMJ UMR 7586, Paris Cedex 05, France

Kristiana Wijaya Faculty of Mathematics and Natural Sciences, Combinatorial Mathematics Research Group, Institut Teknologi Bandung (ITB), Bandung, Indonesia

Tudor Zamfirescu Faculty of Mathematics, University of Dortmund, Dortmund, Germany

Institute of Mathematics "Simion Stoïlow" Roumanian Academy, Bucharest, Roumania

Abdus Salam School of Mathematical Sciences, GC University, Lahore, Pakistan

Rein L. Zeinstra Abdus Salam School of Mathematical Sciences, GC University, Lahore, Pakistan

About the Editors

Pierre Cartier, member of Institut des Hautes Études Scientifiques (IHES) in Bures-sur-Yvette, France, is Emeritus Professor in the University Paris 7. Professor Cartier was a student of the École Normale Supérieure where he obtained his Ph.D. under the supervision of Henri Cartan. He has been a member of the Bourbaki group for over 30 years. In algebraic geometry, a number of objects bear his name: the Cartier duality between Abelian varieties or formal groups, Cartier divisors. He has been a closed collaborator from Grothendieck.

A.D.R. Choudary is director general at the Abdus Salam School of Mathematical Sciences (ASSMS), GC University Lahore, Pakistan. Professor Choudary received his Ph.D. in Mathematics at the University of Bucharest. In 2002, he received the award of "Most Inspirational Educator", and in 2003 received the award of "Distinguished Professor of Research" at Central Washington University, USA. Earlier, he served as a coordinator of the Faculty of Mathematics at the University of Guadalajara, Mexico, and professor of mathematics at Central Washington University, USA. An author of three books, Prof. Choudary has been a visiting scholar at a number of universities and institutes in Europe, the USA, Asia, and Australia. He received several grants from US Department of Education and other similar agencies during his stay in the USA.

Michel Waldschmidt is emeritus professor at Université Pierre et Marie Curie (Paris 6), Paris, France. Professor Waldschmidt studied at the University of Nancy. Then he went to Bordeaux, where he obtained his Ph.D. in 1972 under the supervision of Jean Fresnel. He was immediately offered a temporary position in the University of Orsay by Henri Cartan (1972–1973), and in 1973, he was appointed as a professor at the University Paris VI, where he taught until the end of 2011. He is an expert in number theory, especially Diophantine problems, including transcendence methods.

About the Editors

On Super Edge-Antimagic Total Labeling of Toeplitz Graphs

Martin Bača, Yasir Bashir, Muhammad Faisal Nadeem, and Ayesha Shabbir

1 Introduction

We consider finite undirected graphs without loops and multiple edges. If G is a graph, then $V(G)$ and $E(G)$ stand for the vertex set and the edge set of G, respectively. A *labeling* of a graph is a mapping that carries the graph elements into numbers (usually the positive integers). We deal with labelings with domain either the set of all vertices, or the set of all edges, or the set of all vertices and edges, respectively. We call these labelings a *vertex labeling*, or an *edge labeling*, or a *total labeling*, depending on the graph elements that are being labeled.

For a (p, q)-graph G with p vertices and q edges, a bijective function $f : V(G) \cup E(G) \to \{1, 2, \ldots, p + q\}$ is a *total labeling* of G and the associated edge-weight is $wt_f(uv) = f(u) + f(uv) + f(v)$, for $uv \in E(G)$. An (a, d)*-edge-antimagic total labeling* of G is the total labeling with the property that the edge-weights form an arithmetic progression $\{a, a + d, a + 2d, \ldots, a + (q - 1)d\}$, where $a > 0$ and $d \geq 0$ are two fixed integers. An (a, d)-edge-antimagic total labeling is called *super* if the smallest possible labels appear on the vertices. A graph that admits a (super) (a, d)-edge-antimagic total labeling is called a (super) (a, d)-edge-antimagic total graph.

The research for this article was supported by Slovak VEGA Grant 1/0130/12 and Higher Education Commission Pakistan Grant HEC(FD)/2007/555.

M. Bača (✉)
Department of Applied Mathematics and Informatics, Technical University, Košice, Slovakia
e-mail: martin.baca@tuke.sk

Y. Bashir • M.F. Nadeem • A. Shabbir
Abdus Salam School of Mathematical Sciences, GC University, 68-B, New Muslim Town, Lahore, Pakistan
e-mail: yasirb2@gmail.com; mfaisalnadeem@ymail.com; ashinori@hotmail.com

P. Cartier et al. (eds.), *Mathematics in the 21st Century*, Springer Proceedings in Mathematics & Statistics 98, DOI 10.1007/978-3-0348-0859-0_1

These labelings, introduced by Simanjuntak, Bertault, and Miller in [19], are natural extensions of the concept of a *magic valuation*, studied by Kotzig and Rosa in [15], and the concept of a super edge-magic labeling, defined by Enomoto et al. in [5]. For more information on these labelings, please see [3, 12].

2 Toeplitz Graph

A simple undirected graph T of order p is called *Toeplitz graph* if its adjacency matrix $A(T)$ is Toeplitz. A *Toeplitz matrix* $A(T) = (a_{i,j})$ is a $(p \times p)$ symmetric matrix which has constant values along all diagonals parallel to the main diagonal, i.e., $a_{i,j} = a_{i+1,j+1}$ for each $i, j = 1, 2, \ldots, p-1$. The p distinct diagonals of a $(p \times p)$ symmetric Toeplitz adjacency matrix will be labeled $0, 1, 2, \ldots, p-1$. Diagonal 0 is the main diagonal and it contains only zeros, i.e., $a_{ii} = 0$ for all $i = 1, 2, \ldots, p$, so that there are no loops in the Toeplitz graph. A Toeplitz graph T is uniquely defined by the first row of $A(T)$, a $(0-1)$-sequence. Let $t_1, t_2, \ldots, t_k$ be the diagonals containing ones, $0 < t_1 < t_2 < \cdots < t_k < p$. Then, the corresponding Toeplitz graph will be denoted by $T_p\langle t_1, \ldots, t_k\rangle$. That is, $T_p\langle t_1, \ldots, t_k\rangle$ is the graph with the vertex set $V(T) = \{v_i : i = 1, 2, \ldots, p\}$ in which two vertices u, v of T being connected by an edge if and only if $|u - v| \in \{t_1, t_2, \ldots, t_k\}$. If t_j, $j = 1, 2, \ldots, k$, is the diagonal containing ones, then the diagonal elements a_{i,t_j+i}, $i = 1, 2, \ldots, p - t_j$, determine edges $v_i v_{t_j+i}$ in the Toeplitz graph. Thus, the edge set is $E(T) = \bigcup_{j=1}^{k} \{v_i v_{t_j+i} : i = 1, 2, \ldots, p - t_j\}$, $|V(T)| = p$ and $|E(T)| = pk - \sum_{j=1}^{k} t_j$.

Toeplitz graphs have been introduced by Sierksma and first been investigated by van Dal et al. [4] with respect to their hamiltonicity. Later Heuberger [13] has extended this study in 2002. The properties of Toeplitz graphs, such as bipartiteness, planarity, and colorability, have been studied in [6–9, 14]. For more recent works on Toeplitz graphs, see [16–18]. A Toeplitz graph is not necessarily connected, e.g., see Figs. 1 and 2.

The following result proved by van Dal et al. [4] provides a lower bound on the number of components of a Toeplitz graph.

Theorem 1 ([4]). *$T_p\langle t_1, \ldots, t_k\rangle$ has at least $gcd(t_1, \ldots, t_k)$ components.*

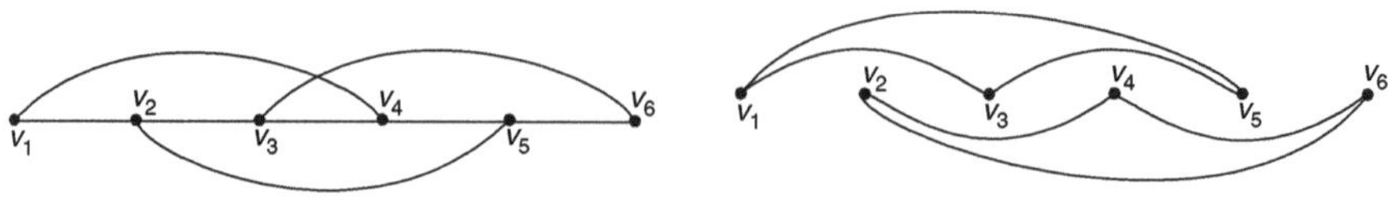

Fig. 1 Toeplitz graphs $T_6\langle 1, 3\rangle$ and $T_6\langle 2, 4\rangle$

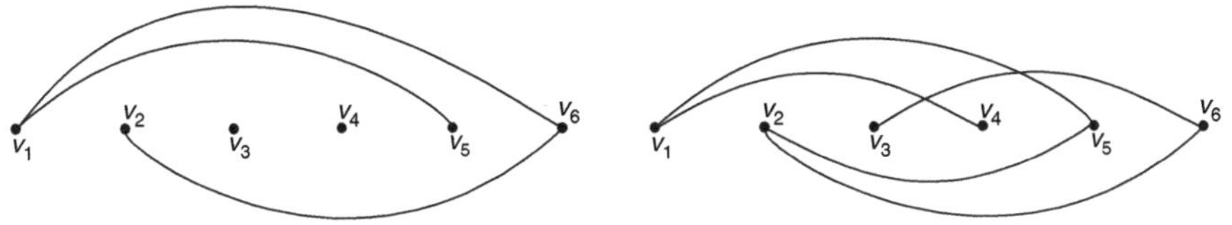

Fig. 2 Toeplitz graphs $T_6\langle 4,5\rangle$ and $T_6\langle 3,4\rangle$

In the paper, we investigate the existence of a super (a,d)-edge-antimagic total labeling for Toeplitz graphs, and for several differences d, we introduce constructions for this labeling.

3 Feasible Values of Difference

We start this section with a necessary condition for a graph to be super (a,d)-edge-antimagic total, which will provide a least upper bound, for a feasible value d.

Lemma 1. *Let $p \geq 3$ and $k \geq 1$ be integers. If a Toeplitz graph $T_p\langle t_1,\dots,t_k\rangle$ is super (a,d)-edge-antimagic total, then $d \leq 1 + \frac{2p-4}{pk-\sum_{j=1}^{k} t_j - 1}$.*

Proof. Assume that $T = T_p\langle t_1,\dots,t_k\rangle$, where $1 \leq t_1 < t_2 < \cdots < t_k < p$, with p vertices and $q = pk - \sum_{j=1}^{k} t_j$ edges has a super (a,d)-edge-antimagic total labeling $f : V(T) \cup E(T) \to \{1,2,\dots,p+q\}$ with the edge-weights $\{a, a+d, a+2d,\dots,a+(q-1)d\}$. The minimum possible edge-weight is the sum of the two smallest possible vertex labels, namely, 1 and 2, and the smallest possible edge label, namely, $p+1$, and so $a = p+4$. On the other hand, the maximum possible edge-weight is no more than the sum of two the largest possible vertex labels, namely, $p-1$ and p, and the largest possible edge label, namely, $p+q$. Thus,

$$a + (q-1)d \leq 3p + q - 1$$

gives

$$d \leq 1 + \frac{2p-4}{pk - \sum_{j=1}^{k} t_j - 1}. \tag{1}$$

□

Denote $A = \frac{2p-4}{pk-\sum_{j=1}^{k} t_j - 1}$ and consider the extremal values of $\sum_{j=1}^{k} t_j$.

Case 1: If $\sum_{j=1}^{k} t_j$ is minimum, i.e., $\sum_{j=1}^{k} t_j = 1 + 2 + \cdots + k = \frac{k(k+1)}{2}$, then $pk - \sum_{j=1}^{k} t_j - 1$ admits the largest value and the fraction A has the smallest possible value

$$A = \frac{2p-4}{pk - \frac{k(k+1)}{2} - 1}. \tag{2}$$

From (1) and (2) it follows that
for $T_p\langle 1\rangle$ the feasible values of difference are $d \in \{0, 1, 2, 3\}$ and
for $T_p\langle 1, 2\rangle$ the feasible values of difference are $d \in \{0, 1, 2\}$.
If $k \geq 3$ and $p = \frac{k+3}{2}$, then for $T_p\langle 1, 2, \ldots, k\rangle$, we have that $d \leq 2$, and
if $k \geq 3$ and $p > \frac{k+3}{2}$, then for $T_p\langle 1, 2, \ldots, k\rangle$, we have that $d \leq 1$.

Case 2: If $\sum_{j=1}^{k} t_j$ is maximum, i.e., $\sum_{j=1}^{k} t_j = (p-k)+(p-k+1)+\cdots+(p-1) = pk - \frac{k(k+1)}{2}$, then the fraction A admits the largest value

$$A = \frac{4p-8}{k(k+1)-2}. \tag{3}$$

For $T_p\langle p-1\rangle$, we have a graph with only one edge and $p-2$ isolates,
for $T_p\langle p-2\rangle$ from (1) it follows that $d \leq 2p-3$, and
for $T_p\langle p-2, p-1\rangle$ from (1) it follows that $d \leq p-1$.
If $k \geq 3$ and $p \leq \frac{k(k+1)+6}{4}$, then from (1) and (3) for $T_p\langle p-k, p-k+1, \ldots, p-1\rangle$, we have that $d \leq 2$, and
if $k \geq 3$ and $p > \frac{k(k+1)+6}{4}$, then from (1) and (3) for $T_p\langle p-k, p-k+1, \ldots,$ p-1$\rangle$, it follows that $d \leq 1 + \frac{4p-8}{k(k+1)-2}$.
Table 1 summarizes the previous facts.

4 Super Edge-Antimagic Total Labelings

In this section, we will study the super edge-antimagicness of Toeplitz graph $T_p\langle t_1, \ldots, t_k\rangle$ for several values of parameter k and for differences $d \in \{0, 1, 2\}$.

In the next theorem, we construct edge-antimagic total labeling with $d = 1$ for arbitrary Toeplitz graph without isolated vertices.

Table 1 Values of d of Toeplitz graph depending on k

for $k = 1$			
Toeplitz graph	$T_p\langle 1\rangle$	...	$T_p\langle p-2\rangle$
Values of d	0,1,2,3	...	0,1,..., 2p-3
for $k = 2$			
Toeplitz graph	$T_p\langle 1,2\rangle$	...	$T_p\langle p-1, p-2\rangle$
Values of d	0,1,2	...	0,1,...,$p-1$
for $k \geq 3$			
Toeplitz graph	$T_p\langle 1,2,\dots k\rangle$, $p > \frac{k+3}{2}$	...	$T_p\langle p-k,\dots,p-1\rangle$, $p > \frac{k(k+1)+6}{4}$
Values of d	0,1	...	0,1, ..., $1+\frac{4p-8}{k(k+1)-2}$

Theorem 2. *$T_p\langle t_1,\dots,t_k\rangle$ with at least $gcd(t_1,\dots,t_k)$ connected components admits a super $(2p+2,1)$-edge-antimagic total labeling.*

Proof. Consider a Toeplitz graph $T_p\langle t_1,\dots,t_k\rangle$ corresponding to the $(p \times p)$ symmetric Toeplitz matrix $A(T) = (a_{i,j})$. Denote the vertices of the Toeplitz graph by $v_1, v_2, \dots, v_p$ such that

$$f(v_i) = i \text{ for } 1 \leq i \leq p.$$

The diagonal elements a_{i,t_j+i}, $i = 1,2,\dots,p-t_j$, determine edges $v_i v_{t_j+i}$ in the Toeplitz graph.

For $1 \leq j \leq k$ and $1 \leq i \leq p-t_j$, we label the edges of the Toeplitz graph in the following way:

$$f(v_i v_{t_j+i}) = p+1-f(v_i)+\sum_{s=1}^{j}(p-t_s).$$

Under the labeling f, the vertices of the Toeplitz graph receive the values from 1 up to p, and we can see that

for $j = 1$ and $1 \leq i \leq p-t_1$, the function f assigns the consecutive labels $p+1, p+2, \dots, 2p-1-t_1, 2p-t_1$ to the edges $v_i v_{t_1+i}$ and

for $j = 2$ and $1 \leq i \leq p-t_2$, the function f assigns the consecutive labels $2p-t_1+1, 2p-t_1+2, \dots, 3p-1-t_1-t_2, 3p-t_1-t_2$ to the edges $v_i v_{t_2+i}$.

By the same manner, we can check the existence of the edge labels for $j = 3,4,\dots,k-2$ and

for $j = k-1$, and $1 \leq i \leq p-t_{k-1}$ the edges $v_i v_{t_{k-1}+i}$ under the function f admit the consecutive labels $(k-1)p+1-\sum_{s=1}^{k-2} t_s, (k-1)p+2-\sum_{s=1}^{k-2} t_s, \dots, kp-\sum_{s=1}^{k-1} t_s$, and

for $j = k$ and $1 \leq i \leq p-t_k$, the edges $v_i v_{t_k+i}$ under the function f admit the consecutive labels $kp+1-\sum_{s=1}^{k-1} t_s, kp+2-\sum_{s=1}^{k-1} t_s, \dots, (k+1)p-\sum_{s=1}^{k} t_s$.

It is easy to see that the labeling f is a bijective function which assigns the set of integers $\left\{1, 2, \ldots, (k+1)p - \sum_{s=1}^{k} t_s\right\}$ to the vertices and edges of the Toeplitz graph. Furthermore, f assigns the numbers $1, 2, \ldots, p$ to the vertices; therefore, it is a super total labeling.

For the edge-weights of $T_p\langle t_1, \ldots, t_k\rangle$, we have:

If $j = 1$, then $wt_f(v_i v_{t_1+i}) = f(v_i) + f(v_i v_{t_1+i}) + f(v_{t_1+i}) = 2p + 1 + i$, and for $1 \le i \le p - t_1$, we obtain the edge-weights from $2p + 2$ up to $3p + 1 - t_1$.

If $j = 2$, then $wt_f(v_i v_{t_2+i}) = 3p + 1 - t_1 + i$, and for $1 \le i \le p - t_2$, we obtain the edge-weights from $3p + 2 - t_1$ up to $4p + 1 - t_1 - t_2$.

If $3 \le j \le k - 1$, then $wt_f(v_i v_{t_j+i}) = f(v_i) + f(v_i v_{t_j+i}) + f(v_{t_j+i}) = (j+1)p + 1 - \sum_{s=1}^{j-1} t_s + i$, and for $1 \le i \le p - t_j$, the edge-weights get values from $(j+1)p + 2 - \sum_{s=1}^{j-1} t_s$ up to $(j+2)p + 1 - \sum_{s=1}^{j} t_s$.

If $j = k$, then $wt_f(v_i v_{t_k+i}) = (k+1)p + 1 - \sum_{s=1}^{k-1} t_s + i$, and for $1 \le i \le p - t_k$, we obtain the edge-weights from $(k+1)p + 2 - \sum_{s=1}^{k-1} t_s$ up to $(k+2)p + 1 - \sum_{s=1}^{k} t_s$.

One can easily verify that the edge-weights are distinct and consecutive from $2p + 2$ up to $(k+2)p + 1 - \sum_{s=1}^{k} t_s$. This implies that Toeplitz graph $T_p\langle t_1, \ldots, t_k\rangle$ has a super $(2p + 2, 1)$-edge-antimagic total labeling. □

Before presenting our next results, we are showing some results of great help.

Let mG be the disjoint union of m copies of a graph G. Figueroa-Centeno, Ichishima, and Muntaner-Batle in [11] proved the following theorem.

Theorem 3 ([11]). *If G is a (super) edge-magic total bipartite or tripartite graph and m is odd, then mG is (super) edge-magic total.*

Bača, Lin, and Muntaner-Batle [1] proved that every path on p vertices has a super edge-magic total labeling. Thus, from Theorem 3, it follows:

Corollary 1. *If m is odd, $m \ge 3$ and $p \ge 2$, then the graph mP_p is super edge-magic total.*

Only a few known results are there on super edge-magicness for an even disjoint union of paths. Figueroa-Centeno, Ichishima, and Muntaner-Batle [11] have shown that the forest $2P_p$, $p > 1$, has a super edge-magic total labeling if and only if $p \ne 2$ or 3. Using an operation, which is, in some sense, a generalization of the Kronecker product of matrices for digraphs, Bača, Lin, and Muntaner-Batle in [2] proved that if $m \equiv 2 \pmod 4$, $m \ge 6$, and $p \ge 4$, then the graph mP_p admits a super edge-magic total labeling.

Figueroa-Centeno et al. have proved the following lemma.

Lemma 2 ([10]). *A (p,q)-graph G is super edge-magic total if and only if there exists a bijective function $f : V(G) \to \{1,2,\ldots,p\}$, such that the set*

$$S = \{f(u) + f(v) : uv \in E(G)\}$$

consists of q consecutive integers. In such a case, f can be extended to a super edge-magic total labeling of G with valence (edge-weight) $a = p + q + s$, where $s = \min(S)$ and

$$S = \{a - (p+1), a - (p+2), \ldots, a - (p+q)\}.$$

Lemma 2 implies that all previous results on super edge-magic total labelings with minimum edge-weight a can be extended to a super $(a - q + 1, 2)$-edge-antimagic total labelings.

Return back to the super edge-antimagicness of Toeplitz graphs for $k = 1$. The graph $T_p\langle t_1\rangle = P_p$, if $t_1 = 1$. Otherwise, it consists of t_1 components, which are not necessarily of equal length. Each of these components is isomorphic to a path and maximum possible length for any of these components is $\left\lfloor \frac{p}{t_1} \right\rfloor$.

Theorem 4. *The graph $T_p\langle t_1\rangle$ has a super $(p + t_1 + 3, 3)$-edge-antimagic total labeling.*

Proof. Consider a total labeling $f : V \cup E \to \{1,2,\ldots,p+q\}$, which is defined as follows:

$$f(v_i) = i \qquad \text{for all } v_i \in V, 1 \le i \le p, \text{ and}$$
$$f(v_i v_{t_1+i}) = p + f(v_i) \qquad \text{for } 1 \le i \le p.$$

For the edge-weights of $T_p\langle t_1\rangle$, we have

$$wt_f(v_i v_{t_1+i}) = f(v_i) + f(v_i v_{t_1+i}) + f(v_{t_1+i}) = p + t_1 + 3i \text{ for } 1 \le i \le p.$$

Thus, the total labeling f has the desired properties. □

Theorem 5. *Let $\frac{p}{t_1}$ be an integer and*

(i) t_1 be odd, $t_1 \ge 3$
(ii) $t_1 \equiv 2 \,(mod\ 4)$, $t_1 \ge 6$

Then, $T_p\langle t_1\rangle$ admits a super (b,d)-edge-antimagic total labeling for $d \in \{0,2\}$.

Proof. If $\frac{p}{t_1}$ is an integer, then $T_p\langle t_1\rangle$ is the disjoint union of t_1 copies of the path on $\frac{p}{t_1}$ vertices, $t_1 P_{\frac{p}{t_1}}$. According to Corollary 1 and the result proved in [2], we obtain a super edge-magicness of $T_p\langle t_1\rangle$. Lemma 2 guarantees the existence of a super $(b,2)$-edge-antimagic total labeling. □

Next theorem gives a result on super edge-antimagicness of Toeplitz graphs for $k = 2$.

Theorem 6. *$T_p\langle t_1, t_1 + 1\rangle$ admits a super edge-magic total labeling and a super $(p + t_1 + 3, 2)$-edge-antimagic total labeling.*

Proof. Consider a graph $T_p\langle t_1, t_1 + 1\rangle$ with the corresponding $(p \times p)$ symmetric Toeplitz matrix $A(T) = (a_{i,j})$ and define a vertex labeling $f : V(T_p\langle t_1, t_1 + 1\rangle) \to \{1, 2, \ldots, p\}$ such that

$$f(v_i) = i \quad \text{for} \quad 1 \leq i \leq p.$$

There are only two consecutive diagonals containing ones, namely, t_1 and $t_1 + 1$. For weights of edges under the vertex labeling f, we have

$$w(v_i v_{t_1+i}) = f(v_i) + f(v_{t_1+i}) = t_1 + 2i \text{ for } 1 \leq i \leq p - t_1 \text{ and}$$
$$w(v_i v_{t_1+1+i}) = f(v_i) + f(v_{t_1+1+i}) = t_1 + 1 + 2i \text{ for } 1 \leq i \leq p - t_1 - 1.$$

So, the vertex labeling f is a bijective function and the weights of edges form the set of $2p-2t_1-1$ consecutive integers, namely, $\{t_1 + 2, t_1 + 3, \ldots, 2p - t_1 - 1, 2p - t_1\}$. With respect to Lemma 2, the vertex labeling f can be extended to a super edge-magic total labeling with the common edge-weight $3p - t_1 + 1$ and also to a super $(p + t_1 + 3, 2)$-edge-antimagic total labeling. This produces the desired result. □

Enomoto et al. in [5] proved the following result.

Theorem 7 ([5]). *If a graph G with p vertices and q edges is super edge-magic total, then $q \leq 2p - 3$.*

With this result in the hand, we are able to prove the following:

Theorem 8. *If $p \geq 3t_1+1$, $t_2 \leq \left\lfloor \frac{p-t_1+1}{2} \right\rfloor$ and $t_3 \leq p-t_1-t_2+2$, then $T_p\langle t_1, t_2, t_3\rangle$ does not admit any super edge-magic total labeling.*

Proof. Suppose that Toeplitz graph $T_p\langle t_1, t_2, t_3\rangle$ admits a super edge-magic total labeling.

Consider the extremal case when $t_2 = t_1 + 1$ and $t_3 = t_1 + 2$. Then with respect to Theorem 7, we have

$$|E(T_p\langle t_1, t_1 + 1, t_1 + 2\rangle)| = 3p - 3t_1 - 3 \leq 2p - 3$$

and $p \leq 3t_1$. Thus, for $p \geq 3t_1 + 1$, the Toeplitz graph $T_p\langle t_1, t_1 + 1, t_1 + 2\rangle$ does not admit any super edge-magic total labeling.

Now, we consider the case when $t_3 = t_2 + 1$. Then according to Theorem 7, we get

$$|E(T_p\langle t_1, t_2, t_2 + 1\rangle)| = 3p - t_1 - 2t_2 - 1 \leq 2p - 3$$

and $t_2 \geq \frac{p-t_1+2}{2}$. Thus, for $p \geq 3t_1 + 1$ and $t_2 \leq \left\lfloor \frac{p-t_1+1}{2} \right\rfloor$, the Toeplitz graph $T_p\langle t_1, t_2, t_2 + 1\rangle$ does not admit any super edge-magic total labeling.

By Theorem 7 every graph $T_p\langle t_1, t_2, t_3\rangle$ with a super edge-magic total labeling satisfies the condition that

$$|E\left(T_p\langle t_1, t_2, t_3\rangle\right)| = 3p - t_1 - t_2 - t_3 \leq 2p - 3$$

and $t_3 \geq p - t_1 - t_2 + 3$.

Thus, for $p \geq 3t_1 + 1$, $t_2 \leq \left\lfloor \frac{p-t_1+1}{2} \right\rfloor$ and $t_3 \leq p - t_1 - t_2 + 2$, the Toeplitz graph $T_p\langle t_1, t_2, t_3\rangle$ does not have any super edge-magic total labeling. □

5 Conclusion

In the foregoing sections, we studied super (a, d)-edge-antimagic total labelings for Toeplitz graph $T_p\langle t_1, \ldots, t_k\rangle$, and we proved the existence of such labelings for $d = 1$ if Toeplitz graph has at least $gcd(t_1, \ldots, t_k)$ connected components. We have shown that if p and t_1 satisfy certain conditions, then the graph $T_p\langle t_1\rangle$ admits a super (b, d)-edge-antimagic total labeling for $d \in \{0, 2\}$. Also we proved the existence of a super edge-magic total labeling and a super $(p + t_1 + 3, 2)$-edge-antimagic total labeling for the graph $T_p\langle t_1, t_2\rangle$ if $t_2 = t_1 + 1$. We have tried to find such labelings also for $t_2 \neq t_1 + 1$ but so far without success. So, we conclude the paper with the following open problem.

Open Problem 1. *For $t_2 \neq t_1 + 1$, determine if there is a super (b, d)-edge-antimagic total labeling of $T_p\langle t_1, t_2\rangle$ for $d \in \{0, 2\}$.*

References

1. Bača, M., Lin, Y., Muntaner-Batle, F.A.: Super edge-antimagic labelings of the path-like trees. Util. Math. **73**, 117–128 (2007)
2. Bača, M., Lin, Y., Muntaner-Batle, F.A.: A note on even disjoint union of paths. AKCE J. Graphs. Comb. **6**(1), 41–46 (2009)
3. Bača, M., Miller, M.: Super Edge-Antimagic Graphs. BrownWalker Press, Boca Raton (2008)
4. van Dal, R., Tijssen, G., Tuza, Z., van der Veen, J., Zamfirescu, C.H., Zamfirescu, T.: Hamiltonian properties of Toeplitz graphs. Discret. Math. **159**, 69–81 (1996)
5. Enomoto, H., Lladó, A.S., Nakamigawa, T., Ringel, G.: Super edge-magic graphs. SUT J. Math. **34**, 105–109 (1998)
6. Euler, R., LeVerge, H., Zamfirescu, T.: A characterization of infinite, bipartite Toeplitz graphs. In: Tung-Hsin, K. (ed.) Combinatorics and Graph Theory'95, Vol. 1. Academia Sinica, pp. 119–130. World Scientific, Singapore (1995)
7. Euler, R.: Characterizing bipartite Toeplitz graphs. Theor. Comput. Sci. **263**, 47–58 (2001)
8. Euler, R.: Coloring planar Toeplitz graphs and the stable set polytope. Discret. Math. **276**, 183–200 (2004)

9. Euler, R., Zamfirescu, T.: On planar Toeplitz graphs. Graphs Comb. (2012). doi:10.1007/s00373-012-1185-8
10. Figueroa-Centeno, R.M., Ichishima, R., Muntaner-Batle, F.A.: The place of super edge-magic labelings among other classes of labelings. Discret. Math. **231**, 153–168 (2001)
11. Figueroa-Centeno, R.M., Ichishima, R., Muntaner-Batle, F.A.: On edge-magic labelings of certain disjoint unions of graphs. Australas. J. Comb. **32**, 225–242 (2005)
12. Gallian, J.: A dynamic survey of graph labeling. Electron. J. Comb. **19**, #DS6 (2012)
13. Heuberger, C.: On Hamiltonian Toeplitz graphs. Discret. Math. **245**, 107–125 (2002)
14. Heuberger, C.: On planarity and colorability of circulant graphs. Discret. Math. **268**, 153–169 (2003)
15. Kotzig, A., Rosa, A.: Magic valuations of finite graphs. Can. Math. Bull. **13**, 451–461 (1970)
16. Malik, S., Qureshi, A.M.: Hamiltonian cycles in directed Toeplitz graphs. Ars Comb. **109**, 511–526 (2013)
17. Malik, S., Zamfirescu, T.: Hamiltonian connectedness in directed Toeplitz graphs. Bull. Math. Soc. Sci. Math. Roum. **53**(2), 145–156 (2010)
18. Nicoloso, S., Pietropaoli, U.: On the chromatic number of Toeplitz graphs. Discret. Appl. Math. (2012). doi:10.1016/j.dam.2011.07.012
19. Simanjuntak, R., Bertault, F., Miller, M.: Two new (a, d)-antimagic graph labelings. In: Proceedings of Eleventh Australasian Workshop on Combinatorial Algorithms, Hunter Valley, pp. 179–189 (2000)

On Ramsey $(2K_2, K_4)$–Minimal Graphs

Edy Tri Baskoro and Kristiana Wijaya

1 Introduction

All graphs in this paper are simple. Let $G(V, E)$ be a graph and $v \in V(G)$. The *degree* of a vertex v, denoted by $d(v)$, is the number of edges incident to the vertex. The *degree sequence* of a graph is the nonincreasing sequence of the degrees of its vertices. If G has n vertices, the degree sequence of G is $(d_1, d_2, \ldots, d_n)$ where $d_i \geq d_{i+1}$ for every $i = 1, 2, \ldots, n-1$.

Let G and H be graphs with order m and n, respectively. The *disjoint union* G and H, denoted by $G \cup H$, is a graph with the vertex set $V(G \cup H) = V(G) \cup V(H)$ and the edge set $E(G \cup H) = E(G) \cup E(H)$. The *join graph* of G and H, denoted by $G + H$, is the graph $V(G) \cup V(H)$ and all edges joining every vertex of G to every vertex of H. Following Borowiecka-Olszewska and Haluszczak [2], we use notation $G \odot H$, for a graph obtained from disjoint graphs G and H by identifying vertices $u \in V(G)$ and $v \in V(H)$. So, the graph $G \odot H$ has $m + n - 1$ vertices. Similarly, we introduce notation $G \ominus H$ for a graph obtained from disjoint graphs G and H by identifying edges $a \in E(G)$ and $e \in E(H)$. The graph $G \ominus H$ has $m + n - 2$ vertices.

For any pair of graphs G and H, notation $F \to (G, H)$ means that in any red-blue coloring on the edges of F, there exists a red copy of G or a blue copy of H in F. A red-blue coloring in F such that neither a red G nor a blue H occurs is called a (G, H)*–coloring*. A graph F is called a *Ramsey* (G, H)*–minimal* if $F \to (G, H)$ but $(F - e) \nrightarrow (G, H)$ for all $e \in E(F)$. The set of all Ramsey (G, H)–minimal graphs is denoted by $\mathfrak{R}(G, H)$.

E.T. Baskoro (✉) • K. Wijaya
Faculty of Mathematics and Natural Sciences, Combinatorial Mathematics Research Group, Institut Teknologi Bandung (ITB), Jalan Ganesa 10, 40132 Bandung, Indonesia
e-mail: ebaskoro@math.itb.ac.id; kristiana.w@students.itb.ac.id

P. Cartier et al. (eds.), *Mathematics in the 21st Century*, Springer Proceedings in Mathematics & Statistics 98, DOI 10.1007/978-3-0348-0859-0_2

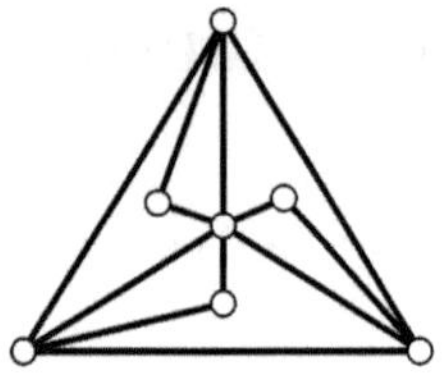

Fig. 1 Graph G in $\mathfrak{R}(2K_2, K_3)$

For a fixed pair of graphs G and H, characterizing all graphs F in $\mathfrak{R}(G, H)$ is a very interesting problem, but it is also a difficult problem, even for small graphs G and H. Burr et al. [3] proved that the set $\mathfrak{R}(mK_2, H)$ is finite for any graph H. In particular, they proved that $\mathfrak{R}(K_2, H) = \{H\}$, for every graph H, $\mathfrak{R}(2K_2, 2K_2) = \{3K_2, C_5\}$, $\mathfrak{R}(2K_2, K_3) = \{2K_3, K_5, G\}$, where G is the graph in Fig. 1. Later, Burr et al. [4] gave a characterization of all graphs in $\mathfrak{R}(tK_2, 2K_2)$ for any $t \geq 2$. Mengersen and Oeckermann [5] gave the proof of $\mathfrak{R}(2K_2, K_{1,2}) = \{2K_{1,2}, C_4, C_5\}$, which was previously mentioned in [4] without proof. In the same paper, they determined all graphs in $\mathfrak{R}(2K_2, K_{1,3})$. Baskoro and Yulianti [1] gave some necessary conditions for graphs in $\mathfrak{R}(2K_2, H)$. They proved the following theorem.

Theorem 1 ([1]). *Let H be a connected graph. Then $2H$ is the only disconnected Ramsey $(2K_2, H)$–minimal graph.* □

Next, the characterization of all graphs which belong to $\mathfrak{R}(2K_2, 2P_n)$ for $n = 4, 5$ was given by Tatanto and Baskoro [6].

In this paper, we give the necessary and sufficient conditions of graphs in $\mathfrak{R}(2K_2, mH)$ for a connected graph H. In particular, we determine all graphs in $\mathfrak{R}(2K_2, K_4)$ with at most 8 vertices. We also give a graph with 9 vertices in $\mathfrak{R}(2K_2, K_4)$. Moreover, we show that a graph obtained from any two disjoint graphs in $\mathfrak{R}(2K_2, K_4)$ by identifying vertices and edges is a member of $\mathfrak{R}(2K_2, 2K_4)$.

2 Properties of Graphs in $\mathfrak{R}(2K_2, H)$

In [1], Baskoro and Yulianti gave necessary conditions for the graphs which belong to $\mathfrak{R}(2K_2, H)$ for any graph H. In this section, we will give the necessary and sufficient conditions for those graphs. Furthermore, if $H = K_4$, then the properties of graphs in $\mathfrak{R}(2K_2, K_4)$ will be discussed. We also give a disconnected graph in $\mathfrak{R}(sK_2, mH)$ for any connected graph H.

Theorem 2. *Let H be any graph. $F \in \mathfrak{R}(2K_2, H)$ if and only if the following conditions are satisfied:*

(i) For every $v \in V(F)$, $F - v \supseteq H$.
(ii) For every K_3 in F, $F - E(K_3) \supseteq H$.
(iii) For every $e \in E(F)$, there exists $v \in V(F)$ or K_3 in F such that $(F-e)-v \not\supseteq H$ or $(F - e) - E(K_3) \not\supseteq H$.

Proof. First, suppose that either (i) is violated by some $v \in V(F)$. Then, color all edges incident to v by red and all the remaining edges by blue. Then, we have a $(2K_2, H)$-coloring of F, a contradiction. Similarly, if (ii) is violated by some K_3, then color the edges of K_3 by red and the remaining edges by blue. By this coloring, we have a $(2K_2, H)$-coloring of F, a contradiction. Furthermore, by the minimality of F, the case (iii) is satisfied.

Conversely, suppose that (i)–(iii) are satisfied. Let us consider any red-blue coloring of F not containing a red $2K_2$. Then either all edges are blue or the red edges form a star or a K_3. In both cases, the existence of a blue H is implied by (i)–(ii). So $F \to (2K_2, H)$. Next, for every $e \in E(F)$, by (iii) there exists a vertex v or a K_3 in F such that $(F - e) - v \not\supseteq H$ or $(F - e) - E(K_3) \not\supseteq H$. Now, define a coloring ϕ of $F - e$ such that $\phi(x)$ is red for all edges x incident to v or all edges $x \in E(K_3)$ and blue for the remaining edges. Then, we obtain that ϕ is a $(2K_2, H)$–coloring of $F - e$. Hence, $F \in \mathfrak{R}(2K_2, H)$. □

Lemma 1. *Let $F \in \mathfrak{R}(2K_2, K_4)$. Then the following conditions are satisfied:*

(i) $\delta(F) \geq 3$ where δ is the minimum degree in F.
(ii) F is not a tree.
(iii) Every vertex $v \in V(F)$ is contained in some K_4 in F.
(iv) Every edge $e \in E(F)$ is contained in some K_4 in F.

Proof. Theorem 2 implies $\delta(F) \geq 3$ and F is not a tree. Suppose now that there exists a vertex $v \in V(F)$ not contained in a K_4 in F. Since $F \in \mathfrak{R}(2K_2, K_4)$, then we have a $(2K_2, K_4)$–coloring of $F - v$. Use this coloring in $F - v$, and color all edges incident to v in F by blue, and we obtain a $(2K_2, K_4)$–coloring of F, a contradiction. Next, suppose that there exists an edge $e \in E(F)$ not contained in a K_4 in F. Since $F \in \mathfrak{R}(2K_2, K_4)$, then we have a $(2K_2, K_4)$–coloring of $F - e$. By using this coloring and color the edge e by blue, we obtain a $(2K_2, K_4)$–coloring of F, a contradiction. □

Theorem 3. *For any integers $s \geq 2$, $m \geq 1$, and any connected graph H, the disconnected graph $(s + m - 1)H$ is in $\mathfrak{R}(sK_2, mH)$.*

Proof. First, we prove that $(s + m - 1)H \to (sK_2, mH)$. Let $F = (s + m - 1)H$. Consider any red-blue coloring of F containing no blue mH. Therefore, there are at most $m - 1$ components of F having blue H. So, we have at least s components of F having no blue H. This means that each of these components will contain a red edge. These red edges together will form sK_2 in F.

Next, we show the minimality. Let $e \in E(F)$. We will prove that $F - e \nrightarrow (sK_2, mH)$. Since $F - e = (s - 1)H \cup (m - 1)H \cup (H - e)$, then define an edge coloring ϕ on $F - e$ such that $\phi(x)$ is blue if $x \in E((m - 1)H \cup s(H - e))$ and red otherwise. Then, it is easy to verify that ϕ is a (sK_2, mH)–coloring. □

Corollary 1. *$2K_4$ is the only disconnected graph in $\mathfrak{R}(2K_2, K_4)$.*

Proof. By Theorem 1. □

Theorem 4. *Let H be a connected graph. Let $F_1, F_2, \ldots, F_m$ be connected graphs in $\mathfrak{R}(2K_2, H)$. Then, graph $F = F_1 \cup F_2 \cup \cdots \cup F_m$ is in $\mathfrak{R}(2K_2, mH)$.*

Proof. Suppose that $F \nrightarrow (2K_2, mH)$, then there exists a $(2K_2, mH)$–coloring of F. It means that there is a $(2K_2, H)$–coloring of F_i for some $i \in \{1, 2, \ldots, m\}$, a contradiction.

Now, we prove that $F - e \nrightarrow (2K_2, mH)$, for any edge e. Let $F - e = F_1 \cup F_2 \cup \cdots \cup (F_i - e) \cup \cdots \cup F_m$, for some $i \in \{1, 2, \ldots, m\}$. Then there exists a $(2K_2, H)$–coloring of $F_i - e$. We use such a coloring in $F_i - e$, and all edges in $(F - e) - (F_i - e)$ are colored by blue. Then, we obtain a $(2K_2, mH)$–coloring of $F - e$. □

3 The Set $\mathfrak{R}(2K_2, K_4)$

We determine all graphs in $\mathfrak{R}(2K_2, K_4)$ with at most 8 vertices. We also give a graph with 9 vertices in $\mathfrak{R}(2K_2, K_4)$.

Theorem 5. *K_6 is the only graph in $\mathfrak{R}(2K_2, K_4)$ with 6 vertices.*

Proof. First, we prove that K_6 satisfies three conditions in Theorem 2. Since $K_6 - v = K_5$, then $K_6 - v \supseteq K_4$. Since $K_6 - E(K_3) = K_3 + \overline{K}_3$, then $K_6 - E(K_3) \supseteq K_4$. Next, for every $e \in E(K_6)$, $K_6 - e = K_4 + \overline{K}_2$. So there exists a K_3 in $K_6 - e$ such that $(K_6 - e) - E(K_3) = K_1 + \overline{K}_3 + \overline{K}_2$ does not contain K_4. Since K_6 is a graph with the maximum number of edges, then K_6 is the only graph in $\mathfrak{R}(2K_2, K_4)$ with 6 vertices. □

Theorem 6. *$\mathfrak{R}(2K_2, K_4)$ contains no connected graphs with 7 vertices.*

Proof. Let F be a connected graph with 7 vertices. If $F \in \mathfrak{R}(2K_2, K_4)$, then F contains a K_4, but it does not contain a K_6. We will show that no graph $F \in \mathfrak{R}(2K_2, K_4)$ on 7 vertices. Since both graphs K_7 and $K_7 - e$ contain K_6, then $K_7, K_7 - e \notin \mathfrak{R}(2K_2, K_4)$, for any edge $e \in E(K_7)$. Hence, F must be a subgraph of $K_7 - e$. Now, we consider $F = K_7 - 2e$. There are two non-isomorphic graphs $K_7 - 2e$, namely, F_a with a degree sequence $(6, 6, 6, 6, 5, 5, 4)$ and F_b with a degree sequence $(6, 6, 6, 5, 5, 5, 5)$.

Now, let us consider the graph F_a with a degree sequence $(6, 6, 6, 6, 5, 5, 4)$. Since F_a contains a K_6, then F_a is not in $\mathfrak{R}(2K_2, K_4)$. Therefore, F must be a subgraph of F_a. So, consider now $F = F_a - e$ which contain no K_6. Then, we obtain such a graph with its degree sequence $(6, 6, 6, 6, 4, 4, 4)$, $(6, 6, 6, 5, 5, 4, 4)$, or $(6, 6, 5, 5, 5, 5, 4)$. We assume $d(v_i) \geq d(v_{i+1})$ for $i = 1, 2, \ldots, 6$. From all graphs, for $V(K_3) = \{v_1, v_2, v_3\}$, we obtain $F - E(K_3)$ does not contain a K_4. So, $F \nrightarrow (2K_2, K_4)$. Therefore, any subgraph F^* with 7 vertices of F will satisfy that $F^* \nrightarrow (2K_2, K_4)$.

Next, we observe the graph F_b with a degree sequence $(6, 6, 6, 5, 5, 5, 5)$. Let $V(F_b) = \{v_1, v_2, \ldots, v_7\}$ where $d(v_i) = 6$ for $i = 1, 2, 3$ and $d(v_i) = 5$

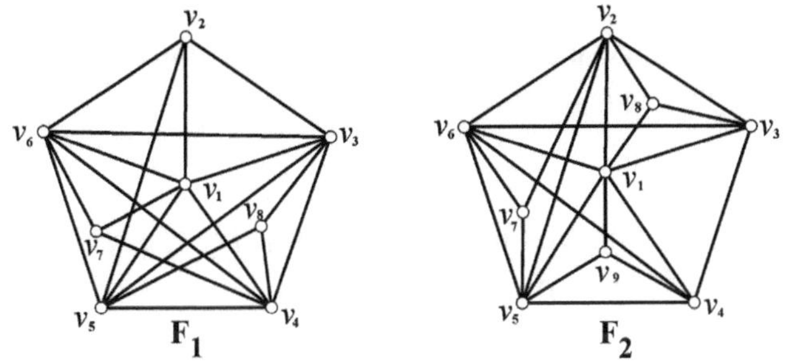

Fig. 2 Graphs in $\mathfrak{R}(2K_2, K_4)$

for $i = 5, 6, 7$. Let ϕ be a coloring of F_b such that $\phi(e)$ is red if $e \in \{v_1v_2, v_1v_3, v_2v_3\}$ and blue otherwise. We obtain a $(2K_2, K_4)$–coloring of F_b. Thus, $F_b \nrightarrow (2K_2, K_4)$. Therefore, any subgraph F^* with 7 vertices of F_b will satisfy that $F^* \nrightarrow (2K_2, K_4)$. This concludes that no graph with 7 vertices is in $\mathfrak{R}(2K_2, K_4)$. □

Now, consider graphs F_1 and F_2 in Fig. 2. Graph F_1 has the vertex set $V(F_1) = \{v_1, v_2, \ldots, v_8\}$ and the edge set $E(F_1) = \{v_iv_j \mid i, j = 1, 2, \ldots, 8,\ i \neq j\} - \{v_1v_8, v_2v_4, v_2v_7, v_2v_8, v_3v_7, v_5v_7, v_6v_8, v_7v_8\}$. Graph F_2 has the vertex set $V(F_2) = \{v_1, v_2, \ldots, v_9\}$ and the edge set $E(F_2) = \{v_iv_j \mid i, j = 1, 2, \ldots, 9,\ i \neq j\} - \{v_iv_7 \mid i = 1, 3, 4\} - \{v_iv_8 \mid i = 4, 5, 6, 7\} - \{v_iv_9 \mid i = 2, 3, 6, 7, 8\} - \{v_2v_4, v_3v_5\}$. We prove that graphs F_1 and F_2 are members of $\mathfrak{R}(2K_2, K_4)$ in the following lemma.

Theorem 7. *The graph F_1 in Fig. 2 is the only graph with 8 vertices in $\mathfrak{R}(2K_2, K_4)$.*

Proof. First, we prove that $F_1 \rightarrow (2K_2, K_4)$. We can see that for every $i \in \{1, 2, \ldots, 8\}$, $F_1 - v_i$ contains a K_4. For every K_3 in F_1, $F_1 - E(K_3)$ contains a K_4. Hence, $F_1 \rightarrow (2K_2, K_4)$. Next, since $F_1 \not\supseteq 2K_4$ and $F_1 \not\supseteq K_6$, then $F_1 \in \mathfrak{R}(2K_2, K_4)$.

Now, suppose there exists a connected graph F with 8 vertices in $\mathfrak{R}(2K_2, K_4)$ but $F \neq F_1$. Let $V(F) = \{v_1, v_2, \ldots, v_8\}$. By Theorem 2, F must contain a K_4,, and we may assume $V(K_4) = \{v_1, v_2, v_3, v_4\}$. By the minimality of F, then F does not contain both $2K_4$ and K_6. By Theorem 2(i), for $i = 1, 2, 3$, $F - v_i$ must contain a K_4. Then (up to isomorphism) the new K_4 in F is formed by the vertex set $\{v_4, v_5, v_6, v_7\}$. So, v_4 is contained in two K_4 in F. Next, there must be a K_4 in $F - v_4$ by Theorem 2(i). Since $\delta(F) \geq 3$ by Lemma 1(i), then (up to isomorphism) the K_4 in $F - v_4$ is formed by the vertex set $\{v_1, v_6, v_7, v_8\}$. Next, by Theorem 2(ii), for $V(K_3) = \{v_1, v_4, v_7\}$, $F - E(K_3)$ must contain a K_4. Then the K_4 in $F - E(K_3)$ can be formed by the vertex set $\{v_2, v_3, v_4, v_6\}$, $\{v_1, v_2, v_5, v_6\}$, or $\{v_3, v_4, v_5, v_6\}$. Otherwise F is the graph F_1 or is not minimal. For all cases, by Theorem 2(ii), $F - E(K_3)$ must contain a K_4, for $V(K_3) = \{v_1, v_4, v_6\}$. But the new K_4 causes F which is not minimal, a contradiction. □

For graphs with 9 vertices, it is not difficult to verify that the graph F_2 in Fig. 2 is a Ramsey $(2K_2, K_4)$-minimal graph. However, characterizing all $(2K_2, K_4)$-minimal graphs is a open problem.

4 Constructing Graphs in $\mathfrak{R}(2K_2, 2K_4)$ by Operations over Graphs in $\mathfrak{R}(2K_2, K_4)$

In this section, we show that a graph obtained from two connected graphs in $\mathfrak{R}(2K_2, K_4)$ by identifying vertices or edges is a member of $\mathfrak{R}(2K_2, 2K_4)$.

Corollary 2. $\{3K_4, 2K_6\} \subseteq \mathfrak{R}(2K_2, 2K_4)$.

Proof. By Theorem 3, we obtain $3K_4 \in \mathfrak{R}(2K_2, 2K_4)$. By Theorem 4, we obtain $2K_6 \in \mathfrak{R}(2K_2, 2K_4)$. □

Theorem 8. *Let $G, H \in \mathfrak{R}(2K_2, K_n)$ be connected graphs and $u \in V(G)$, $v \in V(H)$. If $G \odot H$ is a graph obtained by identifying vertices u and v, then $G \odot H \in \mathfrak{R}(2K_2, 2K_n)$.*

Proof. Let $w \in V(G \odot H)$. Then $(G \odot H) - w$ is either connected or disconnected, depending of the choice of w. If $w = u$ is the identified vertex then $(G \odot H) - w$ is disconnected, that is $(G \odot H) - w = (G - u) \cup (H - v)$. Since $G - u \supseteq K_n$ and $H - v \supseteq K_n$ then $(G \odot H) - w \supseteq 2K_n$. If w is not the identified vertex then $(G \odot H) - w$ is connected, we may assume $(G \odot H) - w = (G - w) \odot H$. Since $G - w \supseteq K_n$ and $H \supseteq K_n$, then $(G \odot H) - w \supseteq 2K_n$. Next, let K_3 in $G \odot H$. Then K_3 is in G or H. Suppose that K_3 is in G, then $(G \odot H) - E(K_3) = (G - E(K_3)) \odot H$. Since $G - E(K_3) \supseteq K_n$ and $H \supseteq K_n$, then $(G \odot H) - E(K_3) \supseteq 2K_n$. So, $G \odot H \to (2K_2, 2K_n)$.

Next, let $e \in E(G \odot H)$, and then $e \in E(G)$ or $e \in E(H)$. We assume $e \in E(G)$. Then, there exists a $(2K_2, K_n)$–coloring ϕ_1 of $G - e$. Now, we define ϕ as a coloring of $(G \odot H) - e$ such that $\phi(a) = \phi_1(a)$ for $a \in E(G - e)$ and blue otherwise. We obtain a $(2K_2, 2K_n)$–coloring of $(G \odot H) - e$. So, $(G \odot H) - e \nrightarrow (2K_2, 2K_n)$. □

Theorem 9. *Let $G, H \in \mathfrak{R}(2K_2, K_n)$ be connected graphs and $a \in E(G)$, $e \in E(H)$. If $(G \ominus H)$ is a graph obtained by identifying edges a and e, then $(G \ominus H) \in \mathfrak{R}(2K_2, 2K_n)$.*

Proof. First, we prove that $(G \ominus H) \to (2K_2, 2K_n)$. Let $w \in V(G \ominus H)$. Then, $(G \ominus H) - w$ is connected. If $w \in V(G)$ is not incident to edge a, then $(G \ominus H) - w = (G - w) \ominus H$. Since $G - w \supseteq K_n$ and $H \supseteq K_n$, then $(G \ominus H) - w \supseteq 2K_n$. If w is incident to edge a, let $a = vw$ then $(G \ominus H) - w = (G - w) \odot (H - w)$ by identifying vertex v. Since $G - w \supseteq K_n$ and $H - w \supseteq K_n$, then $(G \ominus H) - w \supseteq 2K_n$. Let K_3 in $G \ominus H$. Then, this K_3 can contain the edge a or not. If K_3 does not contain a, then $(G \ominus H) - E(K_3) = (G - E(K_3)) \ominus H$. Since $G - E(K_3) \supseteq K_n$ and $H \supseteq K_n$, then $(G \ominus H) - E(K_3) \supseteq 2K_n$. If K_3 contains $a = v_1 w$, then

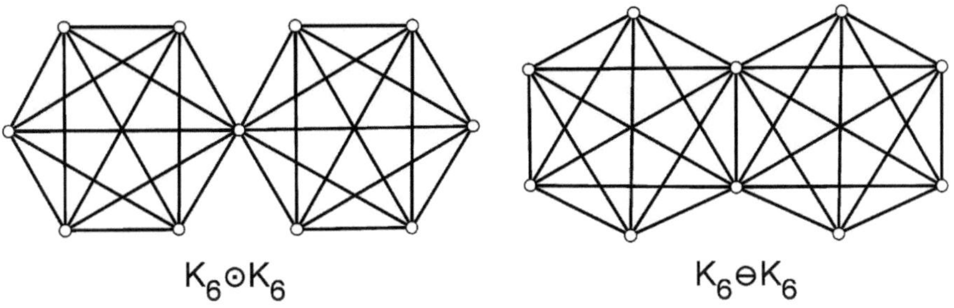

Fig. 3 $K_6 \odot K_6, K_6 \ominus K_6 \in \mathfrak{R}(2K_2, 2K_4)$

$(G \ominus H) - E(K_3) = (G - E(K_3)) \odot (H - e)$ for some $e = v_2 w$ by identifying two vertices $v_1 = v_2$ and w. Since $G - E(K_3) \supseteq K_n$ and $(H - e) \supseteq K_n$, then $(G \ominus H) - E(K_3) \supseteq 2K_n$.

Next, we show the minimality. Let $b \in E(G \ominus H)$. Then the edge b can be the identified edge or not. If b is the identified edge, let $b = vw$, then $(G \ominus H) - b = (G - b) \odot (H - b)$ by identifying two vertices v and w. So, there exists a $(2K_2, K_n)$—coloring of both $G - b$ and $H - b$. Clearly, there exists a $(2K_2, 2K_n)$—coloring of $(G \ominus H) - b$. If b is not the identified edge, we may assume $b \in E(G)$, and then there exists a $(2K_2, K_n)$—coloring ϕ_1 of $G - b$. We define ϕ as a coloring of $(G \ominus H) - b$ such that $\phi(e) = \phi_1(e)$ for $e \in E(G - b)$ and blue otherwise. Then, it is easy to verify that ϕ is a $(2K_2, 2K_n)$—coloring of $(G \ominus H) - b$. □

By Theorems 8 and 9, we have the following corollary.

Corollary 3. $\{K_6 \odot K_6, K_6 \odot F_1, K_6 \odot F_2, F_1 \odot F_1, F_1 \odot F_2, F_2 \odot F_2\} \subseteq \mathfrak{R}(2K_2, 2K_4)$ *and* $\{K_6 \ominus K_6, K_6 \ominus F_1, K_6 \ominus F_2, F_1 \ominus F_1, F_1 \ominus F_2, F_2 \ominus F_2\} \subseteq \mathfrak{R}(2K_2, 2K_4)$. □

References

1. Baskoro, E.T., Yulianti, L.: On Ramsey minimal graphs for $2K_2$ versus P_n. Adv. Appl. Discret. Math. **8**(2), 83–90 (2011)
2. Borowiecka-Olszewska, M., Haluszczak M.: On Ramsey $(K_{1,m}, \mathcal{G})$—minimal graphs. Discret. Math. **313**(19), 1843–1855 (2012)
3. Burr, S.A., Erdös, P., Faudree, R.J., Schelp, R.H.: A class of Ramsey-finite graphs. In: Proceedings of the Ninth Southeastern Conference on Combinatorics, Graph Theory and Computing, Boca Raton, pp. 171–180 (1978)
4. Burr, S.A., Erdös, P., Faudree, R.J., Schelp, R.H.: Ramsey minimal graphs for matchings. In: The Theory and Applications of Graphs (Kalamazoo, 1980), pp. 159–168. Wiley, New York (1981)
5. Mengersen, I., Oeckermann, J.: Matching-star Ramsey sets. Discret. Appl. Math. **95**, 417–424 (1999)
6. Tatanto, D., Baskoro, E.T.: On Ramsey $(2K_2, 2P_n)$—minimal graphs. AIP Conf. Proc. **1450**, 90–95 (2012)

Equilibrium in Choice of Generalized Games

Massimiliano Ferrara and Anton Stefanescu

1 Introduction

The Nash equilibrium (equilibrium point) is the most important solution concept of the noncooperative game theory, and it is defined in terms of the *normal form* of a game, as a strategy combination with the property that no player can gain by unilaterally deviating from it. In the original definition of Nash [8, 9], each player was allowed to choose freely from its own set of individual decisions (strategies), and the players' options were expressed by utility functions defined on the product of the individual strategy spaces. Later, the mode representing a game and the original definition of the equilibrium were extended to cover more general situations met in the noncooperative competitions. This is the case of the social equilibrium of G. Debreu, defined for generalized games where the decision of each player can be chosen only from a subset determined by the selection made by all other players or the case of abstract economies, introduced by W. Shafer and H. Sonnenschein, where the individual preferences are represented as correspondences.

A new concept of equilibrium, called *equilibrium in choice form,* has been introduced in [11]. The formal framework for the definition of equilibria in choice is the *game in choice form,* defined as the family of the sets of individual strategies and a choice profile. Intuitively, a choice profile specifies the desirable outputs of

Based on the paper communicated to the *6th World Conference on 21st Century Mathematics* by the first author.

M. Ferrara
Department SSGES, University "Mediterranea" of Reggio Calabria, Reggio Calabria, Italy

A. Stefanescu (✉)
Faculty of Mathematics and Computer Science, University of Bucharest, Bucharest, Romania
e-mail: astefanescu@fmi.unibuc.ro

P. Cartier et al. (eds.), *Mathematics in the 21st Century*, Springer Proceedings in Mathematics & Statistics 98, DOI 10.1007/978-3-0348-0859-0_3

each player, and, since each output of the game is associated to a game strategy, it can be expressed as a collection of subsets of the set of all game strategies. Particularly, when the players' options are represented by utility functions or by preference relations, a choice profile may be the family of the graphs of the players' best reply mappings.

Then, the set of equilibria in choice coincides with the set of the Nash equilibria so that the definition of the equilibrium in choice captures the main idea of the "best reply" from the definition of the Nash equilibrium, but the new concept is more general, responding to various representations of the players options. Two variants of this concept were discussed in [11] and [12]. The basic one presumes a relaxation of the best reply principle and has obvious counterparts for classical solutions, if this relaxation is accepted. The stronger form of the equilibrium in choice can be considered as a generic notion of noncooperative solution, and several usual versions of such solutions are produced, when the choice profile is designed in different particular ways.

In this paper we rephrase these two kinds of equilibria in the framework of noncooperative games with restricted individual strategies defined, as in Debreu's work, by restriction correspondences. The main existence results refer to equilibria in choice, but applying them to classical models, as those considered by the above-cited authors, we obtain some versions of the known results concerning these models.

2 Equilibrium Concepts in Noncooperative Game Theory

Let us review briefly the main versions of the concept of equilibrium and introduce the general concept of equilibrium in choice.

The main components of the formal representation of any noncooperative competitive situation – called, conventionally, game – are the game mechanism (the game form) and the preference profile.

A game form, denoted by γ, is a system $((X_i)_{i \in N}, G, g)$, where N is the set of the players (here and everywhere in this paper, $N = \{1, 2, \ldots, n\}$). X_i, $i \in N$ are the sets of individual actions, G is the set of the game outcomes, and $g : \prod_{i \in N} X_i \mapsto G$ is the output mapping.

Usually, the possible actions of a player are identified with the plans that the player thinks and that can be followed by the respective actions so that the set X_i will be called "the player i' s strategy set," while the Cartesian product X will be called "the game strategy set."

A preference profile is a collection of individual preferences and, normally, is defined on the set G of game outcomes. If, as usually, g is "onto", then the preferences may be defined on X, since each outcome can be viewed as the result of a game strategy. Hence, a shortened model of a noncooperative game omits G and g from its definition and is defined as $\Gamma = ((X_i)_{i \in N}, (P_i)_{i \in N})$, where $(P_i)_{i \in N}$ is a preference profile.

The first general model of noncooperative games is due to Nash [8, 9], who has considered the cardinal preferences of the players defined as utility functions on the set of game strategies. Thus, by Nash, a n−person noncooperative game is defined as $\Gamma = ((X_i)_{i\in N}, (u_i)_{i\in N})$ (the normal form), where $u_i : X \mapsto R$ is the player i' s utility function (which stands for the individual preference P_i).

Denote by $X_{-i} = \prod_{j\neq i} X_j$ if $i \in N$, and represent the game strategy as (x_{-i}, x_i), where $x_{-i} \in X_{-i}$ and $x_i \in X_i$.

A Nash equilibrium of the game Γ is any game strategy x^* with the property

$$\forall i \in N, u_i\left(x^*\right) \geq u_i\left(x^*_{-i}, x_i\right) \text{ for all } x_i \in X_i. \tag{1}$$

For each $i \in N$, the player i's best reply mapping is the multifunction (correspondence) $\beta_i : X_{-i} \mapsto 2^{X_i}$, defined by

$$\beta_i(x_{-i}) = \{x_i \in X_i | u_i(x_{-i}, x_i) \geq u_i(x_{-i}, y_i), \text{ for all } y_i \in X_i\}.$$

Then, x^* is a Nash equilibrium if and only if it is a fixed point of the game best reply mapping $\beta : X \mapsto 2^X$, defined by

$$\beta(x) = \prod_{i\in N} \beta_i(x_{-i})$$

Also, if $Gr\beta_i$ denotes the graph of β_i, i.e.,

$$Gr\beta_i = \{x_i \in X_i | (x_{-i}, x_i) \in \beta_i(x_{-i}), x_{-i} \in X_{-i}\},$$

then x^* is a Nash equilibrium if and only if

$$x^* \in \cap_{i\in N} Gr\beta_i.$$

A rather more general model of noncooperative games involves the ordinal representation of players' preferences. Thus a preference profile is a family $(r_i)_{i\in N}$ of ordinal preferences (complete, reflexive, and transitive binary relations). In such framework, a Nash equilibrium is defined by the property:

$$\forall i \in N, x^*_i r_i \left(x^*_{-i}, x_i\right) \text{ for all } x_i \in X_i. \tag{2}$$

which, obviously, captures the same idea of the "best response strategy" as the original definition of the equilibrium.

The subsequent evolution of the concept of equilibrium is determined by the mathematical developments in economics.

The generalized game of Debreu [4] alters the representation of the set of game strategies as the Cartesian product of the sets of individual strategies, based on the idea that the choice of an agent in a social system is not entirely free and the actions of all the other agents determine a subset of his own strategy space to which his selection is restricted. Thus, a generalized game is represented by the system $((X_i)_{i\in N}, (\varphi_i)_{i\in N}, (u_i)_{i\in N})$, where $\varphi_i : X_{-i} \mapsto 2^{X_i}$ is the player i's restriction correspondence. Accordingly, the definition of the "social equilibrium" is given by the conditions:

$$\forall i \in N, x_i^* \in \varphi_i\left(x_{-i}^*\right), \text{ and } u_i\left(x^*\right) \geq u_i\left(x_{-i}^*, x_i\right) \text{ for all } x_i \in \varphi_i\left(x_{-i}^*\right). \tag{3}$$

Debreu's existence theorem of social equilibrium was the main mathematical tool used for the proof of the competitive equilibrium existence theorem in the Arrow-Debreu model of an economic system.

Inspired by some earlier results concerning the existence of competitive equilibrium in economies with non-transitive and non-complete preferences, Shafer and Sonnenschein [10] have extended Debreu's model of generalized game to the case when the individual preferences are represented by correspondences.

Their model of abstract economy consists of $((X_i)_{i\in N}, (\varphi_i)_{i\in N}, (\pi_i))$, where $\pi_i : X \mapsto 2^{X_i}$ are the preference correspondences and the equilibrium is defined by:

$$\forall i \in N, x_i^* \in \varphi_i\left(x_{-i}^*\right), \text{ and } \varphi_i\left(x^*\right) \cap \pi_i\left(x^*\right) = \emptyset. \tag{4}$$

Obviously, if $((X_i)_{i\in N}, (u_i)_{i\in N})$ is a normal form game and

$$\pi_i(x) = \{y_i \in X_i | u_i(x_{-i}, y_i) > u_i(x)\},$$

then (4) reduces to (3).

A more general representation of preferences and, therefore, a more general model of noncooperative games have been introduced in [11]. According to the primary approach in noncooperative game theory, when the set of game strategies is the product of individual strategy spaces, a *choice profile* is defined to be any collection $\mathcal{C} = (C_i)_{i\in N}$ of nonempty subsets of X, and a *game in the choice form* is a double family $((X_i)_{i\in N}, (C_i)_{i\in N})$, where $\mathcal{C} = (C_i)_{i\in N}$ is a choice profile.

In this formalism a choice profile is a primary element, and, in order to sustain the definition of the equilibrium, we think of C_i as the set of all game strategies that determines outputs which are desirable for the player i. However, several constructive definitions, starting from different classical representations of a noncooperative game, lead to particular but more precise meanings for this abstract model of the game in choice form.

Now, we adapt these notions to be suitable for the case when the individual choices of any player are restricted by the selections of all other players.

As in the above, let us consider that the game form consists of the family $(X_i)_{i\in N}$ of individual strategies and of family $(\varphi_i)_{i\in N}$ of restriction correspondences. The natural counterparts of the definitions introduced in [11] and [12] are given in the following:

Definition 1. A choice profile under restrictions (c.p.r.) is any family $(C_i)_{i\in N}$ of nonempty sets with $C_i \subseteq Gr\varphi_i$ for each $i \in N$.

Definition 2. A generalized game in the choice form is any system $((X_i)_{i\in N}, (\varphi_i)_{i\in N}, (C_i)_{i\in N})$, where $C = (C_i)_{i\in N}$ is a c.p.r.

In this new framework, the definitions of the two versions of the equilibrium in choice reproduce the definitions stated in [11] with c.p.r. instead of a general choice profile. Thus,

Definition 3. An equilibrium in choice (e.c.) of the generalized game in the choice form $((X_i)_{i\in N}, (\varphi_i)_{i\in N}, (C_i)_{i\in N})$ is any game strategy x^* with the property

$$\forall i \in N, \left(x^*_{-i}, \varphi_i\left(x^*_{-i}\right)\right) \cap C_i \neq \emptyset \Rightarrow x^* \in C_i.$$

If $x^* \in \cap_{i\in N} C_i$, then x^* is said to be a strong equilibrium in choice (s.e.c.).

(Here and in the following, (a, A) stands for the Cartesian product $\{a\} \times A$.)

Following a relatively recent terminology, for any set C in a product space $\prod_{i\in N} X_i$, the sets of the form $C(x_{-i}) = \{y_i \in X_i \mid (x_{-i}, y_i) \in C\}$ and $C(x_i) = \{y_{-i} \in X_{-i} \mid (y_{-i}, x_i) \in C\}$ are called, respectively, the upper sections and lower sections of C.

Remark 1. With the above notation, the definition of the equilibrium in choice can be rephrased as follows: "x^* is an equilibrium in choice iff $x^*_i \in C_i(x^*_{-i})$ for every $i \in N$ for which $\varphi_i(x^*_{-i}) \cap C_i(x^*_{-i}) \neq \emptyset$."

It is easy to see that these definitions enclose, as particular cases, the notions encountered in the above discussion as those related to Debreu's model or to the abstract economy.

Example 1. Let the generalized game be $((X_i)_{i\in N}, (\varphi_i)_{i\in N}, (u_i)_{i\in N})$. If, for each $i \in N$ and $x_{-i} \in X_{-i}$, $u_i(x_{-i}, .)$ reaches its maximum on $\varphi_i(x_{-i})$, set $C_i = \{x \in Gr\varphi_i \mid u_i(x) = \max\limits_{z_i \in \varphi_i(x_{-i})} u_i(x_{-i}, z_i)\}$. Then, x^* is a social equilibrium of the generalized game $((X_i)_{i\in N}, (\varphi_i)_{i\in N}, (u_i)_{i\in N})$ iff it is a strong equilibrium in choice of the generalized game in the choice form $((X_i)_{i\in N}, (\varphi_i)_{i\in N}, (C_i)_{i\in N})$.

Example 2. Let $((X_i)_{i\in N}, (\varphi_i)_{i\in N}, (\pi_i)_{i\in N})$ be an abstract economy. Set $C_i = \{x \in Gr\varphi_i \mid \varphi_i(x_{-i}) \cap \pi_i(x) = \emptyset\},. i \in N$. Then, x^* is an equilibrium of the abstract economy $((X_i)_{i\in N}, (\varphi_i)_{i\in N}, (\pi_i)_{i\in N})$ iff it is a strong equilibrium in choice of the generalized game in the choice form $((X_i)_{i\in N}, (\varphi_i)_{i\in N}, (C_i)_{i\in N})$.

In fact, the concept of strong equilibrium in choice captures the principle of "the best response" common to all previously defined versions of the equilibrium. The weaker form of this concept, called here "equilibrium in choice," relaxes this principle. We can find the correspondents of the equilibrium in choice in the classical models considered in the above examples relaxing the definition conditions of the respective versions of equilibrium.

Definition 4. x^* is a weak social equilibrium (w.s.e.) of the generalized game $((X_i)_{i\in N}, (\varphi_i)_{i\in N}, (u_i)_{i\in N})$ iff

$$u_i(x^*) \geq u_i\left(x_{-i}^*, x_i\right), \forall x_i \in \varphi_i\left(x_{-i}^*\right), \text{ for every } i \in N \text{ for which}$$

$$B_i\left(x_{-i}^*\right) = \left\{ y_i \in \varphi_i\left(x_{-i}^*\right) \mid u_i\left(x_{-i}^*, y_i\right) = \max_{z_i \in \varphi_i\left(x_{-i}^*\right)} u_i\left(x_{-i}^*, z_i\right) \right\} \neq \emptyset.$$

Definition 5. x^* is a weak equilibrium (w.e.) of the abstract economy $((X_i)_{i\in N}, (\varphi_i)_{i\in N}, (\pi_i)_{i\in N})$ iff

$$x_i^* \in \varphi_i\left(x_{-i}^*\right) \text{ and } \varphi_i\left(x_{-i}^*\right) \cap \pi_i(x^*) = \emptyset, \text{ for every } i \in N \text{ for which}$$
$$\left\{y_i \in \varphi_i\left(x_{-i}^*\right) \mid \varphi_i\left(x_{-i}^*\right) \cap \pi_i\left(x_{-i}^*, y_i\right) = \emptyset\right\} \neq \emptyset.$$

Then, if we keep the notation of the Examples 1 and 2, the following statements hold.

Proposition 1. *x^* is a weak social equilibrium of the generalized game $((X_i)_{i\in N}, (\varphi_i)_{i\in N}, (u_i)_{i\in N})$ iff it is an equilibrium in choice of the generalized game in the choice form $((X_i)_{i\in N}, (\varphi_i)_{i\in N}, (C_i)_{i\in N})$, where C_i is as in Example 1.*

Proposition 2. *x^* is a weak equilibrium of the abstract economy $((X_i)_{i\in N}, (\varphi_i)_{i\in N}, (\pi_i)_{i\in N})$ iff it is an equilibrium in choice of the generalized game in the choice form $((X_i)_{i\in N}, (\varphi_i)_{i\in N}, (C_i)_{i\in N})$, where C_i is as in Example 2.*

3 The Main Existence Results

The existence of the s.e.c. is obviously related to the following problem:

"Given n nonempty sets $C_1, C_2, \ldots, C_n$ in the product $\prod_{i=1}^n X_i$ of some topological vector spaces, find conditions under which $\bigcap_{i=1}^n C_i \neq \varnothing$."

The problem is known in the early literature related to fixed point theorems for set-valued mappings and is well represented in the work of Ky Fan [5–7] and other more recent scientists.

Moreover, some applications to the problem of Nash equilibrium were done, even if they appeared as secondary results.

The following two theorems are Fan's intersection theorems for sets with convex sections, rephrased in terms of games in the choice form.

Theorem 1 ([5], Theorem 2). *Let $((X_i)_{i\in N}, (\varphi_i)_{i\in N}, (C_i)_{i\in N})$ be a generalized game in the choice form. Assume that, for each $i \in N$:*

(a) X_i is a nonvoid, compact, and convex set in a locally convex topological vector space.

(b) $C_i \subseteq Gr\varphi_i$ is closed.

(c) *The upper sections* $C_i(x_{-i})$, $x_{-i} \in X_{-i}$ *of* C_i *are nonempty and convex.*

Then, the game has s.e.c.

Theorem 2 ([6], Théorème 1). *Let* $((X_i)_{i\in N}, (\varphi_i)_{i\in N}, (C_i)_{i\in N})$ *be a generalized game in the choice form. Assume that, for each* $i \in N$:

(a) X_i *is a nonvoid, compact, and convex set in a Hausdorff topological vector space.*
(b) *The lower sections* $C_i(x_i)$, $x_i \in X_i$ *of* C_i *are open in* X_{-i}.
(c) *The upper sections* $C_i(x_{-i})$, $x_{-i} \in X_{-i}$ *of* C_i *are nonempty and convex.*

Then, the game has s.e.c.

Note that most intersection theorems assume the convexity of the upper sections of the sets. Our main result below concerns a more general situation.

Theorem 3. *Let* $((X_i)_{i\in N}, (\varphi_i)_{i\in N}, (C_i)_{i\in N})$ *be a generalized game in choice form. Assume that, for each* $i \in N$:

(a) X_i *is a nonvoid, convex, and compact set in a Hausdorff topological vector space* E_i.
(b) C_i *is a nonempty closed subset of* X.
(c) *There exists a sequence* $(G_{ik})_k$ *of subsets of* X *with the properties:*
(c_1) *The upper sections* $G_{ik}(x_{-i})$, $k = 1, 2, \ldots$ *are nonempty and convex, for every* $x_i \in X_{-i}$.
(c_2) *The lower sections* $G_{ik}(x_i)$, $k = 1, 2, \ldots$ *are open in* X_{-i}, *for every* $x_i \in X_i$.
(c_3) $G_{ik} \supseteq G_{ik+1}$, *for every* $k = 1, 2, \ldots$.
(c_4) *For every open set* G *with* $G \supset C_i$, *there exists* k *such that* $G_{ik} \subseteq G$.

Then, the game admits strong equilibria in choice.

Proof. For each k, apply Theorem 2 for the generalized game in the choice form $((X_i)_{i\in N}, (\psi_{ik})_{i\in N}, (G_{ik})_{i\in N},)$, where $\psi_{ik}(x_{-i}) = X_i$, $x_{-i} \in X_{-i}$, $i \in N$. Choose $x^k \in \bigcap_{i\in N} G_{ik}$, $k = 1, 2, \ldots$. Since the sequence $(x^k)_k$ lays in the compact X, it contains a subsequence that converges to some $x^* \in X$. Re-indexing the terms of this subsequence, we can write $x^k \underset{k\to\infty}{\longrightarrow} x^*$, and let us show that x^* is a s.e.c.

Let us suppose, by way of contradiction, that $x^* \notin C_i$ for some $i \in N$. Since C_i is a nonempty compact, a neighborhood V_{x^*} of x^* and an open set G containing C_i can be found such that $V_{x^*} \cap G = \emptyset$. Then, by ($c_4$) and ($c_3$) there is k_1 such that $G_{ik} \subseteq G$ for $k \geq k_1$. On the other hand, V_{x^*} contains all terms of the convergent subsequence beginning with some rank k_2. Hence, $x^k \notin G_{ik}$ for $k \geq \max\{k_1, k_2\}$ contradicting the above assertion. ∎

The existence of the e.c. can be stated in a more general framework as:

"Given n nonempty sets $C_1, C_2, \ldots, C_n$ in the product $\prod_{i=1}^n X_i$ of some topological vector spaces, find conditions under which there exists a point $x^* \in \prod_{i=1}^n X_i$ with the property $\forall i \in N, (x^*_{-i}, \varphi_i(x^*_{-i})) \cap C_i \neq \emptyset \Rightarrow x^* \in C_i$."

This problem has no roots in the previous literature, so that it appears as a new problem which generalized the above-cited one.

The technique of proof of the main results is based on fixed point theorems, as always in the equilibrium theory, but the correspondences involved here are not usual and are specific to the original representation of the games. The proof of the theorem is based on the lemma below. For its proof we refer the reader to [12].

Lemma 1 ([12] Lemma 1). *Let X be a metric space and Y a nonvoid, convex, and compact set in a metrizable topological vector space E. If the correspondence $\gamma : X \mapsto 2^Y$ is closed (i.e., it has a closed graph), then so is the correspondence $\overline{co}\gamma$.*

Theorem 4. *Let $((X_i)_{i\in N}, (\varphi_i)_{i\in N}, (C_i)_{i\in N})$ be a generalized game in choice form. Assume that, for each $i \in N$:*

(a) X_i is a nonvoid, convex, and compact set in a metrizable locally convex space E_i.
(b) φ_i is lower semi-continuous, with nonempty closed values.
(c) C_i is nonempty and closed in X.
(d) The upper sections $C_i(x_{-i})$, $x_{-i} \in X_{-i}$ of C_i are convex.

Then, the game admits equilibria in choice.

Proof. Denote by d_i a distance defining the topology of E_i, $i \in N$. If d is a distance on the product space, let us define the correspondences $\gamma_i : X_{-i} \mapsto 2^{X_i}$, $i = 1,\ldots,n$, by $\gamma_i(x_{-i}) = \{y_i \in \varphi_i(x_{-i}) \mid d((x_{-i}, y_i), C_i) = \min_{z_i\in\varphi_i(x_{-i})} d((x_{-i}, z_i), C_i)\}$.

Obviously, γ_i has nonempty compact values. One can easily verify that it has closed graph, so that it is upper semi-continuous.

Indeed, if $x_{-i}^k \underset{k\to\infty}{\longrightarrow} x_{-i}^0$ and $y_i^k \underset{k\to\infty}{\longrightarrow} y_i^0$, where $y_i^k \in \gamma_i\left(x_{-i}^k\right)$, $k = 1, 2, \ldots$., choose $z_i^0 \in \varphi_i\left(x_{-i}^0\right)$. Since φ_i is lower semi-continuous, there exists a sequence $\left(z_i^k\right)_k$ convergent to z_i^0 with $z_i^k \in \varphi_i\left(x_{-i}^k\right)$, $k = 1, 2, \ldots$. Then, $d\left(\left(x_{-i}^k, y_i^k\right), C_i\right) \le d\left(\left(x_{-i}^k, z_i^k\right), C_i\right)$, for all k, and the continuity of $d(., C_i)$ implies the inequality $d\left(\left(x_{-i}^0, y_i^0\right), C_i\right) \le d\left(\left(x_{-i}^0, z_i^0\right), C_i\right)$. Since $z_i^0 \in \varphi_i\left(x_{-i}^0\right)$ is arbitrary, this means that $d\left(\left(x_{-i}^0, y_i^0\right), C_i\right) = \min_{z_i\in\varphi_i(x_{-i})} d\left(\left(x_{-i}^0, z_i\right), C_i\right)$; hence, $y_i^0 \in \gamma_i\left(x_{-i}^0\right)$.

Define now the correspondences $\psi_i : X_{-i} \mapsto 2^{X_i}$ by

$$\psi_i(x_{-i}) = \overline{co}\gamma_i(x_{-i})$$

Each ψ_i is nonempty, convex, and compact valued, and by Lemma 1 it has closed graph. Define now the correspondence $\psi : X \mapsto 2^X$ by

$$\psi(x) = \prod_{i=1}^{n} \psi_i(x_{-i}).$$

Since each ψ_i is upper semi-continuous and compact valued, so is ψ. Moreover, one can see that it satisfies all assumptions of the Fan-Glicksberg fixed point theorem [5]. To end the proof, we will prove that every fixed point of ψ is an equilibrium in choice. Let x^* be a game strategy such that $x^* \in \psi(x^*)$, i.e., $x_i^* \in \psi_i(x_{-i}^*)$ for $i = 1, \ldots, n$. Suppose that $(x_{-i}^*, \varphi_i(x_{-i}^*)) \cap C_i \neq \emptyset$, for some $i \in N$. Then, $d((x_{-i}^*, y_i), C_i) = 0$ for some $y_i \in \varphi_i(x_{-i}^*)$. Since C_i is closed, this means that $(x_{-i}^*, y_i) \in C_i$ if and only if $y_i \in \gamma_i(x_{-i}^*)$ so that $\gamma_i(x_{-i}^*) = C_i(x_{-i}^*)$. But $x_i^* \in \psi_i(x_{-i}^*) = \overline{co}\gamma_i(x_{-i}^*)$ and since $\gamma_i(x_{-i}^*)$ is the upper section trough x_{-i}^* of C_i, which is closed and convex, it results that $x_i^* \in \gamma_i(x_{-i}^*)$, and hence, $x^* = (x_{-i}^*, x_i^*) \in C_i$. ■

Remark 2. If the assumption (c) of the theorem is reinforced requiring the nonemptyness of the upper sections of C_i, then the conclusion refers to the s.e.c.

4 Applications to the Classical Models

As a corollary of Theorem 1, we can obtain a slight generalization of Debreu's theorem [4] and a new version of the result of Shafer and Sonnenschein [10]. For the proof of the first theorem, the following lemma will be helpful. In fact, this lemma is a part of the "maximum theorem" ([2] and [1]), but for the sake of completeness a brief proof will be done.

Lemma 2. *Let S, T be a Hausdorff topological vector space, $\varphi : S \mapsto 2^T$ be a correspondence, and $f : S \times T \mapsto \mathbb{R}$ be a real-valued function. If φ is lower semi-continuous with nonempty values and f is lower semi-continuous on $S \times T$, then the mapping $\nu : S \mapsto \overline{\mathbb{R}}$ defined by the equality $\nu(s) = \sup_{t \in \varphi(s)} f(s,t)$ is lower semi-continuous on S, for every fixed $t \in T$.*

Proof. It suffices to verify that the set $S_\alpha = \left\{ s \in S \mid \sup_{t \in \varphi(s)} f(s,t) \leq \alpha \right\}$ is closed, for every real α.

Let $(s^k)_k \subset S_\alpha$ be a convergent sequence, $s^k \underset{k \to \infty}{\longrightarrow} s$. If $t \in \varphi(s)$, then the lower semi-continuity of φ implies the existence of a sequence $(t^k)_k \subset T$, with $t^k \underset{k \to \infty}{\longrightarrow} t$ and $t^k \in \varphi(s^k)$, $k = 1, 2, \ldots$. Then, $f(s^k.t^k) \leq \alpha$ for all k, and the lower semi-continuity of f gives us $f(s,t) \leq \alpha$. Since t is arbitrary in $\varphi(s)$, we have that $\nu(s) \leq \alpha$ that means $s \in S_\alpha$. ■

Theorem 5. *Let $((X_i)_{i \in N}, (\varphi_i)_{i \in N}, (u_i)_{i \in N})$ be a generalized game. Assume that, for each $i \in N$:*

(*a*) *X_i is a nonvoid, convex, and compact set of a locally convex topological vector space.*
(*b*) *φ_i is lower semi-continuous and closed with nonempty convex values.*
(*c*) *u_i is continuous on X.*
(*d*) *$u_i(x_{-i}, .)$ is quasi-concave on X_i, for every $x_{-i} \in X_{-i}$.*

Then, the game has social equilibria.

Proof. As it was shown in Example 1, it suffices to prove that the generalized game in the choice form $((X_i)_{i\in N}, (\varphi_i)_{i\in N}, (C_i)_{i\in N})$ with $C_i = \left\{x \in Gr\varphi_i \mid u_i(x) = \max\limits_{z_i \in \varphi_i(x_{-i})} u_i(x_{-i}, z_i)\right\}$ satisfies the assumptions of Theorem 1.

Observe first that $\max\limits_{z_i \in \varphi_i(x_{-i})} u_i(x_{-i}, z_i)$ is well defined (since $u_i(x_{-i}, .)$, is continuous, and φ_i has nonempty compact values) so that C_i is nonempty. Since $C_i = Gr\varphi_i \cap \left\{x \in X \mid u_i(x) \geq \max\limits_{z_i \in \varphi_i(x_{-i})} u_i(x_{-i}, z_i)\right\}$, it follows from Lemma 2 and the assumptions (b) and (c) that it is closed.

For any $x_{-i} \in X_{-i}$, $C_i(x_{-i}) = \left\{y_i \in \varphi_i(x_{-i}) \mid u_i(x_{-i}, y_i) \geq \max\limits_{z_i \in \varphi_i(x_{-i})} u_i(x_{-i}, z_i)\right\} = \varphi_i(x_{-i}) \cap \left\{y_i \in X_i \mid u_i(x_{-i}, y_i) \geq \max\limits_{z_i \in \varphi_i(x_{-i})} u_i(x_{-i}, z_i)\right\}$; hence, this set is convex as the intersection of two convex sets. ■

For an abstract economy, the existence of the equilibria may be also proved by the means of Theorem 1, if a generalized game in the choice form is associated as in Example 2.

Theorem 6. *Assume that the abstract economy* $((X_i)_{i\in N}, (\varphi_i)_{i\in N}, (\pi_i)_{i\in N})$ *satisfies, for each* $i \in N$*, the following conditions:*

(a) X_i *is a nonvoid, convex, and compact set in a locally convex topological vector space.*
(b) φ_i *is lower semi-continuous and closed with nonempty convex values.*
(c) π_i *has an open graph in* $X \times X_i$*.*
(d) If $x \in Gr\varphi_i$*, then* $x_i \notin co\pi_i(x)$*.*
(e) For each $x_{-i} \in X_{-i}$ *the set* $\{y_i \in X_i \mid \varphi_i(x_{-i}) \cap \pi_i(x_{-i}, y_i) = \emptyset\}$ *is convex.*

Then, there exists at least one strong equilibrium.

Proof. Set $C_i = \{x \in Gr\varphi_i \mid \varphi_i(x_{-i}) \cap \pi_i(x) = \emptyset\} = Gr\varphi_i \cap \{x \in X \mid \varphi_i(x_{-i}) \cap \pi_i(x) = \emptyset\}$, and verify the conditions $(b)-(d)$ of Theorem 1.

To prove the closedness of C_i, let $\left(x^k\right)_k$ be a sequence in C_i, convergent to some x^0. Obviously, $x^0 \in Gr\varphi_i$ because φ_i is closed. Suppose, by way of contradiction, that $x^0 \notin C_i$. This means that there is a $y_i^0 \in \varphi_i\left(x_{-i}^0\right) \cap \pi_i\left(x^0\right)$. Because φ_i is lower semi-continuous, a sequence $\left(y_i^k\right)_k$ can be found such that $y_i^k \underset{k\to\infty}{\longrightarrow} y_i^0$, and $y_i^k \in \varphi_i\left(x_i^k\right)$, $k = 1, 2, \ldots$. On the other hand, $y_i^0 \in \pi_i\left(x^0\right)$, i.e., $\left(x^0, y_i^0\right) \in Gr\pi_i$, and by (c), there exists open neighborhood V_{x^0} and $V_{y_i^0}$ of x^0 and y_i^0, respectively, such that $V_{x^0} \times V_{y_i^0} \subseteq Gr\pi_i$. Hence, $\left(x^k, y_i^k\right) \in Gr\pi_i$ for all but finitely many values of k. Thus, $y_i^k \in \varphi_i\left(x_{-i}^k\right) \cap \pi_i\left(x^k\right) \neq \emptyset$, for infinitely many values of k, contradicting the initial assumption of $\left(x^k\right)_k$.

Now, fix $x_{-i} \in X_{-i}$ and prove that the upper section

$$C_i(x_{-i}) = \varphi_i(x_{-i}) \cap \{y_i \in X_i | \varphi_i(x_{-i}) \cap \pi_i(x_{-i}, y_i) = \emptyset\} \tag{5}$$

of C_i is nonempty.

To the contrary, suppose that $C_i(x_{-i}) = \{y_i \in \varphi_i(x_{-i}) | \varphi_i(x_{-i}) \cap \pi_i(x_{-i}, y_i) = \emptyset\} = \emptyset$, for some x_{-i}. This means that $\varphi_i(x_{-i}) \cap \pi_i(x_{-i}, y_i) \neq \emptyset$ for all $y_i \in \varphi_i(x_{-i})$. Hence, the correspondence $\psi_i : \varphi_i(x_{-i}) \mapsto 2^{\varphi_i(x_{-i})}$, defined by

$$\psi_i(y_i) = \varphi_i(x_{-i}) \cap co\pi_i(x_{-i}, y_i)$$

is nonempty and convex valued. Moreover, for each $z_i \in \varphi_i(x_{-i})$, the lower section $\psi_i^{-1}(z_i) = \{y_i \in \varphi_i(x_{-i}) \mid z_i \in \psi_i(y_i)\}$ is open in $\varphi_i(x_{-i})$. For

$$\psi_i^{-1}(z_i) = \begin{cases} \emptyset, & \text{if } z_i \notin \varphi_i(x_{-i}) \\ \varphi_i(x_{-i}) \cap \{y_i \in X_i | (x_{-i}, y_i, z_i) \in Grco\pi_i\}, & \text{if } z_i \in \varphi_i(x_{-i}) \end{cases}.$$

Since $Gr\pi_i$ is open, so is $Grco\pi_i$, and then, if $y_i \in \psi_i^{-1}(z_i)$, an open neighborhood V_{y_i} of y_i can be found, such that $\{x_{-i}\} \times V_{y_i} \times \{z_i\} \subseteq Grco\pi_i$, i.e., $V_{y_i} \subseteq \psi_i^{-1}(z_i)$.

By Browder's fixed point theorem [3], there exists a fixed point of ψ_i, that is, $\overline{y}_i \in \psi_i(\overline{y}_i)$, for some $\overline{y}_i \in \varphi_i(x_{-i})$. Therefore, $\overline{y}_i \in \varphi_i(x_{-i}) \cap co\pi_i(x_{-i}, \overline{y}_i)$, in contradiction with the assumption (d).

Finally, observe that by (5), and the assumption (e), the set $C_i(x_{-i})$ is convex as the intersection of two convex sets. ■

As corollaries of the main existence result for e.c. (Theorem 4), the following theorems establish the existence of the weaker equilibria in generalized games and abstract economies.

As an immediate consequence of Theorem 4 and of Proposition 1, we obtain sufficient conditions for the existence of w.s.e.

Theorem 7. *Let* $((X_i)_{i\in N}, (\varphi_i)_{i\in N}, (u_i)_{i\in N})$ *be a generalized game. Assume that, for each* $i \in N$*:*

(a) X_i *is a nonvoid, convex, and compact set in a metrizable locally convex space.*
(b) φ_i *is closed with convex (possible empty) values.*
(c) The set $\left\{x \in X \mid u_i(x) = \max_{z_i \in \varphi_i(x_{-i})} u_i(x_{-i}, z_i)\right\}$ *is nonempty and closed.*
(d) The set $\left\{y_i \in X_i \mid u_i(x_{-i}, y_i) = \max_{z_i \in \varphi_i(x_{-i})} u_i(x_{-i}, z_i)\right\}$ *is convex (possible empty), for every* $x_{-i} \in X_{-i}$.

Then, the game has weak social equilibria.

Proof. Condition (b) ensures the closedness of $Gr\varphi_i$ and the convexity of the upper sections of the set C_i defined as in Example 1. Conditions (c) and (d) are the transcriptions of the corresponding conditions of Theorem 4.

Now let *us* consider an abstract economy of Shafer-Sonnenschein type.

Theorem 8. *Assume that the abstract economy* $((X_i)_{i\in N}, (P_i)_{i\in N})$ *satisfies, for each* $i \in N$*, the following conditions:*

(a) X_i *is a nonvoid, convex, and compact set in a metrizable locally convex space.*
(b) φ_i *is closed with convex (possible empty) values.*
(c) The set $\{x \in X \mid \varphi_i(x_{-i}) \cap \pi_i(x) = \emptyset\}$ *is nonempty and closed.*
(d) For each $x_{-i} \in X_{-i}$*, the set* $\{y_i \in X_i \mid \varphi_i(x_{-i}) \cap \pi_i(x_{-i}, y_i) = \emptyset\}$ *is convex (possible empty).*

Then, there exists at least one weak equilibrium.

Proof. Similar with the previous one, invoking Example 2 and Proposition 2. ■

References

1. Aubin, J.-P.: Mathematical Methods of Game Theory and Economic Theory. North Holland, Amsterdam (1982)
2. Berge, C., Espaces Topologiques et Fonctions Multivoques. Dunod, Paris (1959)
3. Browder, E.: The fixed point theory of multi-valued mappings in topological vector spaces. Math. Ann. **177**, 283–301 (1968)
4. Debreu, G.: A social equilibrium existence theorem. Proc. Natl. Acad. Sci. U.S.A. **38**, 386–393 (1952)
5. Fan, K.: Fixed-point and minimax theorems in locally convex topological linear spaces. Proc. Natl. Acad. Sci. U.S.A. **38**, 121–126 (1952)
6. Fan, K.: Sur un théorème minimax. C. R. Acad. Sci. Paris **259**, 3925–3928 (1964)
7. Fan, K.: Some properties of convex sets related to fixed point theorems. Math. Ann. **266**, 519–537 (1984)
8. Nash, J.F.: Equilibrium points in n-person games. Proc. Natl. Acad. Sci. U.S.A. **36**, 48–49 (1950)
9. Nash, J.F.: Noncooperative games. Ann. Math. **54**, 286–295 (1951)
10. Shafer, W., Sonnenschein, H.: Equilibrium in abstract economies without ordered preferences. J. Math. Econ. **2**, 345–348 (1975)
11. Stefanescu, M.V., Stefanescu, A.: On the existence of the equilibrium in choice. Math. Rep. **11**(61), N.3, 249–258 (2009)
12. Stefanescu, A., Ferrara, M., Stefanescu, M.V.: Equilibria of the games in choice form. JOTA **155**, 1060–1072 (2012)

On the Motion Induced by a Flat Plate That Applies Oscillating Shear Stresses to an Oldroyd-B Fluid: Applications

Constantin Fetecau, Corina Fetecau, and Dumitru Vieru

1 Introduction

The interest in flows of non-Newtonian fluids substantially increases due to their wide practical applications. Among the many models that have been used to describe the behavior of these fluids, the Oldroyd-B model is amenable to analysis and more importantly experimental corroboration. This model has had some success in describing the response of some polymeric liquids. The Oldroyd-B fluids store energy like a linear elastic solid, and their dissipation is due to two dissipative mechanisms which imply that they arise from a mixture of two viscous fluids. These fluids, which are characterized by three material constants, can describe stress relaxation, creep, and the normal stress differences that develop during simple shear flows. However, they cannot describe either shear thinning or shear thickening, a phenomenon that is exhibited by many polymeric materials. In spite of all this, the Oldroyd-B model is viewed as one of the most successful models for describing the response of a subclass of polymeric liquids. Flows of Oldroyd-B fluids over an infinite plate are extensively studied in the literature, they being some of the most

C. Fetecau (✉)
Department of Mathematics, Technical University of Iasi, 700050 Iasi, Romania

Academy of Romanian Scientists, 050094 Bucuresti, Romania

Abdus Salam School of Mathematical Sciences, GC University, 54600 Lahore, Pakistan
e-mail: c_fetecau@yahoo.com

C. Fetecau
Department of Theoretical Mechanics, Technical University of Iasi, 700050 Iasi, Romania

D. Vieru
Abdus Salam School of Mathematical Sciences, GC University, 54600 Lahore, Pakistan

Department of Theoretical Mechanics, Technical University of Iasi, 700050 Iasi, Romania

P. Cartier et al. (eds.), *Mathematics in the 21st Century*, Springer Proceedings in Mathematics & Statistics 98, DOI 10.1007/978-3-0348-0859-0_4

important motion problems near moving bodies. However, the most part of them corresponds to motion problems for which the velocity is given on the boundary, although in some practical problems, what is specified is the shear stress on the boundary. Furthermore, the "no slip" boundary condition may not be necessarily applicable for flows of polymeric fluids that can slip or slide on the boundary. The first exact solutions for motions of Oldroyd-B fluids in which the shear stress is given on a part of the boundary seem to be those of Waters and King [1]. Later, Tong et al. [2, 3] claim exact solutions corresponding to constant shear stresses on a cylindrical boundary. However, their solutions, as well as those obtained in [4] for the motion over an infinite plate, correspond to a shear stress of the form $f\left(1-e^{-\frac{t}{\lambda}}\right)$ on the boundary.

In the last years, similar solutions seem to be established in the literature for rate-type fluids (see [5–9] and the references therein). However, it is worth pointing out that all these solutions, except their reduced forms for Newtonian and second-grade fluids, correspond to motion problems in which differential expressions of the shear stress are given on the boundary. This is due to the governing equations corresponding to rate-type fluids. These equations, unlike those corresponding to Newtonian and second-grade fluids, contain differential expressions acting on the shear stresses. Due to this fact, one does not exist in the literature exact solutions corresponding to motions of rate-type fluids induced by a solid boundary that applies oscillating shear stresses on the fluid. Such solutions are known for Newtonian and second-grade fluids only [10–16].

The purpose of this work is to provide exact solutions corresponding to the unsteady motions due to an infinite plate that applies oscillating shear stresses to an Oldroyd-B fluid. These solutions, presented as a sum between the steady-state and transient solutions, can easily be particularized to give similar solutions for Maxwell and Newtonian fluids performing the same motion. They can be also used to develop exact solutions for the motion produced by an oscillating plate. For a check of results that have been obtained, the known solution corresponding to the motion induced by an infinite plate that applies a constant shear stress to the fluid is obtained as a limiting case of the cosine solution. Moreover, the steady-state solutions for Newtonian fluids are also recovered as special cases of general solutions. Finally, two applications concerning the motion induced by an infinite plate that is oscillating in its plane are included.

2 Governing Equations

The Cauchy stress tensor $\mathbf{T}$ for an incompressible Oldroyd-B fluid is given by

$$\mathbf{T}=-p\mathbf{I}+\mathbf{S},\ \mathbf{S}+\lambda\left(\frac{d\mathbf{S}}{dt}-\mathbf{LS}-\mathbf{SL}^T\right)=\mu\left[\mathbf{A}+\lambda_r\left(\frac{d\mathbf{A}}{dt}-\mathbf{LA}-\mathbf{AL}^T\right)\right], \quad (1)$$

where $\mathbf{S}$ is the extra-stress tensor, $-p\mathbf{I}$ denotes the indeterminate spherical stress, $\mathbf{L}$ is the velocity gradient, $\mathbf{A}=\mathbf{L}+\mathbf{L}^T$ is the first Rivlin-Ericksen tensor, μ is the dynamic viscosity, λ and $\lambda_r(\leq \lambda)$ are relaxation and retardation times, and d/dt indicate the material time derivative. This model includes as special cases the Maxwell and the linearly viscous fluid models for $\lambda_r = 0$, respectively, $\lambda = \lambda_r = 0$. In the next, we shall study motion problems whose velocity field is of the form

$$\mathbf{v} = \mathbf{v}(y,t) = u(y,t)\mathbf{i}, \tag{2}$$

where $\mathbf{i}$ is the unit vector along the x-direction of a fixed Cartesian coordinate system x, y, and z. For such flows the constraint of incompressibility is automatically satisfied. Assuming that the extra-stress tensor $\mathbf{S}$, as well as the velocity $\mathbf{v}$, is a function of y and t only, it follows from $(1)_2$ and (2) that

$$\left(1+\lambda\frac{\partial}{\partial t}\right)\tau(y,t) = \lambda S_{yy}(y,t)\frac{\partial u(y,t)}{\partial y} + \mu\left(1+\lambda_r\frac{\partial}{\partial t}\right)\frac{\partial u(y,t)}{\partial y},$$

$$\left(1+\lambda\frac{\partial}{\partial t}\right)S_{yy}(y,t) = 0, \tag{3}$$

where $\tau(y,t) = S_{xy}(y,t)$ is the nontrivial shear stress. If the fluid is at rest up to the moment $t = 0$, this is

$$\mathbf{v}(y,0) = \mathbf{0},\ \ \mathbf{S}(y,0) = \mathbf{0}, \tag{4}$$

then the normal stress $S_{yy}(y,t)$ is identically null and Eq. $(3)_1$ becomes

$$\left(1+\lambda\frac{\partial}{\partial t}\right)\tau(y,t) = \mu\left(1+\lambda_r\frac{\partial}{\partial t}\right)\frac{\partial u(y,t)}{\partial y}. \tag{5}$$

Substituting (2) into the balance of linear momentum, in the absence of body forces and a pressure gradient in the flow direction, we obtain the relevant equation

$$\frac{\partial \tau(y,t)}{\partial y} = \rho\frac{\partial u(y,t)}{\partial t} \tag{6}$$

where ρ is the constant density of the fluid. Usually, in the literature, the shear stress $\tau(y,t)$ is eliminated between Eqs. (5) and (6) in order to obtain the governing equation for the velocity $u(y,t)$. However, in order to solve a well-posed shear stress boundary-value problem for rate-type fluids, we follow [17] and eliminate the velocity between the same equations. The surprising result

$$\left(1+\lambda\frac{\partial}{\partial t}\right)\frac{\partial \tau(y,t)}{\partial t} = \nu\left(1+\lambda_r\frac{\partial}{\partial t}\right)\frac{\partial^2 \tau(y,t)}{\partial y^2}, \tag{7}$$

where $\nu = \frac{\mu}{\rho}$ is the kinematic viscosity, shows that for such motions, the shear stress $\tau(y,t)$ satisfies a partial differential equation of the same form as velocity. It allows us both to solve motion problems with a given shear stress on the boundary and to develop new exact solutions for usually boundary-value problems.

3 Motion Due to an Infinite Plate that Applies Oscillating Shears to the Fluid

Let us consider an Oldroyd-B fluid at rest over an infinite flat plate situated in the (x, z) plane. After time $t = 0^+$ the plate applies an oscillating shear stress $f\cos(\omega t)$ or $f\sin(\omega t)$ to the fluid (f is a constant and ω is the frequency of oscillations). Owing to the shear, the fluid is gradually moved and its velocity is of the form (2). The governing equation for shear stress is given by Eq. (7), while the appropriate initial and boundary conditions are

$$u(y,0) = 0,\ \tau(y,0) = 0,\ \left.\frac{\partial \tau(y,t)}{\partial t}\right|_{t=0} = 0;\ y > 0, \tag{8}$$

$$\tau(0,t) = fH(t)\cos(\omega t), \quad \text{or} \quad \tau(0,t) = f\sin(\omega t);\ t \geq 0, \tag{9}$$

$$u(y,t) \longrightarrow 0,\ \tau(y,t) \longrightarrow 0,\ \frac{\partial \tau(y,t)}{\partial y} \longrightarrow 0 \ \text{ as } \ y \longrightarrow \infty, \tag{10}$$

where $H(.)$ is the Heaviside unit step function. In order to obtain exact solutions for the partial differential equations (6) and (7) with the initial and boundary conditions (8)–(10), we shall use the Laplace and Fourier sine transforms. Introducing the following nondimensional variables and functions:

$$y^* = \frac{y}{\sqrt{\nu t_0}}, t^* = \frac{t}{t_0}, \lambda^* = \frac{\lambda}{t_0}, \lambda_r^* = \frac{\lambda_r}{t_0}, \omega^* = \omega t_0,$$
$$u^* = \frac{u}{\sqrt{\frac{\nu}{t_0}}}, \tau^* = \frac{\tau}{f}, \rho^* = \frac{\mu}{f t_0}, \tag{11}$$

where $t_0 > 0$ is a characteristic time, and dropping out the star notations, we attain to the next dimensionless initial and boundary-value problems for $\tau(y,t)$

$$\left(1 + \lambda\frac{\partial}{\partial t}\right)\frac{\partial \tau(y,t)}{\partial t} = \left(1 + \lambda_r\frac{\partial}{\partial t}\right)\frac{\partial^2 \tau(y,t)}{\partial y^2}; y, t > 0, \tag{12}$$

$$\tau(y,0) = 0,\ \left.\frac{\partial \tau(y,t)}{\partial t}\right|_{t=0} = 0;\ y > 0, \tag{13}$$

$$\tau(0,t) = H(t)\cos(\omega t),\ \text{or}\ \tau(0,t) = \sin(\omega t);\ t \geq 0, \tag{14}$$

$$\tau(y,t) \ \text{ and } \ \frac{\partial \tau(y,t)}{\partial y} \longrightarrow 0 \ \text{ as } \ y \longrightarrow \infty, \tag{15}$$

respectively $u(y,t)$

$$\rho\frac{\partial u(y,t)}{\partial t} = \frac{\partial \tau(y,t)}{\partial y}; y,t > 0 \tag{16}$$

$$u(y,0) = 0, \quad \text{if} \quad y > 0 \quad \text{and} \quad u(y,t) \longrightarrow 0 \quad \text{as} \quad y \longrightarrow \infty. \tag{17}$$

3.1 Calculation of the Shear Stress $\tau(y,t)$

Let us denote by $\tau_c(y,t)$ and $\tau_s(y,t)$ the solutions corresponding to the first initial and boundary-value problems and by

$$T(y,t) = \tau_c(y,t) + i\tau_s(y,t) \tag{18}$$

the complex shear stress where i is the imaginary unit. According to Eqs. (12)–(15), the function $T(.,.)$ has to be the solution of the following problem:

$$\left(1+\lambda\frac{\partial}{\partial t}\right)\frac{\partial T(y,t)}{\partial t} = \left(1+\lambda_r\frac{\partial}{\partial t}\right)\frac{\partial^2 T(y,t)}{\partial y^2}; \quad y,t > 0, \tag{19}$$

$$T(y,0) = 0, \quad \left.\frac{\partial T(y,t)}{\partial t}\right|_{t=0} = 0; \quad y > 0, \tag{20}$$

$$T(0,t) = H(t)e^{i\omega t}, \quad \text{if} \quad t \geq 0 \quad \text{and} \quad T(y,t), \ \frac{\partial T(y,t)}{\partial y} \to 0 \quad \text{as} \quad y \to \infty. \tag{21}$$

Applying the Laplace transform to Eqs. (19) and (21) and using (20), we find that

$$(1+\lambda q)q\overline{T}(y,q) = (1+\lambda_r q)\frac{\partial^2 \overline{T}(y,q)}{\partial y^2}; \quad y > 0, \tag{22}$$

$$\overline{T}(0,q) = \frac{1}{q-i\omega} \quad \text{and} \quad \overline{T}(y,q), \ \frac{\partial \overline{T}(y,q)}{\partial y} \to 0 \quad \text{as} \quad y \to \infty, \tag{23}$$

where $\overline{T}(y,q)$ is the Laplace transform of $T(y,t)$ and q is the transform parameter. Now, we apply the Fourier sine transform (see, for instance, [18, Eq. (36) of the subsection 3.3]) to Eq. (22) and use the boundary conditions (23). It results that the Fourier sine transform $\overline{T}_s(\xi,q)$ of $\overline{T}(y,q)$ is given by

$$\overline{T}_s(\xi,q) = \xi\sqrt{\frac{2}{\pi}}\frac{1}{q-i\omega}\frac{\lambda_r q+1}{\lambda q^2+(1+\lambda_r\xi^2)q+\xi^2}, \tag{24}$$

or equivalently

$$\overline{T}_s(\xi,q)=\frac{\xi}{\lambda}\sqrt{\frac{2}{\pi}}\frac{1}{q-i\omega}$$
$$\times\left\{\lambda_r\frac{q+a(\xi)}{[q+a(\xi)]^2-b^2(\xi)}+\frac{1-\lambda_r a(\xi)}{b(\xi)}\frac{b(\xi)}{[q+a(\xi)]^2-b^2(\xi)}\right\},\quad (25)$$

$$\text{where } a(\xi)=\frac{1+\lambda_r\xi^2}{2\lambda}\quad\text{and}\quad b(\xi)=\frac{\sqrt{\left(1+\lambda_r\xi^2\right)^2-4\lambda\xi^2}}{2\lambda}.$$

By applying the inverse Laplace transform to Eq. (25) and using the convolution theorem, we find that

$$T_s(\xi,t)=\frac{\xi}{\lambda}\sqrt{\frac{2}{\pi}}\int_0^t e^{i\omega(t-s)}e^{-a(\xi)s}\left\{\lambda_r ch[b(\xi)s]+\frac{1-\lambda_r a(\xi)}{b(\xi)}sh[b(\xi)s]\right\}ds. \quad (26)$$

Denoting by $I(\xi,t)$ and $J(\xi,t)$ the integrals

$$I(\xi,t)=\int_0^t e^{-[i\omega+a(\xi)]s}ch[b(\xi)s]ds,\quad J(\xi,t)=\int_0^t e^{-[i\omega+a(\xi)]s}sh[b(\xi)s]ds,$$

and making all calculi, it results that

$$I(\xi,t)=\frac{\lambda\left[\lambda\omega^2-\xi^2+i\omega\left(1+\lambda_r\xi^2\right)\right]}{\left(\xi^2-\lambda\omega^2\right)^2+\omega^2\left(1+\lambda_r\xi^2\right)^2}$$
$$\times\left\{[b(\xi)sh[b(\xi)t]-[a(\xi)+i\omega]ch[b(\xi)t]]e^{-i\omega t-a(\xi)t}\right.$$
$$\left.-[a(\xi)+i\omega]\right\},\quad (27)$$

$$J(\xi,t)=\frac{\lambda\left[\lambda\omega^2-\xi^2+i\omega\left(1+\lambda_r\xi^2\right)\right]}{\left(\xi^2-\lambda\omega^2\right)^2+\omega^2\left(1+\lambda_r\xi^2\right)^2}$$
$$\times\left\{[b(\xi)ch[b(\xi)t]+[a(\xi)+i\omega]sh[b(\xi)t]]e^{-i\omega t-a(\xi)t}-b(\xi)\right\}.\quad (28)$$

Lengthy but straightforward computations show that

$$\mathrm{Re}\left\{e^{i\omega t}I(\xi,t)\right\}=\frac{1}{2\gamma\left[\left(\xi^2-\alpha^2\right)^2+\beta^2\right]}\left\{\left[\left(\lambda\omega^2-\xi^2\right)c(\xi)sh[b(\xi)t]\right.\right.$$
$$\left.-\left(\xi^2+\lambda\omega^2\right)\left(1+\lambda_r\xi^2\right)ch[b(\xi)t]\right]e^{-a(\xi)t}$$
$$+\left(\xi^2+\lambda\omega^2\right)\left(1+\lambda_r\xi^2\right)\cos(\omega t)$$
$$\left.+\omega\left[\lambda_r^2\xi^4+2(\lambda-\lambda_r)\xi^2+2\lambda^2\omega^2+1\right]\sin(\omega t)\right\}\quad (29)$$

$$\operatorname{Im}\left\{e^{i\omega t}I(\xi,t)\right\} = \frac{1}{2\gamma\left[(\xi^2-\alpha^2)^2+\beta^2\right]}$$
$$\times\Big\{\big[\omega\big[\lambda_r^2\xi^4+2(\lambda-\lambda_r)\xi^2+2\lambda^2\omega^2+1\big]ch[b(\xi)t]$$
$$+\omega\left(1+\lambda_r\xi^2\right)c(\xi)sh[b(\xi)t]\big]e^{-a(\xi)t}+\left(\xi^2+\lambda\omega^2\right)\left(1+\lambda_r\xi^2\right)\sin(\omega t)$$
$$-\omega\big[\lambda_r^2\xi^4-2(\lambda-\lambda_r)\xi^2+2\lambda^2\omega^2+1\big]\cos(\omega t)\Big\}, \tag{30}$$

$$\operatorname{Re}\left\{e^{i\omega t}J(\xi,t)\right\} = \frac{1}{2\gamma\left[(\xi^2-\alpha^2)^2+\beta^2\right]}$$
$$\times\Big\{\big[\left(\lambda\omega^2-\xi^2\right)c(\xi)ch[b(\xi)t]-\left(\xi^2+\lambda\omega^2\right)\left(1+\lambda_r\xi^2\right)sh[b(\xi)t]\big]e^{-a(\xi)t}$$
$$+\left(\xi^2-\lambda\omega^2\right)c(\xi)\cos(\omega t)+\omega\left(1+\lambda_r\xi^2\right)c(\xi)\sin(\omega t)\Big\}, \tag{31}$$

$$\operatorname{Im}\left\{e^{i\omega t}J(\xi,t)\right\} = \frac{1}{2\gamma\left[(\xi^2-\alpha^2)^2+\beta^2\right]}$$
$$\times\Big\{\Big[\omega\left[\left(1+\lambda_r\xi^2\right)^2+2\lambda\left(\lambda\omega^2-\xi^2\right)\right]sh[b(\xi)t]+\omega\left(1+\lambda_r\xi^2\right)$$
$$\times c(\xi)ch[b(\xi)t]\Big]e^{-a(\xi)t}-\omega\left(1+\lambda_r\xi^2\right)c(\xi)\cos(\omega t)$$
$$-\left(\lambda\omega^2-\xi^2\right)c(\xi)\sin(\omega t)\Big\}, \tag{32}$$

where Re and Im denote the real and imaginary parts of that which follows and

$$\alpha^2=\frac{\omega^2(\lambda-\lambda_r)}{1+\lambda_r^2\omega^2},\quad \beta=\frac{\omega\left(1+\lambda\lambda_r\omega^2\right)}{1+\lambda_r^2\omega^2},\quad \gamma=1+\lambda_r^2\omega^2\quad,$$
$$\textit{and}\quad c(\xi)=\sqrt{(1+\lambda_r\xi^2)^2-4\lambda\xi^2}.$$

Introducing the results from (27) and (28) into Eq. (26) and applying the inverse Fourier sine transform, we obtain the expression of the complex shear stress $T(y,t)$. Finally, by taking the real and imaginary parts of this expression and bearing in mind Eqs. (29)–(32), we find for $\tau_c(y,t)$ and $\tau_s(y,t)$ the simpler forms

$$\tau_c(y,t) = \frac{2}{\gamma\pi} H(t) \int_0^\infty \left\{ \frac{A(\xi)}{c(\xi)} sh[b(\xi)t] + B(\xi) ch[b(\xi)t] \right\} e^{-a(\xi)t} \sin(y\xi) d\xi$$
$$+ \frac{2}{\pi} H(t) \cos(\omega t) \int_0^\infty \frac{\xi\left(\xi^2 - \alpha^2\right)}{\left(\xi^2 - \alpha^2\right)^2 + \beta^2} \sin(y\xi) d\xi$$
$$+ \frac{2\beta}{\pi} H(t) \sin(\omega t) \int_0^\infty \frac{\xi \sin(y\xi)}{\left(\xi^2 - \alpha^2\right)^2 + \beta^2} d\xi, \tag{33}$$

respectively,

$$\tau_s(y,t) = \frac{2}{\pi}\frac{\omega}{\gamma} H(t) \int_0^\infty \left\{ \frac{C(\xi)}{c(\xi)} sh[b(\xi)t] + D(\xi) ch[b(\xi)t] \right\} e^{-a(\xi)t} \sin(y\xi) d\xi$$
$$+ \frac{2}{\pi} H(t) \sin(\omega t) \int_0^\infty \frac{\xi\left(\xi^2 - \alpha^2\right)}{\left(\xi^2 - \alpha^2\right)^2 + \beta^2} \sin(y\xi) d\xi$$
$$- \frac{2\beta}{\pi} H(t) \cos(\omega t) \int_0^\infty \frac{\xi \sin(y\xi)}{\left(\xi^2 - \alpha^2\right)^2 + \beta^2} d\xi, \tag{34}$$

where

$$A(\xi) = \frac{\{\gamma\lambda_r\xi^4 - [1+\lambda\lambda_r\omega^2+2\lambda_r\omega^2(\lambda-\lambda_r)]\xi^2 - (\lambda-\lambda_r)\omega^2\}\xi}{(\xi^2-\alpha^2)^2+\beta^2},\ B(\xi) = \frac{-\gamma\xi^3+(\lambda-\lambda_r)\omega^2\xi}{\left(\xi^2-\alpha^2\right)^2+\beta^2},$$
$$C(\xi) = \frac{\{-[\lambda\gamma+(\lambda-\lambda_r)]\xi+\lambda\ \omega^2(\lambda-\lambda_r)+\gamma\}\xi}{\left(\xi^2-\alpha^2\right)^2+\beta^2},\ D(\xi) = \frac{\left(1+\lambda\lambda_r\omega^2\right)\xi}{\left(\xi^2-\alpha^2\right)^2+\beta^2}.$$

The starting solutions $\tau_c(y,t)$ and $\tau_s(y,t)$, as it was to be expected, are presented as a sum of the steady-state and transient solutions. Such solutions are very important for those who want to eliminate the transients from their rheological measurements. They describe the motion of the fluid some time after its initiation. After that time, when the transients disappear, the starting solutions tend to the steady-state solutions

$$\tau_{cs}(y,t) = \frac{2}{\pi} \cos(\omega t) \int_0^\infty \frac{\xi\left(\xi^2 - \alpha^2\right)}{\left(\xi^2 - \alpha^2\right)^2 + \beta^2} \sin(y\xi) d\xi$$
$$+ \frac{2\beta}{\pi} \sin(\omega t) \int_0^\infty \frac{\xi \sin(y\xi)}{\left(\xi^2 - \alpha^2\right)^2 + \beta^2} d\xi, \tag{35}$$

$$\tau_{ss}(y,t) = \frac{2}{\pi}\sin(\omega t)\int_0^\infty \frac{\xi\left(\xi^2-\alpha^2\right)}{\left(\xi^2-\alpha^2\right)^2+\beta^2}\sin(y\xi)d\xi$$

$$-\frac{2\beta}{\pi}\cos(\omega t)\int_0^\infty \frac{\xi\sin(y\xi)}{\left(\xi^2-\alpha^2\right)^2+\beta^2}d\xi, \tag{36}$$

which are periodic in time and independent of the initial conditions. However, they satisfy the governing equation and boundary conditions.

It is worth pointing out that in the view of the known results (see, for instance, [19, Eqs. (3.731)$_{2,4}$])

$$\int_0^\infty \frac{\xi\sin(y\xi)}{\left(\xi^2-\alpha^2\right)^2+\beta^2}d\xi = \frac{\pi}{2\beta}e^{-my}\sin(ny),$$

$$\int_0^\infty \frac{\xi\left(\xi^2-\alpha^2\right)\sin(y\xi)}{\left(\xi^2-\alpha^2\right)^2+\beta^2}d\xi = \frac{\pi}{2}e^{-my}\cos(ny),$$

where $2m^2 = \sqrt{\alpha^4+\beta^2}-\alpha^2$ and $2n^2 = \sqrt{\alpha^4+\beta^2}+\alpha^2$, the steady-state solutions $\tau_{cs}(y,t)$ and $\tau_{ss}(y,t)$ can be written in the simple forms

$$\tau_{cs}(y,t) = e^{-my}\cos(\omega t - ny),\ \ \tau_{ss}(y,t) = e^{-my}\sin(\omega t - ny). \tag{37}$$

These solutions, as it results from Eqs. (35), (36), or (37), differ by a phase shift. Of course, this is not true for the transient solutions.

3.2 *Calculation of the Velocity* $u(y,t)$

Introducing Eqs. (33) and (34) into Eq. (16), integrating with respect to time from 0 to t, and bearing in mind the conditions (17), we can determine the velocity field corresponding to the two motions. The steady-state components, as they result using Eq. (37), are given by

$$u_{cs}(y,t) = \frac{\sqrt{m^2+n^2}}{\rho\omega}e^{-my}\sin(\omega t - ny + \varphi)$$

$$= \frac{\sqrt{m^2+n^2}}{\rho\omega}e^{-my}\cos\left(\omega t - ny + \varphi + \frac{\pi}{2}\right),$$

$$u_{ss}(y,t) = \frac{\sqrt{m^2+n^2}}{\rho\omega}e^{-my}\cos(\omega t - ny + \varphi), \tag{38}$$

where $tg\varphi = \frac{n}{m}$. The steady-state components of velocity, as expected, also differ by the same phase shift.

3.3 Limiting Case $\omega \to 0$

As a check of our results, let us make $\omega \to 0$ into Eq. (33). In this case, $\alpha = \beta = 0$, $\gamma = 1$, $B(\xi) = \frac{-1}{\xi}$, $A(\xi) = \left(\lambda_r \xi^2 - 1\right) \xi$, and the obtained result

$$\tau(y,t) = H(t)\left\{1 - \frac{2}{\pi}\int_0^\infty [ch[b(\xi)t] + \frac{1-\lambda_r\xi^2}{c(\xi)} sh[b(\xi)t]]e^{-a(\xi)t}\right\}\frac{\sin(y\xi)}{\xi}d\xi, \quad (39)$$

is identical to that obtained in [17, Eq. (24)] (in the dimensional form) and corresponds to the motion induced by an infinite plate that applies a constant shear stress to the fluid.

3.4 Special Cases (Maxwell and Newtonian Fluids)

Finally, it is worth pointing out that making $\lambda_r = 0$ into previous solutions, the corresponding solutions for Maxwell fluids performing the same motions are obtained. Furthermore, by making $\lambda \to 0$ into these last results, the solutions for Newtonian fluids are recovered. The steady-state solutions (37) and (38), for instance, take the simple forms (see [20, Eqs. (20)–(23)] for their dimensional expressions).

$$\tau_{cs}(y,t) = \exp\left(-y\sqrt{\frac{\omega}{2}}\right)\cos\left(\omega t - y\sqrt{\frac{\omega}{2}}\right),$$
$$\tau_{ss}(y,t) = \exp\left(-y\sqrt{\frac{\omega}{2}}\right)\sin\left(\omega t - y\sqrt{\frac{\omega}{2}}\right), \quad (40)$$

$$u_{cs}(y,t) = \frac{1}{\rho\sqrt{\omega}}\exp\left(-y\sqrt{\frac{\omega}{2}}\right)\cos\left(\omega t - y\sqrt{\frac{\omega}{2}} + \frac{3\pi}{4}\right),$$
$$u_{ss}(y,t) = \frac{1}{\rho\sqrt{\omega}}\exp\left(-y\sqrt{\frac{\omega}{2}}\right)\sin\left(\omega t - y\sqrt{\frac{\omega}{2}} + \frac{\pi}{4}\right), \quad (41)$$

while Eq. (39) also reduces to the known result [4, Eq. (4.5)]

$$\tau(y,t) = \operatorname{erfc}\left(\frac{y}{2\sqrt{t}}\right), \quad (42)$$

where erfc(.) is the complementary error function.

4 Applications (Stokes' Problems)

Let us now consider an Oldroyd-B fluid at rest over an infinite flat plate. At time $t = 0^+$, the plate begins to oscillate in its plane and the fluid is gradually moved. Its velocity, as well as the governing equations, is of the same form as before. By eliminating the shear stress $\tau(y,t)$ between Eqs. (5) and (6), we get the governing equation

$$\left(1+\lambda\frac{\partial}{\partial t}\right)\frac{\partial u(y,t)}{\partial t} = \nu\left(1+\lambda_r\frac{\partial}{\partial t}\right)\frac{\partial^2 u(y,t)}{\partial y^2};\, y,t>0, \tag{43}$$

for velocity. This equation, together with the initial and boundary conditions

$$u(y,0)=0,\quad \left.\frac{\partial u(y,t)}{\partial t}\right|_{t=0} = 0;\ \ y>0, \tag{44}$$

$$u(0,t) = UH(t)\cos(\omega t),\qquad \text{or}\qquad u(0,t) = U\sin(\omega t);\ \ t\geq 0, \tag{45}$$

$$u(y,t),\ \frac{\partial u(y,t)}{\partial y} \longrightarrow 0\quad \text{as}\quad y\longrightarrow\infty, \tag{46}$$

leads to an initial and boundary-value problem of the same form as that for the shear stress $\tau(y,t)$. Using the same dimensionless variables and functions as before (for u^* we can also take $\frac{u}{U}$) and bearing in mind the previous results, we find the velocity fields corresponding to the second problem of Stokes

$$\begin{aligned} u_c(y,t) = {} & \frac{2}{\gamma\pi}H(t)\int_0^\infty\left\{\frac{A(\xi)}{c(\xi)}sh[b(\xi)t] + B(\xi)ch[b(\xi)t]\right\}e^{-a(\xi)t}\sin(y\xi)d\xi \\ & +\frac{2}{\pi}H(t)\cos(\omega t)\int_0^\infty \frac{\xi\left(\xi^2-\alpha^2\right)}{\left(\xi^2-\alpha^2\right)^2+\beta^2}\sin(y\xi)d\xi \\ & +\frac{2\beta}{\pi}H(t)\sin(\omega t)\int_0^\infty\frac{\xi\sin(y\xi)}{\left(\xi^2-\alpha^2\right)^2+\beta^2}d\xi, \end{aligned} \tag{47}$$

and

$$\begin{aligned} u_s(y,t) = {} & \frac{2\omega}{\gamma\pi}H(t)\int_0^\infty\left\{\frac{C(\xi)}{c(\xi)}sh[b(\xi)t] + D(\xi)ch[b(\xi)t]\right\}e^{-a(\xi)t}\sin(y\xi)d\xi \\ & +\frac{2}{\pi}H(t)\sin(\omega t)\int_0^\infty \frac{\xi\left(\xi^2-\alpha^2\right)}{\left(\xi^2-\alpha^2\right)^2+\beta^2}\sin(y\xi)d\xi \\ & -\frac{2\beta}{\pi}H(t)\cos(\omega t)\int_0^\infty\frac{\xi\sin(y\xi)}{\left(\xi^2-\alpha^2\right)^2+\beta^2}d\xi. \end{aligned} \tag{48}$$

The steady-state solutions as it results from Eqs. (37) have the simple forms

$$u_{cs}(y,t) = e^{-my}\cos(\omega t - ny),\ \ u_{ss}(y,t) = e^{-my}\sin(\omega t - ny). \tag{49}$$

The Newtonian solutions, as it was to be expected, take the simple forms

$$u_{cs}(y,t) = \exp\left(-y\sqrt{\frac{\omega}{2}}\right)\cos\left(\omega t - y\sqrt{\frac{\omega}{2}}\right),$$
$$u_{ss}(y,t) = \exp\left(-y\sqrt{\frac{\omega}{2}}\right)\sin\left(\omega t - y\sqrt{\frac{\omega}{2}}\right), \tag{50}$$

obtained by Erdogan [21, Eqs. (12) and (17)].

The dimensionless velocity field corresponding to the first problem of Stokes, namely,

$$u(y,t) = H(t)\left\{1 - \frac{2}{\pi}\int_0^\infty\left[ch[b(\xi)t] + \frac{1-\lambda_r\xi^2}{c(\xi)}sh[b(\xi)t]\right]e^{-a(\xi)t}\right\}$$
$$\times\frac{\sin(y\xi)}{\xi}d\xi, \tag{51}$$

is directly obtained from Eq. (39) by putting $u(y,t)$ instead of $\tau(y,t)$. It is identical to that obtained by Christov and Jordan [22] by a different technique.

5 Conclusions

The motion of an Oldroyd-B fluid due to an infinite plate that applies oscillating shear stresses to the fluid is studied by means of integral transforms. Such a problem is not studied in the literature for rate-type fluids, and exact solutions for the nontrivial shear stress $\tau(y,t)$ have been established using a simple but important remark regarding the governing equations. More precisely, the velocity $u(y,t)$ and the shear stress $\tau(y,t)$ corresponding to unsteady unidirectional motions of Oldroyd-B fluids over an infinite plate satisfy partial differential equations of the same form. This remark allows us not only to solve any shear stress boundary-value problem but also to develop new exact solutions using known results from the literature.

In the present work, the starting solutions (33) and (34) are used to develop exact solutions for the motion of Oldroyd-B fluids due to an infinite plate that is oscillating in its plane. These solutions, presented as a sum of steady-state and transient solutions, are important for those that want to eliminate the transients from their experiments. Consequently, an important problem regarding the technical relevance of these solutions is to find the approximate time after which the fluid flows according to the steady-state solutions. This time can be graphically determined by comparing the steady-state and starting solutions.

As a check of general results and based on the above-mentioned remark, the solution corresponding to the first problem of Stokes (see [22, Eq. (6)] or [23, Eq. (4.1)]) is obtained as a limiting case of the present solution (33). All solutions can immediately be particularized to the similar solutions for Maxwell and Newtonian fluids, and some known solutions from the literature are recovered as special cases of the present results. Furthermore, in view of the same remark regarding the governing equations corresponding to the shear stress and velocity, the solution (3.10) from [24] has to be corrected according to the present solution (47). This is not valid for the reduced forms (3.16) and (3.19) corresponding to Maxwell and Newtonian fluids, they being correct.

References

1. Waters, N.D., King, M.J.: Unsteady flow of an elastico-viscous liquid. Rheol. Acta **9**, 345–355 (1970)
2. Tong, D.K., Liu, Y.S.: Exact solutions for the unsteady rotational flow of non-Newtonian fluid in an annular pipe. Int. J. Eng. Sci. **43**, 281–289 (2005)
3. Tong, D.K., Wang, R.H., Yang, H.: Exact solutions for the flow of non-Newtonian fluid with fractional derivative in an annular pipe. Sci. China Ser. G **48**, 485–495 (2005)
4. Fetecau, C., Kannan, K.: A note on an unsteady flow of an Oldroyd-B fluid. Int. J. Math. Math. Sci. **19**, 3185–3194 (2005)
5. Vieru, D., Fetecau, C., Fetecau, C.: Unsteady flow of a generalized Oldroyd-B fluid due to an infiite plate subject to a time-dependent shear stress. Can. J. Phys. **88**, 675–687 (2010)
6. Jamil, M., Fetecau, C.: Some exact solutions for rotating flows of a generalized Burgers' fluid in cylindrical domains. J. Non-Newtonian Fluid Mech. **165**, 1700–1712 (2010)
7. Fetecau, C., Imran, M., Fetecau, C.: Taylor-Couette flow of an Oldroyd-B fluid in an annulus due to a time-dependent couple. Z. Naturforsch. **66a**, 40–46 (2011)
8. Jamil, M., Fetecau, C., Imran, M.: Unsteady helical flows of Oldroyd-B fluids. Commun. Nonlinear Sci. Numer. Simul. **16**, 1378–1386 (2011)
9. Jamil, M., Fetecau, C.: Starting solutions for the motion of a generalized Burgers' fluid between coaxial cylinders. Bound. Value Probl. **2012**, 14 (2012)
10. Bandelli, R., Rajagopal, K.R., Galdi, G.P.: On some unsteady motions of fluids of second grade. Arch. Mech. **47**, 661–676 (1995)
11. Bandelli, R., Rajagopal, K.R.: Start-up flows of second grade fluids in domains with one finite dimension. Int. J. Non-Linear Mech. **30**, 817–839 (1995)
12. Erdogan, M.E.: On unsteady motion of a second grade fluid over a plane wall. Int. J. Non-Linear Mech. **38**, 1045–1051 (2003)
13. Yao, Y., Liu, Y.: Some unsteady flows of second grade fluid over a plane wall. Nonlinear Anal: Real World Appl. **11**, 4302–4311 (2010)
14. Vieru, D., Fetecau, C., Sohail, A.: Flow due to a plate that applies an accelerated shear to a second grade fluid between two parallel walls perpendicular to the plate. Z. Angew. Mat. Phys. **62**, 161–172 (2011)
15. Fetecau, C., Fetecau, C., Rana, M.: General solutions for the unsteady flow of second-grade fluids over an infinite plate that applies arbitrary shear to the fluid. Z. Naturforsch. **66a**, 753–759 (2011)

16. Jamil, M., Rauf, A., Fetecau, C., Khan, N.A.: Helical flows of second grade fluid due to constantly accelerated shear stress. Commun. Nonlinear Sci. Numer. Simul. **16**, 1959–1969 (2011)
17. Fetecau, C., Rubbab, Q., Akhter, S., Fetecau, C.: New method to provide exact solutions for some unidirectional motions of rate type fluids. Therm. Sci. doi:10.2298/TSCI130225130F
18. Sneddon, I.N.: Fourier Transforms. McGraw-Hill Book Company, New York/Toronto/London (1951)
19. Gradshteyn, I.S., Ryzhik, I.M., Jeffrey, A. (eds.): Tables of integrals, series and products, Ed. Academic, San Diego/New York/Boston/London/Sydney/Toronto (1994) (5th edn., translation from Russian)
20. Fetecau, C., Vieru, D., Fetecau, C.: Effects of side walls on the motion of a viscous fluid induced by an infinite plate that applies an oscillating shear stress to the fluid. Cent. Eur. J. Phys. **9**, 816–824 (2011)
21. Erdogan, M.E.: A note on an unsteady flow of a viscous fluid due to an oscillating plane wall. Int. J. Non-Linear Mech. **35**, 1–6 (2000)
22. Christov, C.I., Jordan, P.M.: Comment on "Stokes' first problem for an Oldroyd-B fluid in a porous half space" [Phys. Fluids **17**, 023101 (2005)]. Phys. Fluids **21**, 069101–069102 (2009)
23. Jamil, M.: First problem of Stokes for generalized Burgers' fluids. ISRN Math. Phys. **2012**, Article ID 831063, 17p. doi:10.5402/2012/831063 (2012)
24. Aksel, N., Fetecau, C., Scholle, M.: Starting solutions for some unsteady unidirectional flows of Oldroyd-B fluids. Z. Angew. Math. Phys. **57**, 1–17 (2006)

Basic Properties of the Non-Abelian Global Reciprocity Map

Kâzım İlhan İkeda

Dedicated to my teacher Goro Shimura.

1 Introduction

All through this work, K denotes a global field. That is, K is either a finite extension of $\mathbb{Q}$ or a finite extension of $\mathbb{F}_q(T)$ (i.e., the field of rational functions of a curve defined over a finite field $\mathbb{F}_q$). Let $\mathbb{a}_K$ denote the set of all Archimedean primes of K (so in case K is a function field, then $\mathbb{a}_K = \varnothing$), and let $\mathbb{h}_K$ denote the set of all Henselian (i.e., non-Archimedean) primes of K. For each $\nu \in \mathbb{h}_K \sqcup \mathbb{a}_K$, let K_ν denote the completion of K with respect to the ν-adic absolute value defined on K.

For any field M, let G_M denote the absolute Galois group $\mathrm{Gal}(M^{\mathrm{sep}}/M)$ of M. For any extension A of M, let $(A/M)^{ab}$ denote the maximal Abelian extension of M inside A. If G is a topological group, then the maximal Abelian Hausdorff quotient $G^{ab} = G/G^c$ of G is called the (first) Abelianization of G, where G^c denotes the closure of the (first) commutator subgroup of G. The canonical mapping $\mathrm{red}_{G^c} : G \to G/G^c = G^{ab}$ which is defined by reduction modulo G^c is called the (first) Abelianization map of G. In particular, G_M^{ab} is defined and $G_M^{ab} = \mathrm{Gal}(M^{ab}/M)$, where $M^{ab} = (M^{\mathrm{sep}}/M)^{ab}$.

In case M is a local or a global field, then W_M denotes the absolute Weil group of M, which comes equipped with a continuous homomorphism $\beta_M : W_M \to G_M$

The author would like to thank Pierre Cartier, A. D. Raza Choudary, and Michel Waldschmidt for inviting him to deliver a talk in the 6th World Conference on 21st Century Mathematics 2013, which took place in the Abdus Salam School of Mathematical Sciences, Lahore, on March 6–9, 2013. He would also like to thank the Abdus Salam School of Mathematical Sciences for arranging his stay and for the hospitality he received in Lahore on March 4–11, 2013, which he enjoyed very much. Finally, the author thanks the referee for his or her suggestions, which improved the presentation of this work a lot.

K.İ. İkeda (✉)
Department of Mathematics, Yeditepe University, İnönü Mah., Kayışdağı Cad., 26 Ağustos Yerleşimi, 34755 Ataşehir, Istanbul, Turkey
e-mail: ilhan.ikeda@yeditepe.edu.tr

P. Cartier et al. (eds.), *Mathematics in the 21st Century*, Springer Proceedings in Mathematics & Statistics 98, DOI 10.1007/978-3-0348-0859-0_5

with dense image. The Abelianization W_M^{ab} of W_M is defined, and there exists a continuous homomorphism $\beta_M^{ab} : W_M^{ab} \to G_M^{ab}$ induced naturally from the continuous arrow $\beta_M : W_M \to G_M$. Moreover, if M is a local field, then the absolute Weil group W_M of M has a natural filtration $\{W_M^e\}_{e\in\mathbb{R}_{\geq -1}}$, where W_M^e is the eth ramification subgroup of W_M in upper numbering, which is defined as usual by

$$W_M^e = W_M \cap G_M^e, \tag{1}$$

where G_M^e is the eth upper ramification subgroup of the absolute Galois group G_M of the local field M, for $e \geq -1$. In particular,

$$W_M^0 = I_M, \tag{2}$$

where I_M is the inertia subgroup of W_M.

1.1 An Overview of Abelian Global Class Field Theory

For the basic theory of local and global fields and class field theory, we refer the reader to [11, 15]. The Abelian global class field theory for K establishes a continuous surjective homomorphism

$$(\bullet, K) : \mathbb{J}_K \to W_K^{ab}$$

from the idèle group $\mathbb{J}_K$ of K to the Abelianization W_K^{ab} of the absolute Weil group W_K of K, which is called the Abelian global reciprocity map of K or the global Abelian norm-residue symbol of K. This arrow has kernel $K^\times$ and satisfies certain "functoriality" and "naturality" conditions, which we shall recall now.

First of all, the local and global reciprocity maps $(\bullet, K_\nu)$ and $(\bullet, K)$ are "compatible" for every $\nu \in \mathbb{h}_K \cup \mathbb{a}_K$. That is, the following square

$$\begin{array}{ccc} K_\nu^\times & \xrightarrow{\varepsilon_\nu} & \mathbb{J}_K \\ {\scriptstyle(\bullet,K_\nu)}\downarrow & & \downarrow{\scriptstyle(\bullet,K)} \\ W_{K_\nu}^{ab} & \xrightarrow[e_\nu^{\mathrm{Weil}^{ab}}]{} & W_K^{ab} \end{array} \tag{3}$$

is commutative for each $\nu \in \mathbb{h}_K \cup \mathbb{a}_K$. Next, if $K \subseteq L \subset E$ is any tower of extensions of global fields (note that the inclusions $(L/K)^{ab} \subseteq (E/K)^{ab}$ and $(E/K)^{ab} \subseteq (E/L)^{ab}$ trivially hold), then:

(i) The Abelian global reciprocity map $(\bullet, K)$ of K induces a continuous surjective homomorphism

$$(\bullet, L/K) : \mathbb{J}_K \xrightarrow{(\bullet,K)} W_K^{ab} \xrightarrow{\beta_K^{ab}} G_K^{ab} \to \mathrm{Gal}((L/K)^{ab}/K).$$

The kernel of the surjective homomorphism $(\bullet, L/K)$ is $K^{\times}\mathrm{N}_{L/K}\mathbb{J}_L$. That is, the following sequence

$$1 \to K^{\times}\mathrm{N}_{L/K}\mathbb{J}_L \to \mathbb{J}_K \xrightarrow{(\bullet,L/K)} \mathrm{Gal}((L/K)^{ab}/K) \to 1$$

is exact.

(ii) For every prime ν, the image of the continuous homomorphism defined by the composition

$$K_{\nu}^{\times} \hookrightarrow \mathbb{J}_K \xrightarrow{(\bullet,L/K)} \mathrm{Gal}((L/K)^{ab}/K)$$

is the decomposition group $\mathrm{D}_{\nu}((L/K)^{ab}/K)$ of ν in $\mathrm{Gal}((L/K)^{ab}/K)$, and the image of the continuous homomorphism defined by the composition

$$U_{K_{\nu}} \hookrightarrow \mathbb{J}_K \xrightarrow{(\bullet,L/K)} \mathrm{Gal}((L/K)^{ab}/K)$$

is the inertia group $\mathrm{I}_{\nu}((L/K)^{ab}/K)$ of ν in $\mathrm{Gal}((L/K)^{ab}/K)$. Moreover, for each prime ν, any π_{ν} is mapped to an element in the Frobenius coset modulo $\mathrm{I}_{\nu}((L/K)^{ab}/K)$ in $\mathrm{D}_{\nu}((L/K)^{ab}/K)$.

(iii) The triangle

$$\begin{array}{ccc}
\mathbb{J}_K & \xrightarrow{(\bullet,E/K)} & \mathrm{Gal}((E/K)^{ab}/K) \\
& \searrow^{(\bullet,L/K)} & \downarrow \mathrm{res}_{(L/K)^{ab}} \\
& & \mathrm{Gal}((L/K)^{ab}/K)
\end{array}$$

is commutative, where the right vertical arrow $\mathrm{res}_{(L/K)^{ab}}$ is the "restriction to $(L/K)^{ab}$" map.

(iv) The square

$$\begin{array}{ccc}
\mathbb{J}_L & \xrightarrow{(\bullet,E/L)} & \mathrm{Gal}((E/L)^{ab}/L) \\
\downarrow \mathrm{N}_{L/K} & & \downarrow \mathrm{res}_{(E/K)^{ab}} \\
\mathbb{J}_K & \xrightarrow{(\bullet,E/K)} & \mathrm{Gal}((E/K)^{ab}/K)
\end{array}$$

where the left vertical arrow $\mathrm{N}_{L/K}$ is the idèlic norm map from L to K, and the square

$$\begin{array}{ccc} \mathbb{J}_K & \xrightarrow{(\bullet,E/K)} & \mathrm{Gal}((E/K)^{ab}/K) \\ \downarrow & & \downarrow \mathrm{Ver}_{K\to L} \\ \mathbb{J}_L & \xrightarrow{(\bullet,E/L)} & \mathrm{Gal}((E/L)^{ab}/L) \end{array}$$

where the left vertical arrow is the natural inclusion and the right vertical arrow is the "Verlagerung" (transfer) map from K to L (more precisely from G_K to G_L), are commutative.

(v) (Global existence theorem) For any open subgroup N of finite index in $K^\times\backslash\mathbb{J}_K$, there is a unique Abelian extension $L_N = L$ over K such that the kernel of the Abelian reciprocity homomorphism

$$(\bullet, L/K) : K^\times\backslash\mathbb{J}_K \to \mathrm{Gal}(L/K)$$

relative to the extension L/K is N. Moreover, this assignment

$$N \mapsto L_N$$

defines an inclusion-reversing bijective correspondence

$$\left\{\begin{array}{l}\text{Open subgroups of finite}\\ \text{index in } \mathbb{J}_K \text{ containing } K^\times\end{array}\right\} \rightleftarrows \left\{\begin{array}{l}\text{Finite Abelian exten-}\\ \text{sions of } K \text{ inside } K^{\mathrm{sep}}\end{array}\right\}.$$

(vi) (Ray class groups and ray class fields) Let $\mathfrak{m} = \prod_{\nu\in\mathfrak{h}_K\cup\mathfrak{a}_K} \nu^{e_\nu}$ be a fixed cycle (i.e., modulus) of the global field K, and let $\mathbb{U}_\mathfrak{m}$ be the subgroup of $\mathbb{J}_K$ defined by $\mathfrak{m}$. Then under the canonical surjective homomorphism $\mathbb{J}_K \twoheadrightarrow K^\times\backslash\mathbb{J}_K$, the subgroup $\mathbb{U}_\mathfrak{m}$ of $\mathbb{J}_K$ maps onto a finite-index open subgroup $\overline{\mathbb{U}}_\mathfrak{m}$ of $K^\times\backslash\mathbb{J}_K$. By the global existence theorem, there exists a finite Abelian extension $R_\mathfrak{m}$ of K, called the ray class field of $\mathfrak{m}$, such that the Abelian reciprocity homomorphism $(\bullet, R_\mathfrak{m}/K) : \mathbb{J}_K \to \mathrm{Gal}(R_\mathfrak{m}/K)$ relative to the extension $R_\mathfrak{m}/K$ induces an isomorphism

$$\mathbb{U}_\mathfrak{m}K^\times\backslash\mathbb{J}_K \xrightarrow{\sim} \mathrm{Gal}(R_\mathfrak{m}/K),$$

where $\mathbb{U}_\mathfrak{m}K^\times\backslash\mathbb{J}_K$ is called the ray class group of $\mathfrak{m}$.

(vii) (Splitting of primes) A prime ν in K splits completely in the Abelian extension $(L/K)^{ab}/K$ if and only if $K_\nu^\times \subset \ker(\bullet, L/K)$. Thus,

$$\mathrm{Spl}\left((L/K)^{ab}/K\right) = \{\nu \in \mathfrak{h}_K : K_\nu^\times \subset \ker(\bullet, L/K)\},$$

where $\mathrm{Spl}\left((L/K)^{ab}/K\right)$ denotes the set of all primes in K that split completely in the Abelian extension $(L/K)^{ab}/K$.[1]

1.2 Aim

In [4], we have introduced a certain topological group $\mathscr{I}_K^{\varphi}$ depending only on the global field K and called the *non-Abelian idèle group of* K together with a natural continuous homomorphism

$$\mathsf{NR}_K^{\varphi_K^{\text{Weil}}} : \mathscr{I}_K^{\varphi_K} \to W_K$$

called the *global non-Abelian norm-residue symbol of* K or the *non-Abelian global reciprocity map of* K. Moreover, in [4], we have studied the ℓ-adic representations of the topological group $\mathscr{I}_K^{\varphi_K}$ and observed that the theory of n-dimensional ℓ-adic representations of $\mathscr{I}_K^{\varphi_K}$ is closely related with the Langlands reciprocity principle for $\mathrm{GL}(n)$ over K, for each $n \geq 1$, via the arrow $\mathsf{NR}_K^{\varphi_K^{\text{Weil}}} : \mathscr{I}_K^{\varphi_K} \to W_K$. We remind that, the Langlands reciprocity principle for $\mathrm{GL}(n)$ over K has still not yet been established in case K is assumed to be a number field and $n > 1$ and remains conjectural.

In [4] on the other hand, nothing has been said about the *basic properties* satisfied by the non-Abelian global reciprocity map $\mathsf{NR}_K^{\varphi_K^{\text{Weil}}}$ of K. The aim here therefore, which complements [4], is to study the non-Abelian analogues of the "functoriality" and the "naturality" properties of the map $\mathsf{NR}_K^{\varphi_K^{\text{Weil}}}$ and observe that this arrow indeed deserves its name. That is, we shall introduce the non-Abelian analogues of the "functoriality" and the "naturality" properties of the Abelian global reciprocity map $(\bullet, K)$ of K summarized in Sect. 1.1 and then prove these properties. Moreover, we shall describe the set of primes in K that split in a finite extension L of K, which is one of the main goals of non-Abelian global class field theory.

1.3 Outline

The outline of this paper is as follows. This work heavily depends on [4], so in Sects. 2 and 3, we shall briefly review the theory developed in [4]. More precisely, we shall first recall the construction of the non-Abelian idèle group $\mathscr{I}_K^{\varphi_K}$ of K and

[1] By Čebotarev density theorem, $\mathrm{Spl} : L/K \mapsto \mathrm{Spl}(L/K)$ is an injective and order-reversing mapping from finite Galois extensions L of the global field K into the power set of $\mathfrak{h}_K \cup \mathfrak{a}_K$. The image of the map "Spl" for finite Abelian extensions L of K has a description in terms of the Abelian global reciprocity map $(\bullet, L/K)$ relative to the extension L/K.

then construct the non-Abelian global reciprocity map $\mathsf{NR}_K^{\varphi_K^{\text{Weil}}} : \mathscr{I}_K^{\varphi_K} \to W_K$ of K. In the end of Sect. 3, following [4], we shall state an important and central conjecture on the surjectivity of the non-Abelian global reciprocity map $\mathsf{NR}_K^{\varphi_K^{\text{Weil}}} : \mathscr{I}_K^{\varphi_K} \to W_K$ of K (look at Conjecture 1).

In Sect. 4, which contains the main results of this work, we shall state and prove the functoriality and naturality properties of the non-Abelian global reciprocity map $\mathsf{NR}_K^{\varphi_K^{\text{Weil}}} : \mathscr{I}_K^{\varphi_K} \to W_K$ of the global field K, which constitute the non-Abelian analogues of the basic properties of the Abelian global reciprocity map $(\bullet, K) : \mathbb{J}_K \to W_K^{ab}$ of K sketched in Sect. 1.1. We should mention here that except Sects. 4.8 and 4.9, the results stated in Sect. 4 have formal and straightforward proofs. On the other hand, Sect. 4.8 on the non-Abelian global existence theorem and Sect. 4.9 on the non-Abelian ray class groups and non-Abelian ray class fields are more technical and nontrivial. Finally, we end Sect. 4 by describing the set of primes in K that split in a finite extension L of K.

2 Non-Abelian Idèle Group $\mathscr{I}_K^{\varphi}$ of a Global Field K

All through this work, K denotes a global field. That is, K is a finite extension of $\mathbb{Q}$ or a finite extension of $\mathbb{F}_q(T)$ (i.e., the field of rational functions of a curve defined over a finite field $\mathbb{F}_q$). For the basic theory of local fields and the Abelian local class field theory and for details about global fields and the Abelian global class field theory, we refer the reader to [11, 15]. Let $\mathbb{a}_K$ denote the set of all Archimedean primes of K (so in case K is a function field, then $\mathbb{a}_K = \varnothing$). For each $\nu \in \mathbb{h}_K$, where $\mathbb{h}_K$ denotes the set of all Henselian (= non-Archimedean) primes of K, let K_ν denote the completion of K with respect to the ν-adic absolute value.

Fixing a Lubin-Tate splitting φ_{K_ν} over K_ν, the non-Abelian local reciprocity map

$$\mathbf{\Phi}_{K_\nu}^{(\varphi_{K_\nu})} : G_{K_\nu} \xrightarrow{\sim} \nabla_{K_\nu}^{(\varphi_{K_\nu})}$$

or equivalently the "Weil form" of the non-Abelian local reciprocity map

$$\mathbf{\Phi}_{K_\nu}^{(\varphi_{K_\nu})} : W_{K_\nu} \xrightarrow{\sim} {}_{\mathbb{Z}}\nabla_{K_\nu}^{(\varphi_{K_\nu})}$$

of the local field K_ν is defined. The construction of the topological group $\nabla_{K_\nu}^{(\varphi_{K_\nu})}$, which depends on K_ν and on the choice of the Lubin-Tate splitting φ_{K_ν} over K_ν, introduced here involves the theory of APF-extensions of K_ν and the fields of norm construction of Fontaine and Wintenberger, and ${}_{\mathbb{Z}}\nabla_{K_\nu}^{(\varphi_{K_\nu})}$ is a certain dense subgroup of the topological group $\nabla_{K_\nu}^{(\varphi_{K_\nu})}$ (for details, see [2, 6]). Moreover, the isomorphism $\mathbf{\Phi}_{K_\nu}^{(\varphi_{K_\nu})}$, which is called the *non-Abelian local reciprocity map of* K_ν, is "natural" in the sense that properties such as "existence", "functoriality", and a certain "ramification theoretic" property are all satisfied. The isomorphism $\{\bullet, K_\nu\}_{\varphi_{K_\nu}}$, which is defined to be the inverse $\Phi_{K_\nu}^{(\varphi_{K_\nu})^{-1}}$ of the isomorphism $\Phi_{K_\nu}^{(\varphi_{K_\nu})}$ by

$$\{\bullet, K_\nu\}_{\varphi_{K_\nu}} : \nabla_{K_\nu}^{(\varphi_{K_\nu})} \xrightarrow{\sim} G_{K_\nu}$$

is called the *non-Abelian local norm-residue symbol of* K_ν. For details on non-Abelian local class field theory in the sense of Koch,[2] we refer the reader to the papers [5–7] as well as the influential works of Fesenko [2] and Laubie [10]. Moreover, following Section 8 of [5], together with [7] for a detailed account, for each $\nu \in \mathbb{h}_K$, there exists a certain subgroup ${}_1\nabla_{K_\nu}^{(\varphi_{K_\nu})\underline{0}}$ of ${}_{\mathbb{Z}}\nabla_{K_\nu}^{(\varphi_{K_\nu})}$ satisfying the equality

$$\Phi_{K_\nu}^{(\varphi_{K_\nu})}\left(W_{K_\nu}^{0}\right) = {}_1\nabla_{K_\nu}^{(\varphi_{K_\nu})\underline{0}}, \tag{4}$$

where $W_{K_\nu}^{0} = I_{K_\nu}$ by (1) and (2).

It is then a natural attempt to construct the non-Abelian version of global class field theory of the global field K by "glueing" the non-Abelian local class field theories of respective completions K_ν of K, for $\nu \in \mathbb{h}_K$, following Chevalley-Miyake philosophy of idèles. This program has been carried out in [4] yielding the non-Abelian global reciprocity map of K and the "ultimate" non-Abelian global reciprocity map of K by pushing the idea of Miyake introduced in [12, 13] to the extreme and therefore introducing the non-Abelian idèle group $\mathscr{I}_K^{\varphi_K}$ of K by following the analogy between the non-Abelian local class field theory in the sense of Koch and the Abelian local class field theory of Hasse and taking into account the analogy between the philosophy of Miyake and the philosophy of Chevalley (also look at Iwasawa [8]).

2.1 Digression: Restricted Free Products of Locally Compact Groups

The main reference that we follow very closely and reproduce here is Section 2 of [4]. Let $\{G_i\}_{i\in I}$ be a collection of locally compact topological groups. For all but finitely many $i \in I$, let O_i be a compact open subgroup of G_i. The finite subset of I consisting of all $i \in I$ for which O_i is not defined is denoted by I_∞. For every finite subset S of I satisfying $I_\infty \subseteq S$, define the topological group

$$G_S := \mathop{*}_{i\notin S} O_i * \left(\mathop{*}_{i\in S} G_i\right)$$

[2]Initially, Koch started this theory for metabelian extensions of local fields (look at [9]) using explicit computations with formal Lubin-Tate groups, unlike the more general approach of Fesenko, Laubie, Serbest, and others, which uses APF-extensions and the fields of norm construction of Fontaine and Wintenberger.

as the free product of the topological groups O_i, for $i \in I - S$, and G_i, for $i \in S$, which exists in the category of topological groups (cf. Morris [14]). Then, the *restricted free product* of the collection $\{G_i\}_{i\in I}$ with respect to the collection $\{O_i\}_{i\in I-I_\infty}$, which is denoted by $*'_{i\in I}(G_i : O_i)$, is defined by the injective limit

$$\underset{i\in I}{*}'(G_i : O_i) := \varinjlim_{S} G_S$$

defined over all possible such S, where the connecting morphism

$$\tau_S^T : G_S \to G_T$$

for $S \subseteq T$ is defined naturally by the "universal mapping property of free products"[3] (cf. Hilton-Wu [3] and Morris [14]). The topology on $*'_{i\in I}(G_i : O_i)$ is defined by declaring $X \subseteq *'_{i\in I}(G_i : O_i)$ to be open if $X \cap G_S$ is open in G_S for every S. So, endowed with this topology, $*'_{i\in I}(G_i : O_i)$ is a topological group.

The following proposition (which is Proposition 2.1 of [4]) is a direct consequence of the "universal mapping property of free products".

Proposition 1. *Let $\{G_i\}_{i\in I}$ be a collection of locally compact topological groups, and for all but finitely many $i \in I$, let O_i be a compact open subgroup of G_i. Denote the finite subset of I consisting of all $i \in I$ for which O_i is not defined by I_∞. Assume that for each $i \in I$, a continuous homomorphism*

$$\phi_i : G_i \to H$$

is given. Then, there exists a unique continuous homomorphism

$$\phi_S : G_S \to H$$

defined for each finite subset S of I satisfying $I_\infty \subseteq S$ and a unique continuous homomorphism

$$\phi = \varinjlim_{S} \phi_S : *'_{i\in I}(G_i : O_i) \to H$$

satisfying

$$\phi_S = \phi \circ c_S : G_S \xrightarrow{c_S} *'_{i\in I}(G_i : O_i) \xrightarrow{\phi} H,$$

where $c_S : G_S \to *'_{i\in I}(G_i : O_i)$ is the canonical homomorphism, for every S.

[3] If $\{G_i\}_{i\in I}$ is a collection of topological groups and $*_{i\in I} G_i$ is the free product of this collection together with the canonical embeddings $\iota_{i_o} : G_{i_o} \hookrightarrow *_{i\in I} G_i$, for each $i_o \in I$, then the universal mapping property of free products states that, if for each $i_o \in I$, $\phi_{i_o} : G_{i_o} \to H$ is a continuous homomorphism, then there exists a unique continuous homomorphism $\phi : *_{i\in I} G_i \to H$, such that $\phi \circ \iota_{i_o} = \phi_{i_o}$, for every $i_o \in I$.

Notation 1. As a notation, for a topological group G, the n-fold free product $\overbrace{G * \cdots * G}^{n\text{-copies}}$ of G is denoted by G^{*n}.

2.2 Definition of the Non-Abelian Idèle Group $\mathscr{J}_K^{\underline{\varphi}}$ of K

Now, we introduce the non-Abelian idèle group $\mathscr{J}_K^{\underline{\varphi}}$ of K as follows.

Definition 1. For each $v \in \mathfrak{h}_K$, fix a Lubin-Tate splitting φ_{K_v} and let $\underline{\varphi}_K = \{\varphi_{K_v}\}_{v \in \mathfrak{h}_K}$. If there is no fear of confusion, denote $\underline{\varphi}_K = \underline{\varphi}$. The topological group $\mathscr{J}_K^{\underline{\varphi}}$ defined by the "restricted free product"

$$\mathscr{J}_K^{\underline{\varphi}} := \underset{v \in \mathfrak{h}_K \cup \mathfrak{a}_K}{*}{}' \left({}_{\mathbb{Z}}\nabla_{K_v}^{(\varphi_{K_v})} : {}_1\nabla_{K_v}^{(\varphi_{K_v})^{\underline{0}}} \right)$$

is called the *non-Abelian idèle group of the global field K*. In case K is a number field,

$$\mathscr{J}_K^{\underline{\varphi}} = \underset{v \in \mathfrak{h}_K}{*}{}' \left({}_{\mathbb{Z}}\nabla_{K_v}^{(\varphi_{K_v})} : {}_1\nabla_{K_v}^{(\varphi_{K_v})^{\underline{0}}} \right) * W_{\mathbb{R}}^{*r_1} * W_{\mathbb{C}}^{*r_2},$$

where the finite (= Henselian) part $\mathscr{J}_{K,\mathfrak{h}}^{\underline{\varphi}}$ of $\mathscr{J}_K^{\underline{\varphi}}$ is defined by

$$\mathscr{J}_{K,\mathfrak{h}}^{\underline{\varphi}} := \underset{v \in \mathfrak{h}_K}{*}{}' \left({}_{\mathbb{Z}}\nabla_{K_v}^{(\varphi_{K_v})} : {}_1\nabla_{K_v}^{(\varphi_{K_v})^{\underline{0}}} \right),$$

and the infinite (= Archimedean) part $\mathscr{J}_{K,\mathfrak{a}}^{\underline{\varphi}}$ of $\mathscr{J}_K^{\underline{\varphi}}$ by

$$\mathscr{J}_{K,\mathfrak{a}}^{\underline{\varphi}} := W_{\mathbb{R}}^{*r_1} * W_{\mathbb{C}}^{*r_2}.$$

Here, as usual, r_1 and r_2 denote the number of real and the number of complex conjugate embeddings of the global field K in $\mathbb{C}$.

Remark 1. Note that the non-Abelian idèle group $\mathscr{J}_K^{\underline{\varphi}}$ of K depends only on the global field K and the choice of $\underline{\varphi}$.

Theorem 2.5 of [4] states that the Abelianization $\mathscr{J}_K^{\underline{\varphi}\,ab}$ of the topological group $\mathscr{J}_K^{\underline{\varphi}}$ is canonically isomorphic to the idèle group $\mathbb{J}_K$ of K. Therefore, there exists a continuous surjective homomorphism

$$\mathsf{s}_K : \mathscr{J}_K^{\underline{\varphi}_K} \twoheadrightarrow \mathbb{J}_K$$

defined by the Abelianization map

$$\mathscr{I}_{K}^{\underline{\varphi_K}} \to \mathscr{I}_{K}^{\underline{\varphi ab}}$$

of $\mathscr{I}_{K}^{\underline{\varphi_K}}$.

For each $\nu \in \mathbb{h}_K \cup \mathbb{a}_K$, there exists a natural homomorphism

$$q_\nu : (\mathscr{I}_{K}^{\underline{\varphi}})_\nu := \left\{ \begin{matrix} {}_{\mathbb{Z}}\nabla_{K_\nu}^{(\varphi_{K_\nu})}, & \nu \in \mathbb{h}_K \\ W_{\mathbb{R}}, & \nu \in \mathbb{a}_{K,\mathbb{R}} \\ W_{\mathbb{C}}, & \nu \in \mathbb{a}_{K,\mathbb{C}} \end{matrix} \right\} \to \mathscr{I}_{K}^{\underline{\varphi}},$$

which is defined explicitly via the commutative triangle

$$\begin{array}{ccc} & & (\mathscr{I}_{K}^{\underline{\varphi}})_S \\ & \nearrow^{\iota_\nu^{(S)}} & \\ (\mathscr{I}_{K}^{\underline{\varphi}})_\nu & & \downarrow c_S \\ & \searrow_{q_\nu} & \\ & & \mathscr{I}_{K}^{\underline{\varphi}} \end{array}$$

where S is a finite subset of $\mathbb{h}_K \cup \mathbb{a}_K$ satisfying $\mathbb{a}_K \subseteq S$ and $\nu \in S$. Note that the definition of the continuous homomorphism $q_\nu : (\mathscr{I}_{K}^{\underline{\varphi}})_\nu \to \mathscr{I}_{K}^{\underline{\varphi}}$ does not depend on the choice of S (for details, look at Section 4 of [4]).

3 Non-Abelian Global Reciprocity Map

For $\nu \in \mathbb{h}_K \cup \mathbb{a}_K$, choose an embedding

$$e_\nu : K^{\text{sep}} \hookrightarrow K_\nu^{\text{sep}}.$$

This embedding determines a continuous homomorphism[4] (look at [16] for details)

$$e_\nu^{\text{Weil}} : W_{K_\nu} \to W_K,$$

and, for each $\nu \in \mathbb{h}_K$, a continuous homomorphism

$$\mathsf{NR}_{K_\nu}^{(\varphi_{K_\nu})^{\text{Weil}}} : {}_{\mathbb{Z}}\nabla_{K_\nu}^{(\varphi_{K_\nu})} \xrightarrow[\sim]{\{\bullet, K_\nu\}_{\varphi_{K_\nu}}} W_{K_\nu} \xrightarrow{e_\nu^{\text{Weil}}} W_K.$$

[4]Which is unique if K is a function field and unique up to composition with an inner automorphism of W_K defined by an element of the connected component W_K^o of W_K if K is a number field.

By the "universal mapping property of free products", the following theorem (which is Theorem 3.1 of [4]) follows.

Theorem 1. *There exists a well-defined continuous homomorphism*

$$\mathsf{NR}_{\overline{K}}^{\varphi^{\mathrm{Weil}}} : \mathscr{I}_{\overline{K}}^{\varphi} \to W_K, \tag{5}$$

called the non-Abelian global reciprocity map of K, or the global non-Abelian norm-residue symbol of K, which satisfies

$$\left(\mathsf{NR}_{\overline{K}}^{\varphi^{\mathrm{Weil}}}\right)_S = \mathsf{NR}_{\overline{K}}^{\varphi^{\mathrm{Weil}}} \circ c_S : \left(\mathscr{I}_{\overline{K}}^{\varphi}\right)_S \xrightarrow{c_S} \mathscr{I}_{\overline{K}}^{\varphi} \xrightarrow{\mathsf{NR}_{\overline{K}}^{\varphi^{\mathrm{Weil}}}} W_K,$$

where $c_S : \left(\mathscr{I}_{\overline{K}}^{\varphi}\right)_S \to \mathscr{I}_{\overline{K}}^{\varphi}$ is the canonical homomorphism defined for every finite subset S of $\mathbb{h}_K \cup \mathbb{a}_K$ containing $\mathbb{a}_K$.

Moreover, we have made the following conjecture (look at Conjecture 3.2 in [4]):

Conjecture 1. The homomorphism

$$\mathsf{NR}_{\overline{K}}^{\varphi^{\mathrm{Weil}}} : \mathscr{I}_{\overline{K}}^{\varphi} \to W_K$$

is open, continuous, and surjective.

In this work, we shall assume that Conjecture 1 holds only in Sects. 4.4 and 4.8–4.10.

4 Basic Properties of the Non-Abelian Global Reciprocity Map

In the remaining of this paper, we shall study the basic properties of the non-Abelian global reciprocity map $\mathsf{NR}_{\overline{K}}^{\varphi^{\mathrm{Weil}}} : \mathscr{I}_{\overline{K}}^{\varphi_K} \to W_K$ of the global field K.

4.1 Local-Global Compatibility of the Non-Abelian Norm-Residue Symbols

The *"local-global compatibility"* of $\{\bullet, K_\nu\}_{\varphi_{K_\nu}}$ and $\mathsf{NR}_{\overline{K}}^{\varphi^{Weil}}$ for $\nu \in \mathbb{h}_K$, proved in [4] as Theorem 4.1, states the commutativity of the square

$$\begin{array}{ccc} {}_{\mathbb{Z}}\nabla_{K_\nu}^{(\varphi_{K_\nu})} & \xrightarrow{q_\nu} & \mathscr{I}_K^{\varphi} \\ {\scriptstyle \{\bullet,K_\nu\}_{\varphi_{K_\nu}}}\Big\downarrow & & \Big\downarrow{\scriptstyle \mathrm{NR}_K^{\varphi\,\mathrm{Weil}}} \\ W_{K_\nu} & \xrightarrow[e_\nu^{\mathrm{Weil}}]{} & W_K \end{array}$$

for each $\nu \in \mathbb{h}_K$.

4.2 Relationship with the Abelian Global Reciprocity Map

In this section, we shall study the "behavior" of the non-Abelian global reciprocity map under the Abelianization functor and prove that under Abelianization, the non-Abelian global reciprocity map reduces to the Abelian global reciprocity map.

For $\nu \in \mathbb{h}_K$, define the surjective and continuous homomorphism $\mathsf{a}_{K_\nu} : {}_{\mathbb{Z}}\nabla_{K_\nu}^{(\varphi_{K_\nu})} \to K_\nu^\times$ by the composition

$$\mathsf{a}_{K_\nu} : {}_{\mathbb{Z}}\nabla_{K_\nu}^{(\varphi_{K_\nu})} \xrightarrow{\mathrm{id}_{\mathbb{Z}}\times \mathrm{Pr}_{\tilde{K}_\nu}} \mathbb{Z}\times U_{K_\nu} \xrightarrow{\sim} K_\nu^\times$$

following Remark 4 in [2] combined with the construction of the non-Abelian local norm-residue symbol $\{\bullet, K_\nu\}_{\varphi_{K_\nu}} : {}_{\mathbb{Z}}\nabla_{K_\nu}^{(\varphi_{K_\nu})} \xrightarrow{\sim} W_{K_\nu}$ of K_ν in the "Weil form" as described in [6]. For $\nu \in \mathbb{a}_K$, define the continuous homomorphism

$$\mathsf{a}_{K_\nu} : W_{K_\nu} \to K_\nu^\times$$

to be the natural homomorphism defined by Abelianization $W_{K_\nu}^{ab} \xrightarrow{\sim} K_\nu^\times$ of W_{K_ν}.

Lemma 1. *There exists a unique continuous homomorphism*

$$\mathsf{a}_K : \mathscr{I}_K^{\varphi} \to \mathbb{J}_K$$

which makes the diagram

$$\begin{array}{ccc} (\mathscr{I}_K^{\varphi})_\nu & \xrightarrow{q_\nu} & \mathscr{I}_K^{\varphi} \\ {\scriptstyle \mathsf{a}_{K_\nu}}\Big\downarrow & & \Big\downarrow{\scriptstyle \mathsf{a}_K} \\ K_\nu^\times & \xrightarrow[\varepsilon_\nu]{} & \mathbb{J}_K \end{array}$$

commutative.

Proof. Follows from Proposition 2.1 of [4] applied to the collection of continuous homomorphisms $\varepsilon_\nu \circ \mathsf{a}_K : \left(\mathscr{I}_K^{\varphi}\right)_\nu \xrightarrow{\mathsf{a}_{K_\nu}} K_\nu^\times \xrightarrow{\varepsilon_\nu} \mathbb{J}_K$ defined for each $\nu \in \mathbb{h}_K \cup \mathbb{a}_K$. □

Lemma 2. *For each $\nu \in \mathbb{h}_K \cup \mathbb{o}_K$, there exists a continuous homomorphism*

$$q_\nu^{ab} : \left(\mathscr{I}_K^{\varphi}\right)_\nu^{ab} \to \mathscr{I}_K^{\varphi ab},$$

called the Abelianization of the natural homomorphism $q_\nu : \left(\mathscr{I}_K^{\varphi}\right)_\nu \to \mathscr{I}_K^{\varphi}$ that makes the diagram

$$\begin{array}{ccc}
(\mathscr{I}_K^{\varphi})_\nu & \xrightarrow{q_\nu} & \mathscr{I}_K^{\varphi} \\
\Big\downarrow {\scriptstyle \mathrm{red}_{(\mathscr{I}_K^{\varphi})_\nu^c}} & & \Big\downarrow {\scriptstyle \mathrm{red}_{\mathscr{I}_K^{\varphi c}}} \\
(\mathscr{I}_K^{\varphi})_\nu^{ab} & \xrightarrow[q_\nu^{ab}]{} & \mathscr{I}_K^{\varphi ab}
\end{array}$$

commutative.

Proof. The proof is trivial. □

Theorem 2. *The non-Abelian global reciprocity map*

$$\mathsf{NR}_K^{\varphi^{\mathrm{Weil}}} : \mathscr{I}_K^{\varphi} \to W_K$$

of K sits in the following commutative diagram

$$\begin{array}{ccc}
\mathscr{I}_K^{\varphi} & \xrightarrow{\mathsf{NR}_K^{\varphi^{\mathrm{Weil}}}} & W_K \\
\Big\downarrow {\scriptstyle \mathrm{a}_K} & & \Big\downarrow {\scriptstyle \mathrm{red}_{W_K^c}} \\
\mathbb{J}_K & \xrightarrow[(\bullet, K)]{} & W_K^{ab}.
\end{array}$$

Proof. It suffices to prove the equality

$$(\mathrm{red}_{W_K^c} \circ \mathsf{NR}_K^{\varphi^{\mathrm{Weil}}})_S \circ \iota_\nu^{(S)} = ((\bullet, K) \circ \mathsf{a}_K)_S \circ \iota_\nu^{(S)},$$

where S is any finite subset of $\mathbb{h}_K \cup \mathbb{o}_K$ satisfying $\mathbb{o}_K \subseteq S$ and $\nu \in S$. In fact, for such an S and for any $\nu \in S$, in case $\nu \in \mathbb{h}_K$, then

$$\begin{aligned}
\left(\mathrm{red}_{W_K^c} \circ \mathsf{NR}_K^{\varphi^{\mathrm{Weil}}}\right)_S \circ \iota_\nu^{(S)} &= \mathrm{red}_{W_K^c} \circ \mathsf{NR}_K^{\varphi^{\mathrm{Weil}}} \circ c_S \circ \iota_\nu^{(S)} \\
&= \mathrm{red}_{W_K^c} \circ \left(\mathsf{NR}_K^{\varphi^{\mathrm{Weil}}} \circ q_\nu\right) \\
&= \mathrm{red}_{W_K^c} \circ \left(e_\nu^{\mathrm{Weil}} \circ \{\bullet, K_\nu\}_{\varphi_\nu}\right).
\end{aligned}$$

Moreover, the following diagram

$$\begin{array}{ccccc}
{}_{\mathbb{Z}}\nabla_{K_\nu}^{(\varphi_{K_\nu})} & \xrightarrow[\sim]{\{\bullet, K_\nu\}_{\varphi_{K_\nu}}} & W_{K_\nu} & \xrightarrow{e_\nu^{\mathrm{Weil}}} & W_K \\
\Big\downarrow{\scriptstyle \mathsf{a}_{K_\nu}} & & \Big\downarrow{\scriptstyle \mathrm{red}_{W_{K_\nu}^c}} & & \Big\downarrow{\scriptstyle \mathrm{red}_{W_K^c}} \\
K_\nu^\times & \xrightarrow[(\bullet, K_\nu)]{\sim} & W_{K_\nu}^{ab} & \xrightarrow[e_\nu^{\mathrm{Weil}^{ab}}]{} & W_K^{ab}
\end{array} \tag{6}$$

commutes, as the rectangle

$$\begin{array}{ccc}
W_{K_\nu} & \xrightarrow{e_\nu^{\mathrm{Weil}}} & W_K \\
\Big\downarrow{\scriptstyle \mathrm{red}_{W_{K_\nu}^c}} & & \Big\downarrow{\scriptstyle \mathrm{red}_{W_K^c}} \\
W_{K_\nu}^{ab} & \xrightarrow[e_\nu^{\mathrm{Weil}^{ab}}]{} & W_K^{ab}
\end{array}$$

is naturally commutative, and the diagram

$$\begin{array}{ccc}
{}_{\mathbb{Z}}\nabla_{K_\nu}^{(\varphi_{K_\nu})} & \xrightarrow[\sim]{\{\bullet, K_\nu\}_{\varphi_{K_\nu}}} & W_{K_\nu} \\
\Big\downarrow{\scriptstyle \mathsf{a}_{K_\nu}} & & \Big\downarrow{\scriptstyle \mathrm{red}_{W_{K_\nu}^c}} \\
K_\nu^\times & \xrightarrow[(\bullet, K_\nu)]{\sim} & W_{K_\nu}^{ab}
\end{array}$$

which relates the Abelian with the non-Abelian local reciprocity maps is commutative by Remark 4 in [2] combined with the construction of the non-Abelian local norm-residue symbol $\{\bullet, K_\nu\}_{\varphi_{K_\nu}} : {}_{\mathbb{Z}}\nabla_{K_\nu}^{(\varphi_{K_\nu})} \xrightarrow{\sim} W_{K_\nu}$ of K_ν in the "Weil form" as described in [6]. Therefore,

$$\begin{aligned}
\left(\mathrm{red}_{W_K^c} \circ \mathsf{NR}_K^{\varphi^{\mathrm{Weil}}}\right)_S \circ \iota_\nu^{(S)} &= \mathrm{red}_{W_K^c} \circ \left(e_\nu^{\mathrm{Weil}} \circ \{\bullet, K_\nu\}_{\varphi_\nu}\right) \\
&= \left(e_\nu^{\mathrm{Weil}^{ab}} \circ (\bullet, K_\nu)\right) \circ \mathsf{a}_{K_\nu} \\
&= ((\bullet, K) \circ \varepsilon_\nu) \circ \mathsf{a}_{K_\nu},
\end{aligned}$$

by the commutativity of the diagram (6) and by the compatibility diagram (3) of the local and the global Abelian reciprocity maps. Now, by Lemma 1,

$$
\begin{aligned}
\left(\mathrm{red}_{W_K^c} \circ \mathsf{NR}_{\underline{K}}^{\underline{\varphi}^{\mathrm{Weil}}}\right)_S \circ \iota_\nu^{(S)} &= ((\bullet, K) \circ \varepsilon_\nu) \circ \mathfrak{a}_{K_\nu} \\
&= (\bullet, K) \circ (\mathfrak{a}_K \circ q_\nu) \\
&= (\bullet, K) \circ \mathfrak{a}_K \circ c_S \circ \iota_\nu^{(S)} \\
&= ((\bullet, K) \circ \mathfrak{a}_K)_S \circ \iota_\nu^{(S)},
\end{aligned}
$$

which completes the proof. The case $\nu \in \mathbb{a}_K$ can be proved similarly. □

4.3 Non-Abelian Idèles in Field Extensions

In this section, we shall study the relationship between the non-Abelian idèle group $\mathscr{J}_K^{\underline{\varphi}_K}$ of K and the non-Abelian idèle group $\mathscr{J}_L^{\underline{\varphi}_L}$ of L, where L is a finite extension of the global field K.

Remark 2. Let L be a finite extension of the global field K. Fixing $\underline{\varphi}_K = \{\varphi_{K_\nu}\}_{\nu \in \mathbb{h}_K}$ uniquely determines $\underline{\varphi}_L = \{\varphi_{L_\mu}\}_{\mu \in \mathbb{h}_L}$ via Koch-de Shalit process applied to compatible extensions of K_ν for each $\nu \in \mathbb{h}_K$ (for details, look at [6]).

Thus, let L be a finite extension of the global field K. The absolute Weil group W_L of L is the open subgroup of W_K defined by $W_L = \beta_K^{-1}(G_L)$, where the absolute Weil group W_K of the global field K comes equipped with a continuous homomorphism $\beta_K : W_K \to G_K$ with dense image. Moreover, the open subgroup W_L of W_K is equipped with a continuous homomorphism $\beta_L : W_L \to G_L$, which sits in the commutative square

$$
\begin{array}{ccc}
W_L & \xrightarrow{\beta_L} & G_L \\
{\scriptstyle \gamma_{L/K}}\downarrow & & \downarrow{\scriptstyle \mathrm{id}_{G_L}} \\
W_K & \xrightarrow[\beta_K]{} & G_K
\end{array}
\tag{7}
$$

and with dense image, where the left vertical arrow

$$\gamma_{L/K} : W_L \hookrightarrow W_K$$

is the natural embedding, that is, the identity map defined by the inclusion mapping $W_L := \beta_K^{-1}(G_L) \subset W_K$. Let $\nu \in \mathbb{h}_K \cup \mathbb{a}_K$ and $\mu \in \mathbb{h}_L \cup \mathbb{a}_L$ so that $\mu \mid \nu$. Then, L_μ is a finite extension of K_ν, and W_{L_μ} is an open subgroup of W_{K_ν} defined by $W_{L_\mu} = \beta_{K_\nu}^{-1}\left(G_{L_\mu}\right)$, where the absolute Weil group W_{K_ν} of the local field K_ν comes equipped with a continuous homomorphism $\beta_{K_\nu} : W_{K_\nu} \to G_{K_\nu}$ with dense image. The open subgroup W_{L_μ} of W_{K_ν} is equipped with a continuous homomorphism $\beta_{L_\mu} : W_{L_\mu} \to G_{L_\mu}$ with dense image, and the square

$$\begin{array}{ccc} W_{L_\mu} & \xrightarrow{\beta_{L_\mu}} & G_{L_\mu} \\ {\scriptstyle \gamma_{L_\mu/K_\nu}}\downarrow & & \downarrow{\scriptstyle \mathrm{id}_{G_{L_\mu}}} \\ W_{K_\nu} & \xrightarrow[\beta_{K_\nu}]{} & G_{K_\nu} \end{array} \tag{8}$$

commutes, where the left vertical arrow

$$\gamma_{L_\mu/K_\nu} : W_{L_\mu} \hookrightarrow W_{K_\nu}$$

is the natural embedding, namely, the identity mapping defined by the inclusion $W_{L_\mu} := \beta_{K_\nu}^{-1}(G_{L_\mu}) \subset W_{K_\nu}$.

For any place $\nu \in \mathbb{h}_K \cup \mathbb{a}_K$ of K, the fixed embedding

$$e_\nu : K^{\mathrm{sep}} \hookrightarrow K_\nu^{\mathrm{sep}} \tag{9}$$

uniquely determines a continuous homomorphism

$$e_\nu^{\mathrm{Galois}} : G_{K_\nu} \to G_K$$

defined by the restriction to K^{sep} and a continuous homomorphism

$$e_\nu^{\mathrm{Weil}} : W_{K_\nu} \to W_K$$

so that the following square

$$\begin{array}{ccc} W_{K_\nu} & \xrightarrow{e_\nu^{\mathrm{Weil}}} & W_K \\ {\scriptstyle \beta_{K_\nu}}\downarrow & & \downarrow{\scriptstyle \beta_K} \\ G_{K_\nu} & \xrightarrow[e_\nu^{\mathrm{Galois}}]{} & G_K \end{array} \tag{10}$$

is commutative (for details, look at Proposition 1.6.1 of Tate [16]). Moreover, for any finite extension L/K, the fixed embedding (9) uniquely determines an embedding

$$e_\mu : L^{\mathrm{sep}} \hookrightarrow L_\mu^{\mathrm{sep}} \tag{11}$$

which is defined to be the unique arrow that makes the square

$$\begin{array}{ccc} L^{sep} & \overset{\exists!\, e_\mu}{\dashrightarrow} & L_\mu^{sep} \\ \| & & \| \\ K^{sep} & \xrightarrow[e_\nu]{} & K_\nu^{sep} \end{array}$$

commutative, for every $\mu \in \mathbb{h}_L \cup \mathbb{a}_L$ satisfying $\mu \mid \nu$. Therefore, the embedding (11) uniquely determines a continuous homomorphism

$$e_{\mu}^{\text{Galois}} : G_{L_\mu} \to G_L$$

defined by the restriction to L^{sep}. The relationship between the continuous homomorphisms e_{ν}^{Galois} and e_{μ}^{Galois} is given by the following lemma.

Lemma 3. *Let L be a finite extension of the global field K. For any finite or infinite prime ν of K and for any prime μ of L lying above ν, the following square*

$$\begin{array}{ccc} G_{L_\mu} & \xrightarrow{e_{\mu}^{\text{Galois}}} & G_L \\ {\scriptstyle \text{id}_{G_{L_\mu}}} \downarrow & & \downarrow {\scriptstyle \text{id}_{G_L}} \\ G_{K_\nu} & \xrightarrow[e_{\nu}^{\text{Galois}}]{} & G_K \end{array} \tag{12}$$

is commutative.

Proof. In fact, for any $\sigma \in G_{L_\mu}$,

$$\begin{aligned} \text{id}_{G_L} \circ e_{\mu}^{\text{Galois}}(\sigma) &= \sigma \mid_{L^{\text{sep}}} \\ &= \sigma \mid_{K^{\text{sep}}} \\ &= e_{\nu}^{\text{Galois}} \circ \text{id}_{G_{L_\mu}}(\sigma) \end{aligned}$$

which completes the proof of the commutativity of the square (12). □

The embedding (11) also determines a continuous homomorphism

$$e_{\mu}^{\text{Weil}} : W_{L_\mu} \to W_L$$

defined by

$$e_{\mu}^{\text{Weil}}(w) = e_{\nu}^{\text{Weil}} \circ \gamma_{L_\mu/K_\nu}(w),$$

for every $w \in W_{L_\mu}$. In fact, for each $w \in W_{L_\mu}$, the commutativity of the square (10) yields

$$\begin{aligned} \left(\beta_K \circ e_{\nu}^{\text{Weil}}\right) \circ \gamma_{L_\mu/K_\nu}(w) &= \left(e_{\nu}^{\text{Galois}} \circ \beta_{K_\nu}\right) \circ \gamma_{L_\mu/K_\nu}(w) \\ &= e_{\nu}^{\text{Galois}} \circ \left(\beta_{K_\nu} \circ \gamma_{L_\mu/K_\nu}\right)(w) \\ &= e_{\nu}^{\text{Galois}} \circ \left(\text{id}_{G_{L_\mu}} \circ \beta_{L_\mu}\right)(w) \end{aligned}$$

where the last equality follows from the commutative square (8). Therefore,

$$\begin{aligned}\left(\beta_K \circ e_\nu^{\mathrm{Weil}}\right) \circ \gamma_{L_\mu/K_\nu}(w) &= e_\nu^{\mathrm{Galois}} \circ \left(\mathrm{id}_{G_{L_\mu}} \circ \beta_{L_\mu}\right)(w)\\ &= \left(e_\nu^{\mathrm{Galois}} \circ \mathrm{id}_{G_{L_\mu}}\right) \circ \beta_{L_\mu}(w)\\ &= \left(\mathrm{id}_{G_L} \circ e_\mu^{\mathrm{Galois}}\right) \circ \beta_{L_\mu}(w)\end{aligned}$$

by Lemma 3. Therefore, for each $w \in W_{L_\mu}$,

$$e_\nu^{\mathrm{Weil}} \circ \gamma_{L_\mu/K_\nu}(w) \in \beta_K^{-1}(G_L) = W_L.$$

The relationship between the continuous homomorphisms e_ν^{Weil} and e_μ^{Weil} is described in the following lemma.

Lemma 4. *Let L be a finite extension of the global field K. For any finite or infinite prime ν of K and for any prime μ of L lying above ν, the following square*

$$\begin{array}{ccc} W_{L_\mu} & \xrightarrow{e_\mu^{\mathrm{Weil}}} & W_L \\ {\scriptstyle \gamma_{L_\mu/K_\nu}}\downarrow & & \downarrow{\scriptstyle \gamma_{L/K}} \\ W_{K_\nu} & \xrightarrow[e_\nu^{\mathrm{Weil}}]{} & W_K \end{array} \tag{13}$$

is commutative.

Proof. For any $w \in W_{L_\mu}$, the following equalities

$$\begin{aligned}e_\nu^{\mathrm{Weil}} \circ \gamma_{L_\mu/K_\nu}(w) &= e_\mu^{\mathrm{Weil}}(w)\\ &= \gamma_{L/K} \circ e_\mu^{\mathrm{Weil}}(w)\end{aligned}$$

are immediate from the definition of $e_\mu^{\mathrm{Weil}} : W_{L_\mu} \to W_L$, completing the proof of the commutativity of the square (13). □

Moreover, the continuous homomorphisms $e_\mu^{\mathrm{Galois}} : G_{L_\mu} \to G_L$ and $e_\mu^{\mathrm{Weil}} : W_{L_\mu} \to W_L$ make the following square

$$\begin{array}{ccc} W_{L_\mu} & \xrightarrow{e_\mu^{\mathrm{Weil}}} & W_L \\ {\scriptstyle \beta_{L_\mu}}\downarrow & & \downarrow{\scriptstyle \beta_L} \\ G_{L_\mu} & \xrightarrow[e_\mu^{\mathrm{Galois}}]{} & G_L \end{array} \tag{14}$$

commutative. In fact,

$$\begin{aligned}\beta_L \circ e_\mu^{\mathrm{Weil}} &= \left(\beta_K \circ \gamma_{L/K}\right) \circ e_\mu^{\mathrm{Weil}} \\ &= \beta_K \circ e_\nu^{\mathrm{Weil}} \circ \gamma_{L_\mu/K_\nu},\end{aligned}$$

where the equalities follow from the commutativity of the squares (7) and (13). Therefore,

$$\begin{aligned}\beta_L \circ e_\mu^{\mathrm{Weil}} &= \left(\beta_K \circ e_\nu^{\mathrm{Weil}}\right) \circ \gamma_{L_\mu/K_\nu} \\ &= e_\nu^{\mathrm{Galois}} \circ \left(\beta_{K_\nu} \circ \gamma_{L_\mu/K_\nu}\right) \\ &= e_\nu^{\mathrm{Galois}} \circ \left(\mathrm{id}_{G_{L_\mu}} \circ \beta_{L_\mu}\right)\end{aligned}$$

by the commutative squares (10) and (8). Thus, by Lemma 3,

$$\begin{aligned}\beta_L \circ e_\mu^{\mathrm{Weil}} &= \left(e_\nu^{\mathrm{Galois}} \circ \mathrm{id}_{G_{L_\mu}}\right) \circ \beta_{L_\mu} \\ &= \mathrm{id}_{G_L} \circ \left(e_\mu^{\mathrm{Galois}} \circ \beta_{L_\mu}\right) \\ &= e_\mu^{\mathrm{Galois}} \circ \beta_{L_\mu},\end{aligned}$$

which completes the proof of the commutativity of the square (14).

Now, for any $\nu \in \mathbb{h}_K$ and for any $\mu \in \mathbb{h}_L$ satisfying $\mu \mid \nu$, the following square

$$\begin{array}{ccc} W_{L_\mu} & \xrightarrow[\sim]{\boldsymbol{\Phi}_{L_\mu}^{(\varphi_{L_\mu})}} & {}_{\mathbb{Z}}\nabla_{L_\mu}^{(\varphi_{L_\mu})} \\ {\scriptstyle \gamma_{L_\mu/K_\nu}}\big\downarrow & & \big\downarrow{\scriptstyle \mathscr{N}_{L_\mu/K_\nu}^{\infty}} \\ W_{K_\nu} & \xrightarrow[\sim]{\boldsymbol{\Phi}_{K_\nu}^{(\varphi_{K_\nu})}} & {}_{\mathbb{Z}}\nabla_{K_\nu}^{(\varphi_{K_\nu})} \end{array} \tag{15}$$

is commutative (look at pp. 39 of [6] for details), where the right vertical arrow

$$\mathscr{N}_{L_\mu/K_\nu}^{\infty} : {}_{\mathbb{Z}}\nabla_{L_\mu}^{(\varphi_{L_\mu})} \to {}_{\mathbb{Z}}\nabla_{K_\nu}^{(\varphi_{K_\nu})}$$

is the continuous homomorphism defined in pp. 39 of [6].

For each $\mu \in \mathbb{h}_L$, define a continuous homomorphism

$${}_{\mathbb{Z}}\nabla_{L_\mu}^{(\varphi_{L_\mu})} \to \mathscr{J}_K^{\underline{\varphi}_K}$$

by the composition

$$_{\mathbb{Z}}\nabla_{L_\mu}^{(\varphi_{L_\mu})} \xrightarrow{\mathscr{N}^{\infty}_{L_\mu/K_\nu}} {}_{\mathbb{Z}}\nabla_{K_\nu}^{(\varphi_{K_\nu})} \xrightarrow{q_\nu} \mathscr{I}_K^{\underline{\varphi}_K},$$

and for each $\mu \in \mathbb{a}_L$, define a continuous homomorphism

$$W_{L_\mu} \to \mathscr{I}_K^{\underline{\varphi}_K}$$

by the composition

$$W_{L_\mu} \underset{\gamma_{L_\mu/K_\nu}}{\hookrightarrow} W_{K_\nu} \xrightarrow{q_\nu} \mathscr{I}_K^{\underline{\varphi}_K}.$$

By Proposition 2.1 of [4], the following proposition follows at once.

Proposition 2. *There exists a unique continuous homomorphism*

$$\mathscr{N}^{\infty}_{L/K} : \mathscr{I}_L^{\underline{\varphi}_L} \to \mathscr{I}_K^{\underline{\varphi}_K},$$

called the "norm" homomorphism between the non-Abelian idèle group $\mathscr{I}_L^{\underline{\varphi}_L}$ of L and the non-Abelian idèle group $\mathscr{I}_K^{\underline{\varphi}_K}$ of K, which satisfies

$$\left(\mathscr{N}^{\infty}_{L/K}\right)_S = \mathscr{N}^{\infty}_{L/K} \circ c_S : \left(\mathscr{I}_L^{\underline{\varphi}_L}\right)_S \xrightarrow{c_S} \mathscr{I}_L^{\underline{\varphi}_L} \xrightarrow{\mathscr{N}^{\infty}_{L/K}} \mathscr{I}_K^{\underline{\varphi}_K},$$

where $c_S : \left(\mathscr{I}_L^{\underline{\varphi}_L}\right)_S \to \mathscr{I}_L^{\underline{\varphi}_L}$ is the canonical homomorphism defined for every finite subset S of $\mathbb{h}_L \cup \mathbb{a}_L$ containing $\mathbb{a}_L$.

Note that this homomorphism is transitive in the tower of finite extensions of the global field K.

Proposition 3. *Let $K \subseteq L \subseteq E$ be a tower of finite extensions of the global field K. Then, the equality*

$$\mathscr{N}^{\infty}_{E/K} = \mathscr{N}^{\infty}_{L/K} \circ \mathscr{N}^{\infty}_{E/L}$$

holds.

Proof. It suffices to prove that

$$\left(\mathscr{N}^{\infty}_{E/K}\right)_S \circ \iota_\lambda^{(S)} = \left(\mathscr{N}^{\infty}_{L/K} \circ \mathscr{N}^{\infty}_{E/L}\right)_S \circ \iota_\lambda^{(S)},$$

where $\lambda \in \mathbb{h}_E \cup \mathbb{a}_E$ and S a finite subset of $\mathbb{h}_E \cup \mathbb{a}_E$ such that $\lambda \in S$ and $\mathbb{a}_E \subset S$. So, let λ be any prime of E and S any such subset of $\mathbb{h}_E \cup \mathbb{a}_E$. The composite homomorphism

$$\left(\mathscr{N}_{E/K}^{\infty}\right)_S = \mathscr{N}_{E/K}^{\infty} \circ c_S : \left(\mathscr{I}_E^{\varphi_E}\right)_S \xrightarrow{c_S} \mathscr{I}_E^{\varphi_E} \xrightarrow{\mathscr{N}_{E/K}^{\infty}} \mathscr{I}_K^{\varphi_K}$$

satisfies

$$\left(\mathscr{N}_{E/K}^{\infty}\right)_S \circ \iota_\lambda^{(S)} = \left(\mathscr{N}_{E/K}^{\infty}\right)_\lambda$$

and

$$\left(\mathscr{N}_{E/K}^{\infty}\right)_S \circ \iota_\lambda^{(S)} = \left(\mathscr{N}_{E/K}^{\infty} \circ c_S\right) \circ \iota_\lambda^{(S)} = \mathscr{N}_{E/K}^{\infty} \circ \left(c_S \circ \iota_\lambda^{(S)}\right) = \mathscr{N}_{E/K}^{\infty} \circ q_\lambda.$$

Therefore,

$$\left(\mathscr{N}_{E/K}^{\infty}\right)_S \circ \iota_\lambda^{(S)} = \mathscr{N}_{E/K}^{\infty} \circ q_\lambda = \left(\mathscr{N}_{E/K}^{\infty}\right)_\lambda.$$

If $\lambda \in \mathbb{h}_E$, then

$$\left(\mathscr{N}_{E/K}^{\infty}\right)_S \circ \iota_\lambda^{(S)} = \mathscr{N}_{E/K}^{\infty} \circ q_\lambda = \left(\mathscr{N}_{E/K}^{\infty}\right)_\lambda = q_\nu \circ \mathscr{N}_{E_\lambda/K_\nu}^{\infty},$$

where $\nu \in \mathbb{h}_K$ is the finite prime of K given by $\nu = \lambda \cap O_K$. Moreover,

$$\begin{aligned}
\left(\mathscr{N}_{L/K}^{\infty} \circ \mathscr{N}_{E/L}^{\infty}\right)_S \circ \iota_\lambda^{(S)} &= \left(\mathscr{N}_{L/K}^{\infty} \circ \mathscr{N}_{E/L}^{\infty}\right) \circ q_\lambda \\
&= \mathscr{N}_{L/K}^{\infty} \circ \left(\mathscr{N}_{E/L}^{\infty} \circ q_\lambda\right) \\
&= \mathscr{N}_{L/K}^{\infty} \circ \left(\mathscr{N}_{E/L}^{\infty}\right)_\lambda \\
&= \mathscr{N}_{L/K}^{\infty} \circ \left(q_\mu \circ \mathscr{N}_{E_\lambda/L_\mu}^{\infty}\right),
\end{aligned}$$

where $\mu \in \mathbb{h}_L$ is the finite prime of L defined by $\mu = \lambda \cap O_L$. So it follows that

$$\begin{aligned}
\left(\mathscr{N}_{L/K}^{\infty} \circ \mathscr{N}_{E/L}^{\infty}\right)_S \circ \iota_\lambda^{(S)} &= \mathscr{N}_{L/K}^{\infty} \circ \left(q_\mu \circ \mathscr{N}_{E_\lambda/L_\mu}^{\infty}\right) \\
&= \left(\mathscr{N}_{L/K}^{\infty} \circ q_\mu\right) \circ \mathscr{N}_{E_\lambda/L_\mu}^{\infty} \\
&= \left(\mathscr{N}_{L/K}^{\infty}\right)_\mu \circ \mathscr{N}_{E_\lambda/L_\mu}^{\infty} \\
&= q_\nu \circ \mathscr{N}_{L_\mu/K_\nu}^{\infty} \circ \mathscr{N}_{E_\lambda/L_\mu}^{\infty},
\end{aligned}$$

where $\nu \in \mathbb{h}_K$ is the finite prime of K given by $\nu = \lambda \cap O_K = \mu \cap O_K$. Now, by the transitivity rule proved in pp. 39 of [6],

$$\mathscr{N}^{\infty}_{L_\mu/K_\nu} \circ \mathscr{N}^{\infty}_{E_\lambda/L_\mu} = \mathscr{N}^{\infty}_{E_\lambda/K_\nu}.$$

Therefore, it follows that

$$\left(\mathscr{N}^{\infty}_{E/K}\right)_S \circ \iota^{(S)}_\lambda = \left(\mathscr{N}^{\infty}_{L/K} \circ \mathscr{N}^{\infty}_{E/L}\right)_S \circ \iota^{(S)}_\lambda$$

for $\lambda \in \mathbb{h}_E$. If $\lambda \in \mathbb{a}_E$, then

$$\left(\mathscr{N}^{\infty}_{E/K}\right)_S \circ \iota^{(S)}_\lambda = \mathscr{N}^{\infty}_{E/K} \circ q_\lambda = \left(\mathscr{N}^{\infty}_{E/K}\right)_\lambda = q_\nu \circ \gamma_{E_\lambda/K_\nu},$$

where $\nu \in \mathbb{a}_K$ is the infinite prime of K defined by $\nu = \lambda\mid_K$. Also,

$$\begin{aligned}\left(\mathscr{N}^{\infty}_{L/K} \circ \mathscr{N}^{\infty}_{E/L}\right)_S \circ \iota^{(S)}_\lambda &= \left(\mathscr{N}^{\infty}_{L/K} \circ \mathscr{N}^{\infty}_{E/L}\right) \circ q_\lambda \\ &= \mathscr{N}^{\infty}_{L/K} \circ \left(\mathscr{N}^{\infty}_{E/L} \circ q_\lambda\right) \\ &= \mathscr{N}^{\infty}_{L/K} \circ \left(\mathscr{N}^{\infty}_{E/L}\right)_\lambda \\ &= \mathscr{N}^{\infty}_{L/K} \circ \left(q_\mu \circ \gamma_{E_\lambda/L_\mu}\right),\end{aligned}$$

where $\mu \in \mathbb{a}_L$ is the infinite prime of L defined by $\mu = \lambda\mid_L$. Thus, it follows that

$$\begin{aligned}\left(\mathscr{N}^{\infty}_{L/K} \circ \mathscr{N}^{\infty}_{E/L}\right)_S \circ \iota^{(S)}_\lambda &= \mathscr{N}^{\infty}_{L/K} \circ \left(q_\mu \circ \gamma_{E_\lambda/L_\mu}\right) \\ &= \left(\mathscr{N}^{\infty}_{L/K} \circ q_\mu\right) \circ \gamma_{E_\lambda/L_\mu} \\ &= \left(\mathscr{N}^{\infty}_{L/K}\right)_\mu \circ \gamma_{E_\lambda/L_\mu} \\ &= q_\nu \circ \gamma_{L_\mu/K_\nu} \circ \gamma_{E_\lambda/L_\mu},\end{aligned}$$

where $\nu \in \mathbb{a}_K$ is the infinite prime of K defined by $\nu = \lambda\mid_K = \mu\mid_K$. Now, by the transitivity rule

$$\gamma_{L_\mu/K_\nu} \circ \gamma_{E_\lambda/L_\mu} = \gamma_{E_\lambda/K_\nu},$$

it follows that

$$\left(\mathscr{N}^{\infty}_{E/K}\right)_S \circ \iota^{(S)}_\lambda = \left(\mathscr{N}^{\infty}_{L/K} \circ \mathscr{N}^{\infty}_{E/L}\right)_S \circ \iota^{(S)}_\lambda,$$

for $\lambda \in \mathbb{a}_E$, which completes the proof. □

Theorem 3. *Let L be a finite extension of the global field K. Then, the following square*

$$\begin{array}{ccc}
\mathcal{J}_L^{\varphi_L} & \xrightarrow{\mathsf{NR}_L^{\varphi_L}{}^{\mathrm{Weil}}} & W_L \\
{\scriptstyle \mathcal{N}_{L/K}^{\infty}}\downarrow & & \downarrow{\scriptstyle \gamma_{L/K}} \\
\mathcal{J}_K^{\varphi_K} & \xrightarrow[\mathsf{NR}_K^{\varphi_K}{}^{\mathrm{Weil}}]{} & W_K
\end{array}$$

is commutative.

Proof. It suffices to prove that

$$\left(\mathsf{NR}_K^{\varphi_K}{}^{\mathrm{Weil}} \circ \mathcal{N}_{L/K}^{\infty}\right)_S \circ \iota_\mu^{(S)} = \left(\gamma_{L/K} \circ \mathsf{NR}_L^{\varphi_L}{}^{\mathrm{Weil}}\right)_S \circ \iota_\mu^{(S)},$$

where $\mu \in \mathbb{h}_L \cup \mathbb{a}_L$ and S is any finite subset of $\mathbb{h}_L \cup \mathbb{a}_L$ satisfying $\mu \in S$ and $\mathbb{a}_L \subset S$. So, let μ be any prime of L and S any such subset of $\mathbb{h}_L \cup \mathbb{a}_L$. Then, clearly the following identities

$$\begin{aligned}
\left(\mathsf{NR}_K^{\varphi_K}{}^{\mathrm{Weil}} \circ \mathcal{N}_{L/K}^{\infty}\right)_S \circ \iota_\mu^{(S)} &= \left(\mathsf{NR}_K^{\varphi_K}{}^{\mathrm{Weil}} \circ \mathcal{N}_{L/K}^{\infty}\right) \circ q_\mu \\
&= \mathsf{NR}_K^{\varphi_K}{}^{\mathrm{Weil}} \circ \left(\mathcal{N}_{L/K}^{\infty} \circ q_\mu\right) \\
&= \mathsf{NR}_K^{\varphi_K}{}^{\mathrm{Weil}} \circ \left(\mathcal{N}_{L/K}^{\infty}\right)_\mu
\end{aligned}$$

hold. Now, first assume that $\mu \in \mathbb{h}_L$. Then,

$$\begin{aligned}
\left(\mathsf{NR}_K^{\varphi_K}{}^{\mathrm{Weil}} \circ \mathcal{N}_{L/K}^{\infty}\right)_S \circ \iota_\mu^{(S)} &= \mathsf{NR}_K^{\varphi_K}{}^{\mathrm{Weil}} \circ \left(\mathcal{N}_{L/K}^{\infty}\right)_\mu \\
&= \mathsf{NR}_K^{\varphi_K}{}^{\mathrm{Weil}} \circ \left(q_\nu \circ \mathcal{N}_{L_\mu/K_\nu}^{\infty}\right) \\
&= \left(\mathsf{NR}_K^{\varphi_K}{}^{\mathrm{Weil}} \circ q_\nu\right) \circ \mathcal{N}_{L_\mu/K_\nu}^{\infty} \\
&= \left(e_\nu^{\mathrm{Weil}} \circ \{\bullet, K_\nu\}_{\varphi_{K_\nu}}\right) \circ \mathcal{N}_{L_\mu/K_\nu}^{\infty} \\
&= e_\nu^{\mathrm{Weil}} \circ \left(\{\bullet, K_\nu\}_{\varphi_{K_\nu}} \circ \mathcal{N}_{L_\mu/K_\nu}^{\infty}\right),
\end{aligned}$$

where $\nu \in \mathbb{h}_K$ is the finite prime of K defined by $\nu = \mu \cap O_K$. Now, the diagram (15) (equivalently, the square (7.4) of [6]), namely, the following square

$$\begin{array}{ccc} W_{L_\mu} & \xleftarrow[\sim]{\{\bullet, L_\mu\}_{\varphi_{L_\mu}}} & {}_{\mathbb{Z}}\nabla_{L_\mu}^{(\varphi_{L_\mu})} \\ {\scriptstyle \gamma_{L_\mu/K_\nu}} \downarrow & & \downarrow {\scriptstyle \mathcal{N}_{L_\mu/K_\nu}^{\infty}} \\ W_{K_\nu} & \xleftarrow[\{\bullet, K_\nu\}_{\varphi_{K_\nu}}]{\sim} & {}_{\mathbb{Z}}\nabla_{K_\nu}^{(\varphi_{K_\nu})} \end{array}$$

is commutative. Therefore,

$$\begin{aligned} \left(\mathsf{NR}_K^{\varphi_K^{\text{Weil}}} \circ \mathcal{N}_{L/K}^{\infty}\right)_S \circ \iota_\mu^{(S)} &= e_\nu^{\text{Weil}} \circ \left(\{\bullet, K_\nu\}_{\varphi_{K_\nu}} \circ \mathcal{N}_{L_\mu/K_\nu}^{\infty}\right) \\ &= e_\nu^{\text{Weil}} \circ \left(\gamma_{L_\mu/K_\nu} \circ \{\bullet, L_\mu\}_{\varphi_{L_\mu}}\right) \\ &= \left(e_\nu^{\text{Weil}} \circ \gamma_{L_\mu/K_\nu}\right) \circ \{\bullet, L_\mu\}_{\varphi_{L_\mu}}. \end{aligned}$$

Now, by commutative square (13) of Lemma 4, continuing the computation,

$$\begin{aligned} \left(\mathsf{NR}_K^{\varphi_K^{\text{Weil}}} \circ \mathcal{N}_{L/K}^{\infty}\right)_S \circ \iota_\mu^{(S)} &= \left(e_\nu^{\text{Weil}} \circ \gamma_{L_\mu/K_\nu}\right) \circ \{\bullet, L_\mu\}_{\varphi_{L_\mu}} \\ &= \left(\gamma_{L/K} \circ e_\mu^{\text{Weil}}\right) \circ \{\bullet, L_\mu\}_{\varphi_{L_\mu}} \\ &= \gamma_{L/K} \circ \left(e_\mu^{\text{Weil}} \circ \{\bullet, L_\mu\}_{\varphi_{L_\mu}}\right) \\ &= \gamma_{L/K} \circ \left(\mathsf{NR}_L^{\varphi_L^{\text{Weil}}} \circ q_\mu\right), \end{aligned}$$

where the last equality follows from the local-global compatibility of the non-Abelian norm-residue symbols. Therefore,

$$\begin{aligned} \left(\mathsf{NR}_K^{\varphi_K^{\text{Weil}}} \circ \mathcal{N}_{L/K}^{\infty}\right)_S \circ \iota_\mu^{(S)} &= \gamma_{L/K} \circ \left(\mathsf{NR}_L^{\varphi_L^{\text{Weil}}} \circ q_\mu\right) \\ &= \left(\gamma_{L/K} \circ \mathsf{NR}_L^{\varphi_L^{\text{Weil}}}\right) \circ q_\mu \\ &= \left(\gamma_{L/K} \circ \mathsf{NR}_L^{\varphi_L^{\text{Weil}}}\right) \circ c_S \circ \iota_\mu^{(S)} \\ &= \left(\gamma_{L/K} \circ \mathsf{NR}_L^{\varphi_L^{\text{Weil}}}\right)_S \circ \iota_\mu^{(S)}. \end{aligned}$$

Now, if $\mu \in \infty_L$, then

$$\left(\mathsf{NR}_K^{\varphi_K^{\mathrm{Weil}}} \circ \mathcal{N}_{L/K}^{\infty}\right)_S \circ \iota_\mu^{(S)} = \mathsf{NR}_K^{\varphi_K^{\mathrm{Weil}}} \circ \left(\mathcal{N}_{L/K}^{\infty}\right)_\mu$$
$$= \mathsf{NR}_K^{\varphi_K^{\mathrm{Weil}}} \circ \left(q_\nu \circ \gamma_{L_\mu/K_\nu}\right),$$

by the equality $\left(\mathcal{N}_{L/K}^{\infty}\right)_\mu = q_\nu \circ \gamma_{L_\mu/K_\nu}$, where $\nu \in \infty_K$ is the infinite prime of K defined by $\nu = \mu\mid_K$. Thus,

$$\left(\mathsf{NR}_K^{\varphi_K^{\mathrm{Weil}}} \circ \mathcal{N}_{L/K}^{\infty}\right)_S \circ \iota_\mu^{(S)} = \mathsf{NR}_K^{\varphi_K^{\mathrm{Weil}}} \circ \left(q_\nu \circ \gamma_{L_\mu/K_\nu}\right)$$
$$= \left(\mathsf{NR}_K^{\varphi_K^{\mathrm{Weil}}} \circ q_\nu\right) \circ \gamma_{L_\mu/K_\nu}$$
$$= e_\nu^{\mathrm{Weil}} \circ \gamma_{L_\mu/K_\nu},$$

as $\left(\mathsf{NR}_K^{\varphi_K^{\mathrm{Weil}}}\right)_\nu = \mathsf{NR}_K^{\varphi_K^{\mathrm{Weil}}} \circ q_\nu = e_\nu^{\mathrm{Weil}}$. Therefore,

$$\left(\mathsf{NR}_K^{\varphi_K^{\mathrm{Weil}}} \circ \mathcal{N}_{L/K}^{\infty}\right)_S \circ \iota_\mu^{(S)} = e_\nu^{\mathrm{Weil}} \circ \gamma_{L_\mu/K_\nu}$$
$$= \gamma_{L/K} \circ e_\mu^{\mathrm{Weil}}, \tag{16}$$

where the last equality follows from Lemma 4. Note that

$$\left(\mathsf{NR}_L^{\varphi_L^{\mathrm{Weil}}}\right)_\mu = \mathsf{NR}_L^{\varphi_L^{\mathrm{Weil}}} \circ q_\mu = e_\mu^{\mathrm{Weil}}. \tag{17}$$

Thus, substituting (17) into (16),

$$\left(\mathsf{NR}_K^{\varphi_K^{\mathrm{Weil}}} \circ \mathcal{N}_{L/K}^{\infty}\right)_S \circ \iota_\mu^{(S)} = \gamma_{L/K} \circ e_\mu^{\mathrm{Weil}}$$
$$= \gamma_{L/K} \circ \left(\mathsf{NR}_L^{\varphi_L^{\mathrm{Weil}}} \circ q_\mu\right)$$
$$= \left(\gamma_{L/K} \circ \mathsf{NR}_L^{\varphi_L^{\mathrm{Weil}}}\right) \circ q_\mu$$
$$= \left(\gamma_{L/K} \circ \mathsf{NR}_L^{\varphi_L^{\mathrm{Weil}}}\right) \circ c_S \circ \iota_\mu^{(S)}$$
$$= \left(\gamma_{L/K} \circ \mathsf{NR}_L^{\varphi_L^{\mathrm{Weil}}}\right)_S \circ \iota_\mu^{(S)},$$

where $\nu \in \infty_K$ is the infinite prime of K defined by $\nu = \mu\mid_K$. This completes the proof. □

4.4 Relative Non-Abelian Global Reciprocity Maps

Recall that the absolute Weil group W_K of the global field K comes equipped with a continuous homomorphism $\beta_K : W_K \to G_K$ with dense image. Let L be a finite Galois extension of K. Then, $W_L := \beta_K^{-1}(G_L)$ is an open subgroup of W_K, and there exists an isomorphism of topological groups

$$W_K/W_L \xrightarrow[\sim]{\beta_{L/K}} G_K/G_L \xrightarrow[\sim]{\mathrm{res}_L^*} \mathrm{Gal}(L/K),$$

where the left arrow is defined by

$$\beta_{L/K} : w \pmod{W_L} \mapsto \beta_K(w) \pmod{G_L}$$

for every $w \in W_K$, and res_L^* is the isomorphism induced from the surjective homomorphism $\mathrm{res}_L : G_K \to \mathrm{Gal}(L/K)$.

The following two lemmas are trivial but useful.

Lemma 5. *Let L be a finite Galois extension of the global field K. Then, the following square*

$$\begin{array}{ccc} W_K & \xrightarrow{\mathrm{red}_{W_L}} & W_K/W_L \\ {\scriptstyle \beta_K}\downarrow & & \downarrow{\scriptstyle \beta_{L/K}} \\ G_K & \xrightarrow[\mathrm{red}_{G_L}]{} & G_K/G_L \end{array} \tag{18}$$

is commutative, where the top and the bottom horizontal arrows are the reduction modulo W_L and the reduction modulo G_L morphisms, respectively.

Lemma 6. *Let $K \subseteq L \subseteq E$ be a tower of finite Galois extensions of the global field K. Then, the following diagram*

$$\begin{array}{ccc} W_L & \xrightarrow{\mathrm{red}_{W_E}} & W_L/W_E \\ {\scriptstyle \gamma_{L/K}}\downarrow & & \downarrow{\scriptstyle \beta_{E/L}} \\ W_K & & G_L/G_E \\ {\scriptstyle \mathrm{red}_{W_E}}\downarrow & & \downarrow{\scriptstyle \mathrm{res}_E^*} \\ W_K/W_E & & \mathrm{Gal}(E/L) \\ {\scriptstyle \beta_{E/K}}\downarrow & & \downarrow{\scriptstyle \mathrm{id}_{\mathrm{Gal}(E/L)}} \\ G_K/G_E & \xrightarrow[\mathrm{res}_E^*]{} & \mathrm{Gal}(E/K) \end{array} \tag{19}$$

is commutative.

Proof. The proof follows from the equality

$$\beta_L(w) \mid_E = \beta_K(w) \mid_E,$$

for every $w \in W_L$. □

For each finite Galois extension L/K, there exists a continuous homomorphism defined by the composition

$$\mathsf{NR}_{L/K}^{\varphi_K^{\text{Weil}}} : \mathscr{I}_K^{\varphi_K} \xrightarrow{\mathsf{NR}_K^{\varphi_K^{\text{Weil}}}} W_K \xrightarrow[\text{reduction modulo } W_L]{\text{red}_{W_L}} W_K/W_L \xrightarrow[\sim]{\beta_{L/K}} G_K/G_L \xrightarrow[\sim]{\text{res}_L^*} \text{Gal}(L/K)$$

and called the *non-Abelian global reciprocity map relative to the extension L/K*, or the *global non-Abelian norm-residue symbol relative to the extension L/K* in this work.

Remark 3. This continuous homomorphism is furthermore a surjection if we assume that Conjecture 1 holds.

Notation 2. Keeping the notation introduced in [4], let $\mathscr{N}_K^{\varphi_K}$ denote the kernel $\ker\left(\mathsf{NR}_K^{\varphi_K^{\text{Weil}}}\right)$ of the global non-Abelian norm-residue symbol $\mathsf{NR}_K^{\varphi_K^{\text{Weil}}} : \mathscr{I}_K^{\varphi_K} \to W_K$ of K.

Theorem 4. *Assume that Conjecture 1 holds. The global non-Abelian norm-residue symbol*

$$\mathsf{NR}_{L/K}^{\varphi_K^{\text{Weil}}} : \mathscr{I}_K^{\varphi_K} \to \text{Gal}(L/K),$$

relative to the finite Galois extension L/K is surjective with open kernel

$$\ker\left(\mathsf{NR}_{L/K}^{\varphi_K^{\text{Weil}}}\right) = \mathscr{N}_K^{\varphi_K} \mathscr{N}_{L/K}^{\infty}\left(\mathscr{I}_L^{\varphi_L}\right)$$

and induces a topological group isomorphism

$$\mathsf{NR}_{L/K}^{\varphi_K^{\text{Weil}*}} : \mathscr{I}_K^{\varphi_K}/\mathscr{N}_K^{\varphi_K} \mathscr{N}_{L/K}^{\infty}\left(\mathscr{I}_L^{\varphi_L}\right) \xrightarrow{\sim} \text{Gal}(L/K).$$

Proof. As we have assumed that Conjecture 1 holds, the surjectivity of the arrow $\mathsf{NR}_{L/K}^{\varphi_K^{\text{Weil}}} : \mathscr{I}_K^{\varphi_K} \to \text{Gal}(L/K)$ is clear. Moreover, the kernel $\ker\left(\mathsf{NR}_{L/K}^{\varphi_K^{\text{Weil}}}\right)$ of the continuous surjective homomorphism $\mathsf{NR}_{L/K}^{\varphi_K^{\text{Weil}}} : \mathscr{I}_K^{\varphi_K} \to \text{Gal}(L/K)$ is

$$\ker\left(\mathsf{NR}_{L/K}^{\varphi_K^{\text{Weil}}}\right) = \left(\mathsf{NR}_K^{\varphi_K^{\text{Weil}}}\right)^{-1}(W_L),$$

which proves that $\ker\left(\mathsf{NR}^{\varphi_K^{\text{Weil}}}_{L/K}\right)$ is open, as W_L is an open subgroup of W_K. Now, the inclusion $\mathcal{N}_K^{\varphi_K}\mathcal{N}_{L/K}^{\infty}\left(\mathcal{J}_L^{\varphi_L}\right) \subseteq \left(\mathsf{NR}_K^{\varphi_K^{\text{Weil}}}\right)^{-1}(W_L)$ follows from Theorem 3. In order to prove the reverse inclusion, let $x \in \mathcal{J}_K^{\varphi_K}$ such that $\mathsf{NR}_K^{\varphi_K^{\text{Weil}}}(x) \in W_L$. As Conjecture 1 is assumed to be true, there exists $x' \in \mathcal{J}_L^{\varphi_L}$ such that

$$\gamma_{L/K}\mathsf{NR}_L^{\varphi_L^{\text{Weil}}}(x') = \mathsf{NR}_K^{\varphi_K^{\text{Weil}}}(x).$$

Thus, again by Theorem 3,

$$\begin{aligned}\mathsf{NR}_K^{\varphi_K^{\text{Weil}}}(x) &= \gamma_{L/K}\mathsf{NR}_L^{\varphi_L^{\text{Weil}}}(x') \\ &= \mathsf{NR}_K^{\varphi_K^{\text{Weil}}}\left(\mathcal{N}_{L/K}^{\infty}(x')\right).\end{aligned}$$

Therefore, there exists $\xi \in \mathcal{N}_K^{\varphi_K}$ such that

$$\xi\mathcal{N}_{L/K}^{\infty}(x') = x.$$

Hence, the inclusion $\left(\mathsf{NR}_K^{\varphi_K^{\text{Weil}}}\right)^{-1}(W_L) \subseteq \mathcal{N}_K^{\varphi_K}\mathcal{N}_{L/K}^{\infty}\left(\mathcal{J}_L^{\varphi_L}\right)$ follows as well. So the equalities

$$\begin{aligned}\ker\left(\mathsf{NR}^{\varphi_K^{\text{Weil}}}_{L/K}\right) &= \left(\mathsf{NR}_K^{\varphi_K^{\text{Weil}}}\right)^{-1}(W_L) \\ &= \mathcal{N}_K^{\varphi_K}\mathcal{N}_{L/K}^{\infty}\left(\mathcal{J}_L^{\varphi_L}\right)\end{aligned}$$

hold. Finally, as W_L is an open subgroup of W_K, Conjecture 1 implies that the isomorphism

$$\mathsf{NR}^{\varphi_K^{\text{Weil}*}}_{L/K} : \mathcal{J}_K^{\varphi_K}/\mathcal{N}_K^{\varphi_K}\mathcal{N}_{L/K}^{\infty}\left(\mathcal{J}_L^{\varphi_L}\right) \xrightarrow{\sim} \text{Gal}(L/K)$$

induced from the non-Abelian global reciprocity map

$$\mathsf{NR}^{\varphi_K^{\text{Weil}}}_{L/K} : \mathcal{J}_K^{\varphi_K} \to \text{Gal}(L/K)$$

relative to the extension L/K is a topological isomorphism, which completes the proof. □

Remark 4. Again assume that Conjecture 1 holds. Let L/K be a finite separable (not necessarily Galois) extension. Then, following the same lines of reasoning of the proof of Theorem 4, the equality

$$\left(\mathsf{NR}_K^{\varphi_K^{\mathrm{Weil}}}\right)^{-1}(W_L) = \mathcal{N}_K^{\varphi_K}\mathcal{N}_{L/K}^{\infty}\left(\mathcal{I}_L^{\varphi_L}\right)$$

follows immediately from Theorem 3.

4.5 Relationship with the Relative Abelian Global Reciprocity Maps

Let L/K be a finite Galois extension. In this section, we shall study the "behavior" of the non-Abelian global reciprocity map relative to the extension L/K under the Abelianization functor and prove that under Abelianization, the non-Abelian global reciprocity map relative to the extension L/K reduces to the Abelian global reciprocity map relative to the Abelian extension $(L/K)^{ab}/K$.

Theorem 5. *The non-Abelian global reciprocity map*

$$\mathsf{NR}_{L/K}^{\varphi_K^{\mathrm{Weil}}} : \mathcal{I}_K^{\varphi} \to \mathrm{Gal}(L/K)$$

relative to the extension L/K sits in the following commutative diagram

$$\begin{array}{ccc}
\mathcal{I}_K^{\varphi} & \xrightarrow{\mathsf{NR}_{L/K}^{\varphi_K^{\mathrm{Weil}}}} & \mathrm{Gal}(L/K) \\
{\scriptstyle \mathrm{a}_K}\big\downarrow & & \big\downarrow{\scriptstyle \mathrm{res}_{(L/K)^{ab}}} \\
\mathbb{J}_K & \xrightarrow[(\bullet,L/K)]{} & \mathrm{Gal}((L/K)^{ab}/K).
\end{array}$$

Proof. By Theorem 2, the diagram

$$\begin{array}{ccc}
\mathcal{I}_K^{\varphi} & \xrightarrow{\mathsf{NR}_K^{\varphi^{\mathrm{Weil}}}} & W_K \\
{\scriptstyle \mathrm{a}_K}\big\downarrow & & \big\downarrow{\scriptstyle \mathrm{red}_{W_K^c}} \\
\mathbb{J}_K & \xrightarrow[(\bullet,K)]{} & W_K^{ab}
\end{array}$$

is commutative. Moreover, the diagram

$$
\begin{array}{ccccccc}
W_K & \xrightarrow{\mathrm{red}_{W_L}} & W_K/W_L & \xrightarrow{\beta_{L/K}} & G_K/G_L & \xrightarrow{\mathrm{res}_L^*} & \mathrm{Gal}(L/K) \\
\Big\downarrow{\scriptstyle \mathrm{red}_{W_K^c}} & & & & & & \Big\downarrow{\scriptstyle \mathrm{res}_{(L/K)^{ab}}} \\
W_K^{ab} & & \xrightarrow[\beta_K^{ab}]{\qquad\qquad} & & G_K^{ab} & \longrightarrow & \mathrm{Gal}((L/K)^{ab}/K)
\end{array}
$$

is trivially commutative, because $\mathrm{res}_{(L/K)^{ab}}(\beta_K(w)\mid_L) = \beta_K(w)\mid_{(L/K)^{ab}}$, for every $w \in W_K$. Thus, the commutativity of the diagram

$$
\begin{array}{ccccccccc}
\mathscr{J}_K^{\varphi} & \xrightarrow{\mathrm{NR}_K^{\varphi_K^{\mathrm{Weil}}}} & W_K & \xrightarrow{\mathrm{red}_{W_L}} & W_K/W_L & \xrightarrow{\beta_{L/K}} & G_K/G_L & \xrightarrow{\mathrm{res}_L^*} & \mathrm{Gal}(L/K) \\
\Big\downarrow{\scriptstyle \mathrm{a}_K} & & \Big\downarrow{\scriptstyle \mathrm{red}_{W_K^c}} & & & & & & \Big\downarrow{\scriptstyle \mathrm{res}_{(L/K)^{ab}}} \\
\mathbb{J}_K & \xrightarrow[(\bullet,K)]{} & W_K^{ab} & & \xrightarrow[\beta_K^{ab}]{\qquad\qquad} & & G_K^{ab} & \longrightarrow & \mathrm{Gal}((L/K)^{ab}/K)
\end{array}
$$

(with the curved arrow $\mathrm{NR}_{L/K}^{\varphi_K^{\mathrm{Weil}}} : \mathscr{J}_K^{\varphi} \to \mathrm{Gal}(L/K)$ along the top and the curved arrow $(\bullet,L/K) : \mathbb{J}_K \to \mathrm{Gal}((L/K)^{ab}/K)$ along the bottom)

follows, completing the proof. □

4.6 Decomposition and Inertia Groups

Let L/K be a finite Galois extension. For $\nu \in \mathbb{h}_K \cup \mathbb{a}_K$, let $D_\nu := \mathrm{D}_\nu(L/K)$ and $I_\nu := \mathrm{I}_\nu(L/K)$ denote, respectively, the decomposition and the inertia groups of ν in $\mathrm{Gal}(L/K)$ determined by the continuous homomorphism $e_\nu^{\mathrm{Weil}} : W_{K_\nu} \to W_K$. That is, the subgroups D_ν and I_ν of $\mathrm{Gal}(L/K)$ are defined by

$$D_\nu = \mathrm{res}_L^* \circ \beta_{L/K} \circ \mathrm{red}_{W_L} \circ e_\nu^{\mathrm{Weil}}(W_{K_\nu})$$

and

$$I_\nu = \mathrm{res}_L^* \circ \beta_{L/K} \circ \mathrm{red}_{W_L} \circ e_\nu^{\mathrm{Weil}}\left(W_{K_\nu}^0\right).$$

Recall that for $\nu \in \mathbb{a}_K$, the group $W_{K_\nu}^0$ is defined by $W_{K_\nu}^0 = W_{K_\nu}$.

Theorem 6. *For every prime $\nu \in \mathbb{h}_K \cup \mathbb{a}_K$:*

(i) *The image of the continuous homomorphism defined by the composition*

$$\left(\mathscr{I}_K^{\underline{\varphi}_K}\right)_\nu := \left\{ \begin{array}{ll} {}_{\mathbb{Z}}\nabla_{K_\nu}^{(\varphi_{K_\nu})}, & \nu \in \text{h}_K \\ W_{\mathbb{R}}, & \nu \in \text{a}_{K,\mathbb{R}} \\ W_{\mathbb{C}}, & \nu \in \text{a}_{K,\mathbb{C}} \end{array} \right\} \xrightarrow{q_\nu} \mathscr{I}_K^{\underline{\varphi}_K} \xrightarrow{\mathsf{NR}_{L/K}^{\underline{\varphi}_K^{\text{Weil}}}} \text{Gal}(L/K)$$

is the decomposition group D_ν of ν in Gal(L/K) *determined by the continuous homomorphism $e_\nu^{\text{Weil}} : W_{K_\nu} \to W_K$;*

(ii) *The image of the continuous homomorphism defined by the composition*

$$\left(\mathscr{I}_K^{\underline{\varphi}_K}\right)_\nu^0 := \left\{ \begin{array}{ll} {}_{1}\nabla_{K_\nu}^{(\varphi_{K_\nu})^{\underline{0}}}, & \nu \in \text{h}_K \\ W_{\mathbb{R}}^0, & \nu \in \text{a}_{K,\mathbb{R}} \\ W_{\mathbb{C}}^0, & \nu \in \text{a}_{K,\mathbb{C}} \end{array} \right\} \xrightarrow{q_\nu} \mathscr{I}_K^{\underline{\varphi}_K} \xrightarrow{\mathsf{NR}_{L/K}^{\underline{\varphi}_K^{\text{Weil}}}} \text{Gal}(L/K)$$

is the inertia group I_ν of ν in Gal(L/K) *determined by the continuous homomorphism $e_\nu^{\text{Weil}} : W_{K_\nu} \to W_K$.*

Proof. First, assume that $\nu \in \text{h}_K$. Then, clearly

$$\begin{aligned} \mathsf{NR}_{L/K}^{\underline{\varphi}_K^{\text{Weil}}} \circ q_\nu &= \left(\text{res}_L^* \circ \beta_{L/K} \circ \text{red}_{W_L} \circ \mathsf{NR}_K^{\underline{\varphi}_K^{\text{Weil}}} \right) \circ q_\nu \\ &= \text{res}_L^* \circ \beta_{L/K} \circ \text{red}_{W_L} \circ \left(\mathsf{NR}_K^{\underline{\varphi}_K^{\text{Weil}}} \circ q_\nu \right) \\ &= \text{res}_L^* \circ \beta_{L/K} \circ \text{red}_{W_L} \circ \left(e_\nu^{\text{Weil}} \circ \{\bullet, K_\nu\}_{\varphi_{K_\nu}} \right) \end{aligned}$$

by the local-global compatibility of the non-Abelian norm-residue symbols discussed in Sect. 4.1. Thus, the image $\mathsf{NR}_{L/K}^{\underline{\varphi}_K^{\text{Weil}}} \circ q_\nu \left({}_{\mathbb{Z}}\nabla_{K_\nu}^{(\varphi_{K_\nu})} \right)$ of $\mathsf{NR}_{L/K}^{\underline{\varphi}_K^{\text{Weil}}} \circ q_\nu$ is given by

$$\begin{aligned} \mathsf{NR}_{L/K}^{\underline{\varphi}_K^{\text{Weil}}} \circ q_\nu \left({}_{\mathbb{Z}}\nabla_{K_\nu}^{(\varphi_{K_\nu})} \right) &= \text{res}_L^* \circ \beta_{L/K} \circ \text{red}_{W_L} \circ e_\nu^{\text{Weil}} \circ \{\bullet, K_\nu\}_{\varphi_{K_\nu}} \left({}_{\mathbb{Z}}\nabla_{K_\nu}^{(\varphi_{K_\nu})} \right) \\ &= \text{res}_L^* \circ \beta_{L/K} \circ \text{red}_{W_L} \circ e_\nu^{\text{Weil}} (W_{K_\nu}) \\ &= D_\nu \end{aligned}$$

which is the decomposition group of ν in Gal(L/K) determined by $e_\nu^{\text{Weil}} : W_{K_\nu} \to W_K$, and the image $\mathsf{NR}_{L/K}^{\underline{\varphi}_K^{\text{Weil}}} \circ q_\nu \left({}_{1}\nabla_{K_\nu}^{(\varphi_{K_\nu})^{\underline{0}}} \right)$ of $\mathsf{NR}_{L/K}^{\underline{\varphi}_K^{\text{Weil}}} \circ q_\nu$ is given by

$$\begin{aligned} \mathsf{NR}_{L/K}^{\underline{\varphi}_K^{\text{Weil}}} \circ q_\nu \left({}_{1}\nabla_{K_\nu}^{(\varphi_{K_\nu})^{\underline{0}}} \right) &= \text{res}_L^* \circ \beta_{L/K} \circ \text{red}_{W_L} \circ e_\nu^{\text{Weil}} \circ \{\bullet, K_\nu\}_{\varphi_{K_\nu}} \left({}_{1}\nabla_{K_\nu}^{(\varphi_{K_\nu})^{\underline{0}}} \right) \\ &= \text{res}_L^* \circ \beta_{L/K} \circ \text{red}_{W_L} \circ e_\nu^{\text{Weil}} (W_{K_\nu}^0) \\ &= I_\nu \end{aligned}$$

which is the inertia group of ν in $\text{Gal}(L/K)$ determined by $e_\nu^{\text{Weil}} : W_{K_\nu} \to W_K$. Next, assume that $\nu \in \infty_K$. Then,

$$
\begin{aligned}
\mathsf{NR}_{L/K}^{\varphi_K^{\text{Weil}}} \circ q_\nu &= \left(\text{res}_L^* \circ \beta_{L/K} \circ \text{red}_{W_L} \circ \mathsf{NR}_K^{\varphi_K^{\text{Weil}}}\right) \circ q_\nu \\
&= \text{res}_L^* \circ \beta_{L/K} \circ \text{red}_{W_L} \circ \left(\mathsf{NR}_K^{\varphi_K^{\text{Weil}}} \circ q_\nu\right) \\
&= \text{res}_L^* \circ \beta_{L/K} \circ \text{red}_{W_L} \circ e_\nu^{\text{Weil}},
\end{aligned}
$$

and the image of $\mathsf{NR}_{L/K}^{\varphi_K^{\text{Weil}}} \circ q_\nu$ is given by

$$
\mathsf{NR}_{L/K}^{\varphi_K^{\text{Weil}}} \circ q_\nu \left(W_{K_\nu}\right) = \text{res}_L^* \circ \beta_{L/K} \circ \text{red}_{W_L} \circ e_\nu^{\text{Weil}} \left(W_{K_\nu}\right) = D_\nu = I_\nu,
$$

which completes the proof. □

Remark 5. Alternatively, we can prove the Archimedean case of Theorem 6 as follows.

If $\nu \in \infty_{K,\mathbb{C}}$, that is, ν is a complex Archimedean prime of K, then the decomposition and the inertia groups D_ν and I_ν of ν in $\text{Gal}(L/K)$ determined by the continuous homomorphism $e_\nu^{\text{Weil}} : W_{\mathbb{C}} := \mathbb{C}^\times \to W_K$ are the same. Moreover, $D_\nu = I_\nu$ is the trivial subgroup $\langle \text{id}_L \rangle$ of $\text{Gal}(L/K)$, and the image of $\mathsf{NR}_{L/K}^{\varphi_K^{\text{Weil}}} \circ q_\nu$ is given by

$$
\begin{aligned}
\mathsf{NR}_{L/K}^{\varphi_K^{\text{Weil}}} \circ q_\nu(W_{\mathbb{C}}) &= \left\{\beta_K\left(e_\nu^{\text{Weil}}(w)\right) |_L : w \in W_{\mathbb{C}}\right\} \\
&= \left\{e_\nu^{\text{Galois}}(\beta_{\mathbb{C}}(w)) |_L : w \in W_{\mathbb{C}}\right\} = \langle \text{id}_L \rangle = D_\nu = I_\nu
\end{aligned}
$$

where the equalities in the last line follow from the commutativity of the square (10) given by

$$
\begin{array}{ccc}
W_{\mathbb{C}} & \xrightarrow{e_\nu^{\text{Weil}}} & W_K \\
{\scriptstyle \beta_{\mathbb{C}}}\downarrow & & \downarrow{\scriptstyle \beta_K} \\
G_{\mathbb{C}} & \xrightarrow[e_\nu^{\text{Galois}}]{} & G_K
\end{array}
$$

and from the fact that $\beta_{\mathbb{C}} : W_{\mathbb{C}} \to G_{\mathbb{C}}$ is the trivial mapping.

If $\nu \in \infty_{K,\mathbb{R}}$, that is, ν is a real Archimedean prime of K, then the decomposition and the inertia groups D_ν and I_ν of ν in $\text{Gal}(L/K)$ determined by the continuous homomorphism $e_\nu^{\text{Weil}} : W_{\mathbb{R}} := \mathbb{C}^\times \cup j\mathbb{C}^\times \to W_K$ are the same, and $D_\nu = I_\nu$ is the subgroup $\langle e_{\nu}^{\mathrm{Galois}}(c)\mid_{L}\rangle$ of $\mathrm{Gal}(L/K)$, which is of order at most 2. Here, $c:\mathbb{C}\to\mathbb{C}$ denotes the $\mathbb{R}$-automorphism of $\mathbb{C}$ defined by the complex conjugation $c: z\mapsto \overline{z}$, for $z\in\mathbb{C}$. (For example, if K is a totally real number field and L/K is a totally real extension, then $e_{\nu}^{\mathrm{Galois}}(c)\mid_{L}=\mathrm{id}_L$.) Therefore, the image of $\mathsf{NR}_{L/K}^{\varphi_K^{\mathrm{Weil}}}\circ q_{\nu}$ is

$$\begin{aligned}\mathsf{NR}_{L/K}^{\varphi_K^{\mathrm{Weil}}}\circ q_{\nu}(W_{\mathbb{R}}) &= \left\{\beta_K\left(e_{\nu}^{\mathrm{Weil}}(w)\right)\mid_{L}: w\in W_{\mathbb{R}}\right\}\\ &= \left\{e_{\nu}^{\mathrm{Galois}}(\beta_{\mathbb{R}}(w))\mid_{L}: w\in W_{\mathbb{R}}\right\}\\ &= \left\{\mathrm{id}_L, e_{\nu}^{\mathrm{Galois}}(c)\mid_{L}\right\} = D_{\nu} = I_{\nu},\end{aligned}$$

as $\beta_{\mathbb{R}}(\mathbb{C}^{\times}) = \mathrm{id}_{\mathbb{C}}$ and $\beta_{\mathbb{R}}(j\mathbb{C}^{\times}) = c$ (look at 1.4.3 of [16] for details), which completes the proof.

4.7 Basic Functorial Properties

Theorem 7. *Let $K\subseteq L\subseteq E$ be a tower of finite Galois extensions of the global field K. Then, the triangle*

$$\begin{array}{ccc} \mathscr{I}_K^{\varphi_K} & \xrightarrow{\ \mathsf{NR}_{E/K}^{\varphi_K^{\mathrm{Weil}}}\ } & \mathrm{Gal}(E/K)\\ & \searrow^{\mathsf{NR}_{L/K}^{\varphi_K^{\mathrm{Weil}}}} & \Big\downarrow \mathrm{res}_L\\ & & \mathrm{Gal}(L/K)\end{array}$$

where the right vertical arrow is the restriction map to L, is commutative.

Proof. First note that for the tower $K\subseteq L\subseteq E$ of finite Galois extensions of K, the following rectangle

$$\begin{array}{ccccc} W_K/W_E & \xrightarrow[\sim]{\ \beta_{E/K}\ } & G_K/G_E & \xrightarrow[\sim]{\ \mathrm{res}_E^*\ } & \mathrm{Gal}(E/K)\\ \Big\downarrow & & \Big\downarrow & & \Big\downarrow \mathrm{res}_L\\ W_K/W_L & \xrightarrow[\beta_{L/K}]{\sim} & G_K/G_L & \xrightarrow[\mathrm{res}_L^*]{\sim} & \mathrm{Gal}(L/K)\end{array}$$

is commutative. Thus, it suffices to prove the commutativity of the triangle

$$
\begin{array}{ccc}
\mathscr{I}_K^{\varphi_K} & \xrightarrow{\mathrm{red}_{W_E}\circ \mathrm{NR}_K^{\varphi_K\text{-Weil}}} & W_K/W_E \\
& \searrow^{\mathrm{red}_{W_L}\circ \mathrm{NR}_K^{\varphi_K\text{-Weil}}} & \downarrow \\
& & W_K/W_L
\end{array}
$$

which is clear. □

Theorem 8. *Let $K \subseteq L \subseteq E$ be a tower of finite Galois extensions of the global field K. Then, the square*

$$
\begin{array}{ccc}
\mathscr{I}_L^{\varphi_L} & \xrightarrow{\mathrm{NR}_{E/L}^{\varphi_L\text{-Weil}}} & \mathrm{Gal}(E/L) \\
\Big\downarrow{\scriptstyle \mathscr{N}_{L/K}^{\infty}} & & \Big\downarrow{\scriptstyle \mathrm{id}_{\mathrm{Gal}(E/L)}} \\
\mathscr{I}_K^{\varphi_K} & \xrightarrow[\mathrm{NR}_{E/K}^{\varphi_K\text{-Weil}}]{} & \mathrm{Gal}(E/K)
\end{array}
$$

is commutative.

Proof. It suffices to prove that

$$\left(\mathrm{NR}_{E/K}^{\varphi_K\text{-Weil}} \circ \mathscr{N}_{L/K}^{\infty}\right)_S \circ \iota_{\mu}^{(S)} = \left(\mathrm{NR}_{E/L}^{\varphi_L\text{-Weil}}\right)_S \circ \iota_{\mu}^{(S)},$$

where $\mu \in \mathbb{h}_L \cup \mathbb{a}_L$ and S is any finite subset of $\mathbb{h}_L \cup \mathbb{a}_L$ satisfying $\mu \in S$ and $\mathbb{a}_L \subset S$. So, let μ be any prime of L and S any such subset of $\mathbb{h}_L \cup \mathbb{a}_L$. Then, clearly the following identities

$$
\begin{aligned}
\left(\mathrm{NR}_{E/K}^{\varphi_K\text{-Weil}} \circ \mathscr{N}_{L/K}^{\infty}\right)_S \circ \iota_{\mu}^{(S)} &= \left(\mathrm{NR}_{E/K}^{\varphi_K\text{-Weil}} \circ \mathscr{N}_{L/K}^{\infty}\right) \circ q_{\mu} \\
&= \mathrm{NR}_{E/K}^{\varphi_K\text{-Weil}} \circ \left(\mathscr{N}_{L/K}^{\infty} \circ q_{\mu}\right) \\
&= \mathrm{NR}_{E/K}^{\varphi_K\text{-Weil}} \circ \left(\mathscr{N}_{L/K}^{\infty}\right)_{\mu}
\end{aligned}
$$

hold. Now, first assume that $\mu \in \mathbb{h}_L$. Then,

$$\begin{aligned}
\left(\mathsf{NR}_{E/K}^{\varphi_K^{\mathrm{Weil}}} \circ \mathcal{N}_{L/K}^{\infty}\right)_S \circ \iota_\mu^{(S)} &= \mathsf{NR}_{E/K}^{\varphi_K^{\mathrm{Weil}}} \circ \left(\mathcal{N}_{L/K}^{\infty}\right)_\mu \\
&= \mathsf{NR}_{E/K}^{\varphi_K^{\mathrm{Weil}}} \circ \left(q_\nu \circ \mathcal{N}_{L_\mu/K_\nu}^{\infty}\right) \\
&= \left(\mathsf{NR}_{E/K}^{\varphi_K^{\mathrm{Weil}}} \circ q_\nu\right) \circ \mathcal{N}_{L_\mu/K_\nu}^{\infty} \\
&= \left(\mathrm{res}_E^* \circ \beta_{E/K} \circ \mathrm{red}_{W_E} \circ \mathsf{NR}_K^{\varphi_K^{\mathrm{Weil}}} \circ q_\nu\right) \circ \mathcal{N}_{L_\mu/K_\nu}^{\infty} \\
&= \left(\mathrm{res}_E^* \circ \beta_{E/K} \circ \mathrm{red}_{W_E} \circ e_\nu^{\mathrm{Weil}} \circ \{\bullet, K_\nu\}_{\varphi_{K_\nu}}\right) \circ \mathcal{N}_{L_\mu/K_\nu}^{\infty} \\
&= \mathrm{res}_E^* \circ \beta_{E/K} \circ \mathrm{red}_{W_E} \circ e_\nu^{\mathrm{Weil}} \circ \left(\{\bullet, K_\nu\}_{\varphi_{K_\nu}} \circ \mathcal{N}_{L_\mu/K_\nu}^{\infty}\right),
\end{aligned}$$

where $\nu \in \mathbb{h}_K$ is the finite prime of K defined by $\nu = \mu \cap O_K$. Now, the diagram (15) (equivalently, the square (7.4) of [6]), that is, the following square

$$\begin{array}{ccc}
W_{L_\mu} & \xleftarrow[\sim]{\{\bullet, L_\mu\}_{\varphi_{L_\mu}}} & {}_{\mathbb{Z}}\nabla_{L_\mu}^{(\varphi_{L_\mu})} \\
\Big\downarrow {\scriptstyle \gamma_{L_\mu/K_\nu}} & & \Big\downarrow {\scriptstyle \mathcal{N}_{L_\mu/K_\nu}^{\infty}} \\
W_{K_\nu} & \xleftarrow[\{\bullet, K_\nu\}_{\varphi_{K_\nu}}]{\sim} & {}_{\mathbb{Z}}\nabla_{K_\nu}^{(\varphi_{K_\nu})}
\end{array}$$

is commutative. Therefore,

$$\begin{aligned}
\left(\mathsf{NR}_{E/K}^{\varphi_K^{\mathrm{Weil}}} \circ \mathcal{N}_{L/K}^{\infty}\right)_S \circ \iota_\mu^{(S)} &= \mathrm{res}_E^* \circ \beta_{E/K} \circ \mathrm{red}_{W_E} \circ e_\nu^{\mathrm{Weil}} \circ \left(\{\bullet, K_\nu\}_{\varphi_{K_\nu}} \circ \mathcal{N}_{L_\lambda/K_\nu}^{\infty}\right) \\
&= \mathrm{res}_E^* \circ \beta_{E/K} \circ \mathrm{red}_{W_E} \circ e_\nu^{\mathrm{Weil}} \circ \left(\gamma_{L_\mu/K_\nu} \circ \{\bullet, L_\mu\}_{\varphi_{L_\mu}}\right) \\
&= \mathrm{res}_E^* \circ \beta_{E/K} \circ \mathrm{red}_{W_E} \circ \left(e_\nu^{\mathrm{Weil}} \circ \gamma_{L_\mu/K_\nu}\right) \circ \{\bullet, L_\mu\}_{\varphi_{L_\mu}}.
\end{aligned}$$

Now, by commutative square (13) of Lemma 4, continuing the computation,

$$\begin{aligned}
\left(\mathsf{NR}_{E/K}^{\varphi_K^{\mathrm{Weil}}} \circ \mathcal{N}_{L/K}^{\infty}\right)_S \circ \iota_\mu^{(S)} &= \mathrm{res}_E^* \circ \beta_{E/K} \circ \mathrm{red}_{W_E} \circ \left(e_\nu^{\mathrm{Weil}} \circ \gamma_{L_\mu/K_\nu}\right) \circ \{\bullet, L_\mu\}_{\varphi_{L_\mu}} \\
&= \mathrm{res}_E^* \circ \beta_{E/K} \circ \mathrm{red}_{W_E} \circ \left(\gamma_{L/K} \circ e_\mu^{\mathrm{Weil}}\right) \circ \{\bullet, L_\mu\}_{\varphi_{L_\mu}} \\
&= \mathrm{res}_E^* \circ \beta_{E/K} \circ \mathrm{red}_{W_E} \circ \gamma_{L/K} \circ \left(e_\mu^{\mathrm{Weil}} \circ \{\bullet, L_\mu\}_{\varphi_{L_\mu}}\right)
\end{aligned}$$

$$= \mathrm{res}_E^* \circ \beta_{E/K} \circ \mathrm{red}_{W_E} \circ \gamma_{L/K} \circ \left(\mathsf{NR}_L^{\varphi_L^{\mathrm{Weil}}} \circ q_\mu\right)$$

$$= \mathrm{res}_E^* \circ \beta_{E/K} \circ \mathrm{red}_{W_E} \circ \gamma_{L/K} \circ \left(\mathsf{NR}_L^{\varphi_L^{\mathrm{Weil}}} \circ c_S \circ \iota_\mu^{(S)}\right)$$

$$= \mathrm{res}_E^* \circ \beta_{E/L} \circ \mathrm{red}_{W_E} \circ \left(\mathsf{NR}_L^{\varphi_L^{\mathrm{Weil}}} \circ c_S \circ \iota_\mu^{(S)}\right),$$

where the last equality follows from Lemma 6. Thus,

$$\begin{aligned}\left(\mathsf{NR}_{E/K}^{\varphi_K^{\mathrm{Weil}}} \circ \mathscr{N}_{L/K}^{\infty}\right)_S \circ \iota_\mu^{(S)} &= \mathrm{res}_E^* \circ \beta_{E/L} \circ \mathrm{red}_{W_E} \circ \left(\mathsf{NR}_L^{\varphi_L^{\mathrm{Weil}}} \circ c_S \circ \iota_\mu^{(S)}\right)\\ &= \left(\mathrm{res}_E^* \circ \beta_{E/L} \circ \mathrm{red}_{W_E} \circ \mathsf{NR}_L^{\varphi_L^{\mathrm{Weil}}}\right) \circ c_S \circ \iota_\mu^{(S)}\\ &= \mathsf{NR}_{E/L}^{\varphi_L^{\mathrm{Weil}}} \circ c_S \circ \iota_\mu^{(S)}\\ &= \left(\mathsf{NR}_{E/L}^{\varphi_L^{\mathrm{Weil}}}\right)_S \circ \iota_\mu^{(S)}.\end{aligned}$$

Now, if $\mu \in \mathbb{O}_L$, then

$$\begin{aligned}\left(\mathsf{NR}_{E/K}^{\varphi_K^{\mathrm{Weil}}} \circ \mathscr{N}_{L/K}^{\infty}\right)_S \circ \iota_\mu^{(S)} &= \mathsf{NR}_{E/K}^{\varphi_K^{\mathrm{Weil}}} \circ \left(\mathscr{N}_{L/K}^{\infty}\right)_\mu\\ &= \mathsf{NR}_{E/K}^{\varphi_K^{\mathrm{Weil}}} \circ \left(q_\nu \circ \gamma_{L_\mu/K_\nu}\right),\end{aligned}$$

by the equality $\left(\mathscr{N}_{L/K}^{\infty}\right)_\mu = q_\nu \circ \gamma_{L_\mu/K_\nu}$, where $\nu \in \mathbb{O}_K$ is the infinite prime of K defined by $\nu = \mu\mid_K$. Then, by the definition of the non-Abelian global reciprocity map $\mathsf{NR}_{E/K}^{\varphi_K^{\mathrm{Weil}}}$ relative to the extension E/K,

$$\begin{aligned}\left(\mathsf{NR}_{E/K}^{\varphi_K^{\mathrm{Weil}}} \circ \mathscr{N}_{L/K}^{\infty}\right)_S \circ \iota_\mu^{(S)} &= \mathsf{NR}_{E/K}^{\varphi_K^{\mathrm{Weil}}} \circ \left(q_\nu \circ \gamma_{L_\mu/K_\nu}\right)\\ &= \left(\mathsf{NR}_{E/K}^{\varphi_K^{\mathrm{Weil}}} \circ q_\nu\right) \circ \gamma_{L_\mu/K_\nu}\\ &= \left(\mathrm{res}_E^* \circ \beta_{E/K} \circ \mathrm{red}_{W_E} \circ \mathsf{NR}_K^{\varphi_K^{\mathrm{Weil}}} \circ q_\nu\right) \circ \gamma_{L_\mu/K_\nu}\\ &= \left(\mathrm{res}_E^* \circ \beta_{E/K} \circ \mathrm{red}_{W_E} \circ e_\nu^{\mathrm{Weil}}\right) \circ \gamma_{L_\mu/K_\nu},\end{aligned}$$

as $\left(\mathsf{NR}_K^{\varphi_K^{\text{Weil}}}\right)_\nu = \mathsf{NR}_K^{\varphi_K^{\text{Weil}}} \circ q_\nu = e_\nu^{\text{Weil}}$. Therefore,

$$\begin{aligned}\left(\mathsf{NR}_{E/K}^{\varphi_K^{\text{Weil}}} \circ \mathscr{N}_{L/K}^{\infty}\right)_S \circ \iota_\mu^{(S)} &= \left(\text{res}_E^* \circ \beta_{E/K} \circ \text{red}_{W_E} \circ e_\nu^{\text{Weil}}\right) \circ \gamma_{L_\mu/K_\nu} \\ &= \text{res}_E^* \circ \beta_{E/K} \circ \text{red}_{W_E} \circ \left(e_\nu^{\text{Weil}} \circ \gamma_{L_\mu/K_\nu}\right) \\ &= \text{res}_E^* \circ \beta_{E/K} \circ \text{red}_{W_E} \circ \left(\gamma_{L/K} \circ e_\mu^{\text{Weil}}\right), \end{aligned} \tag{20}$$

where the last equality follows from Lemma 4. Note that

$$\left(\mathsf{NR}_L^{\varphi_L^{\text{Weil}}}\right)_\mu = \mathsf{NR}_L^{\varphi_L^{\text{Weil}}} \circ q_\mu = e_\mu^{\text{Weil}}. \tag{21}$$

Thus, substituting (21) into (20),

$$\begin{aligned}\left(\mathsf{NR}_{E/K}^{\varphi_K^{\text{Weil}}} \circ \mathscr{N}_{L/K}^{\infty}\right)_S \circ \iota_\mu^{(S)} &= \text{res}_E^* \circ \beta_{E/K} \circ \text{red}_{W_E} \circ \left(\gamma_{L/K} \circ e_\mu^{\text{Weil}}\right) \\ &= \text{res}_E^* \circ \beta_{E/K} \circ \text{red}_{W_E} \circ \gamma_{L/K} \circ \left(\mathsf{NR}_L^{\varphi_L^{\text{Weil}}} \circ q_\mu\right) \\ &= \text{res}_E^* \circ \beta_{E/L} \circ \text{red}_{W_E} \circ \left(\mathsf{NR}_L^{\varphi_L^{\text{Weil}}} \circ q_\mu\right), \end{aligned}$$

where the last equality follows from Lemma 6. Thus, using the definition of the non-Abelian global reciprocity map $\mathsf{NR}_{E/L}^{\varphi_L^{\text{Weil}}}$ relative to the extension E/L,

$$\begin{aligned}\left(\mathsf{NR}_{E/K}^{\varphi_K^{\text{Weil}}} \circ \mathscr{N}_{L/K}^{\infty}\right)_S \circ \iota_\mu^{(S)} &= \left(\text{res}_E^* \circ \beta_{E/L} \circ \text{red}_{W_E} \circ \mathsf{NR}_L^{\varphi_L^{\text{Weil}}}\right) \circ q_\mu \\ &= \mathsf{NR}_{E/L}^{\varphi_L^{\text{Weil}}} \circ q_\mu \\ &= \mathsf{NR}_{E/L}^{\varphi_L^{\text{Weil}}} \circ c_S \circ \iota_\mu^{(S)} \\ &= \left(\mathsf{NR}_{E/L}^{\varphi_L^{\text{Weil}}}\right)_S \circ \iota_\mu^{(S)}, \end{aligned}$$

where $\nu \in \mathbb{O}_K$ is the infinite prime of K defined by $\nu = \mu \mid_K$. This completes the proof. □

4.8 Non-Abelian Global Existence Theorem

In this section, under the assumption that Conjecture 1 holds, we shall prove the following theorem, which is the non-Abelian generalization of the existence theorem of Abelian global class field theory. As usual, K denotes a global field.

Theorem 9 (Non-Abelian global existence theorem). *There exists an inclusion-reversing bijective correspondence*

$$\left\{\begin{array}{l}\textit{Finite Galois exten-}\\ \textit{sions of } K \textit{ inside } K^{\mathrm{sep}}\end{array}\right\} \rightleftarrows \left\{\begin{array}{l}\textit{Open normal subgroups of finite}\\ \textit{index in } \mathscr{I}_K^{\underline{\varphi}_K} \textit{ containing } \mathscr{N}_K^{\underline{\varphi}_K}\end{array}\right\}$$

defined by

$$L \mapsto \mathscr{N}_K^{\underline{\varphi}_K} \mathscr{N}_{L/K}^{\infty}\left(\mathscr{I}_L^{\underline{\varphi}_L}\right),$$

for each finite Galois extension L of K inside K^{sep}, where $\mathscr{N}_K^{\underline{\varphi}_K} = \ker\left(\mathsf{NR}_K^{\underline{\varphi}_K\ \mathrm{Weil}}\right)$.

In order to prove the injectivity part of the non-Abelian global existence theorem, we shall first prove the following lemma:

Lemma 7. *Let L/K be a finite Galois extension and E/K any finite separable extension.*

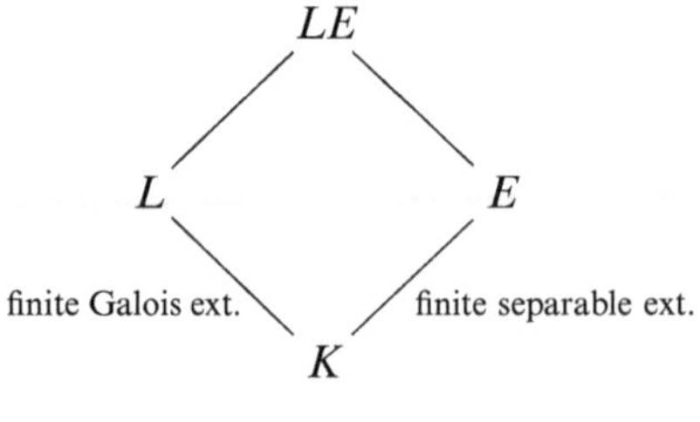

Then,

$$\mathscr{N}_E^{\underline{\varphi}_E} \mathscr{N}_{LE/E}^{\infty}\left(\mathscr{I}_{LE}^{\underline{\varphi}_{LE}}\right) = \left(\mathscr{N}_{E/K}^{\infty}\right)^{-1}\left(\mathscr{N}_K^{\underline{\varphi}_K} \mathscr{N}_{L/K}^{\infty}\left(\mathscr{I}_L^{\underline{\varphi}_L}\right)\right),$$

and the rectangle

$$\begin{array}{ccc}
\mathscr{I}_E^{\underline{\varphi}_E} / \mathscr{N}_E^{\underline{\varphi}_E} \mathscr{N}_{LE/E}^{\infty}(\mathscr{I}_{LE}^{\underline{\varphi}_{LE}}) & \xrightarrow[\sim]{\mathsf{NR}_{LE/E}^{\underline{\varphi}_E\ \mathrm{Weil}^*}} & \mathrm{Gal}(LE/E) \\
\Big\downarrow {\scriptstyle (\mathrm{red}_{\mathscr{N}_K^{\underline{\varphi}_K} \mathscr{N}_{L/K}^{\infty}(\mathscr{I}_L^{\underline{\varphi}_L})} \circ \mathscr{N}_{E/K}^{\infty})^*} & & \Big\downarrow {\scriptstyle \mathrm{res}_L} \\
\mathscr{I}_K^{\underline{\varphi}_K} / \mathscr{N}_K^{\underline{\varphi}_K} \mathscr{N}_{L/K}^{\infty}(\mathscr{I}_L^{\underline{\varphi}_L}) & \xrightarrow[\mathsf{NR}_{L/K}^{\underline{\varphi}_K\ \mathrm{Weil}^*}]{\sim} & \mathrm{Gal}(L/K)
\end{array}$$

is commutative.

Proof. In fact, the following diagram

$$\begin{array}{ccccccc}
W_E & \xrightarrow{\mathrm{red}_{W_{LE}}} & W_E/W_{LE} & \xrightarrow[\sim]{\beta_{LE/E}} & G_E/G_{LE} & \xrightarrow[\sim]{\mathrm{res}^*_{LE}} & \mathrm{Gal}(LE/E) \\
{\scriptstyle \gamma_{E/K}}\big\downarrow & & & & & & \big\downarrow{\scriptstyle \mathrm{res}_L} \\
W_K & \xrightarrow{\mathrm{red}_{W_L}} & W_K/W_L & \xrightarrow[\sim]{\beta_{L/K}} & G_K/G_L & \xrightarrow[\sim]{\mathrm{res}^*_{L}} & \mathrm{Gal}(L/K)
\end{array}$$

is commutative, because for each $w \in W_E$, the equalities

$$\mathrm{res}_L(\mathrm{res}_{LE}(\beta_E(w))) = \mathrm{res}_L(\beta_E(w)) = \mathrm{res}_L\left(\beta_K \circ \gamma_{E/K}(w)\right)$$

hold by (7). Thus, by Theorem 3, the diagram

$$\begin{array}{ccccccccc}
\multicolumn{9}{c}{\overbrace{\hspace{30em}}^{\mathrm{NR}^{\varphi_E\,\mathrm{Weil}}_{LE/E}}} \\
\mathscr{I}_E^{\varphi_E} & \xrightarrow{\mathrm{NR}^{\varphi_E\,\mathrm{Weil}}_E} & W_E & \xrightarrow{\mathrm{red}_{W_{LE}}} & W_E/W_{LE} & \xrightarrow[\sim]{\beta_{LE/E}} & G_E/G_{LE} & \xrightarrow[\sim]{\mathrm{res}^*_{LE}} & \mathrm{Gal}(LE/E) \\
{\scriptstyle \mathscr{N}^{\infty}_{E/K}}\big\downarrow & & \big\downarrow{\scriptstyle \gamma_{E/K}} & & & & & & \big\downarrow{\scriptstyle \mathrm{res}_L} \\
\mathscr{I}_K^{\varphi_K} & \xrightarrow{\mathrm{NR}^{\varphi_K\,\mathrm{Weil}}_K} & W_K & \xrightarrow{\mathrm{red}_{W_L}} & W_K/W_L & \xrightarrow[\sim]{\beta_{L/K}} & G_K/G_L & \xrightarrow[\sim]{\mathrm{res}^*_{L}} & \mathrm{Gal}(L/K) \\
\multicolumn{9}{c}{\underbrace{\hspace{30em}}_{\mathrm{NR}^{\varphi_K\,\mathrm{Weil}}_{L/K}}}
\end{array} \tag{22}$$

is commutative. Now, if the equality

$$\mathscr{N}_E^{\varphi_E}\mathscr{N}^{\infty}_{LE/E}\left(\mathscr{I}_{LE}^{\varphi_{LE}}\right) = \left(\mathscr{N}^{\infty}_{E/K}\right)^{-1}\left(\mathscr{N}_K^{\varphi_K}\mathscr{N}^{\infty}_{L/K}\left(\mathscr{I}_L^{\varphi_L}\right)\right)$$

holds, then the composite homomorphism

$$\mathscr{I}_E^{\varphi_E} \xrightarrow{\mathscr{N}^{\infty}_{E/K}} \mathscr{I}_K^{\varphi_K} \xrightarrow{\mathrm{red}_{\mathscr{N}_K^{\varphi_K}\mathscr{N}^{\infty}_{L/K}(\mathscr{I}_L^{\varphi_L})}} \mathscr{I}_K^{\varphi_K}/\mathscr{N}_K^{\varphi_K}\mathscr{N}^{\infty}_{L/K}\left(\mathscr{I}_L^{\varphi_L}\right)$$

naturally induces the well-defined group homomorphism

$$\mathscr{I}_E^{\varphi_E}/\mathscr{N}_E^{\varphi_E}\mathscr{N}^{\infty}_{LE/E}\left(\mathscr{I}_{LE}^{\varphi_{LE}}\right) \xrightarrow{\left(\mathrm{red}_{\mathscr{N}_K^{\varphi_K}\mathscr{N}^{\infty}_{L/K}\left(\mathscr{I}_L^{\varphi_L}\right)}\circ\mathscr{N}^{\infty}_{E/K}\right)^*} \mathscr{I}_K^{\varphi_K}/\mathscr{N}_K^{\varphi_K}\mathscr{N}^{\infty}_{L/K}\left(\mathscr{I}_L^{\varphi_L}\right)$$

as $\ker\left(\mathrm{red}_{\mathscr{N}_K^{\varphi_K}\mathscr{N}^{\infty}_{L/K}\left(\mathscr{I}_L^{\varphi_L}\right)}\circ\mathscr{N}^{\infty}_{E/K}\right) = \mathscr{N}_E^{\varphi_E}\mathscr{N}^{\infty}_{LE/E}\left(\mathscr{I}_{LE}^{\varphi_{LE}}\right)$, and the commutative rectangle (22) yields the commutativity of the diagram

$$\begin{array}{ccc} \mathcal{I}_E^{\varphi_E} / \mathcal{N}_E^{\varphi_E}\mathcal{N}_{LE/E}^{\infty}(\mathcal{I}_{LE}^{\varphi_{LE}}) & \xrightarrow[\sim]{\mathsf{NR}_{LE/E}^{\varphi_E}{}^{\mathrm{Weil}*}} & \mathrm{Gal}(LE/E) \\ \Big\downarrow (\mathrm{red}_{\mathcal{N}_K^{\varphi_K}\mathcal{N}_{L/K}^{\infty}(\mathcal{I}_L^{\varphi_L})} \circ \mathcal{N}_{E/K}^{\infty})^* & & \Big\downarrow \mathrm{res}_L \\ \mathcal{I}_K^{\varphi_K} / \mathcal{N}_K^{\varphi_K}\mathcal{N}_{L/K}^{\infty}(\mathcal{I}_L^{\varphi_L}) & \xrightarrow[\mathsf{NR}_{L/K}^{\varphi_K}{}^{\mathrm{Weil}*}]{\sim} & \mathrm{Gal}(L/K). \end{array}$$

Thus, it remains to prove that the equality

$$\mathcal{N}_E^{\varphi_E}\mathcal{N}_{LE/E}^{\infty}\left(\mathcal{I}_{LE}^{\varphi_{LE}}\right) = \left(\mathcal{N}_{E/K}^{\infty}\right)^{-1}\left(\mathcal{N}_K^{\varphi_K}\mathcal{N}_{L/K}^{\infty}\left(\mathcal{I}_L^{\varphi_L}\right)\right)$$

holds. In order to do so, let

$$\alpha \in \mathcal{N}_E^{\varphi_E}\mathcal{N}_{LE/E}^{\infty}\left(\mathcal{I}_{LE}^{\varphi_{LE}}\right) = \ker\left(\mathsf{NR}_{LE/E}^{\varphi_E}{}^{\mathrm{Weil}}\right),$$

where the equality follows from Theorem 4. Then,

$$\mathsf{NR}_{L/K}^{\varphi_K}{}^{\mathrm{Weil}}\left(\mathcal{N}_{E/K}^{\infty}(\alpha)\right) = \mathrm{res}_L \circ \mathsf{NR}_{LE/E}^{\varphi_E}{}^{\mathrm{Weil}}(\alpha) = \mathrm{id}_L$$

by the commutativity of the diagram (22) and by the choice of α. Therefore, it follows that

$$\alpha \in \left(\mathcal{N}_{E/K}^{\infty}\right)^{-1}\left(\ker\left(\mathsf{NR}_{L/K}^{\varphi_K}{}^{\mathrm{Weil}}\right)\right) = \left(\mathcal{N}_{E/K}^{\infty}\right)^{-1}\left(\mathcal{N}_K^{\varphi_K}\mathcal{N}_{L/K}^{\infty}\left(\mathcal{I}_L^{\varphi_L}\right)\right),$$

where the equality follows from Theorem 4. For the reverse inclusion, assume that

$$\alpha \in \left(\mathcal{N}_{E/K}^{\infty}\right)^{-1}\left(\ker\left(\mathsf{NR}_{L/K}^{\varphi_K}{}^{\mathrm{Weil}}\right)\right) = \left(\mathcal{N}_{E/K}^{\infty}\right)^{-1}\left(\mathcal{N}_K^{\varphi_K}\mathcal{N}_{L/K}^{\infty}\left(\mathcal{I}_L^{\varphi_L}\right)\right),$$

where the equality follows from Theorem 4. Therefore,

$$\mathsf{NR}_{L/K}^{\varphi_K}{}^{\mathrm{Weil}}\left(\mathcal{N}_{E/K}^{\infty}(\alpha)\right) = \mathrm{res}_L \circ \mathsf{NR}_{LE/E}^{\varphi_E}{}^{\mathrm{Weil}}(\alpha) = \mathrm{id}_L$$

by the commutativity of the diagram (22) and by the choice of α. Thus, $\mathsf{NR}_{LE/E}^{\varphi_E}{}^{\mathrm{Weil}}(\alpha) \in \mathrm{Gal}(LE/E)$ such that

$$\mathsf{NR}_{LE/E}^{\varphi_E}{}^{\mathrm{Weil}}(\alpha)\Big|_L = \mathrm{id}_L \text{ and } \mathsf{NR}_{LE/E}^{\varphi_E}{}^{\mathrm{Weil}}(\alpha)\Big|_E = \mathrm{id}_E,$$

proving that $\mathsf{NR}_{LE/E}^{\varphi_E}{}^{\mathrm{Weil}}(\alpha) = \mathrm{id}_{LE}$. So, it follows that

$$\alpha \in \ker\left(\mathsf{NR}_{LE/E}^{\varphi_E\ \mathrm{Weil}}\right) = \mathscr{N}_E^{\varphi_E}\mathscr{N}_{LE/E}^{\infty}\left(\mathscr{I}_{LE}^{\varphi_{LE}}\right),$$

where the equality follows from Theorem 4, which completes the proof. □

Now, proof of the injectivity part of the non-Abelian global existence theorem follows. In fact, let L and E be any two finite Galois extensions of K inside K^{sep} such that

$$\mathscr{N}_K^{\varphi_K}\mathscr{N}_{L/K}^{\infty}\left(\mathscr{I}_L^{\varphi_L}\right) = \mathscr{N}_K^{\varphi_K}\mathscr{N}_{E/K}^{\infty}\left(\mathscr{I}_E^{\varphi_E}\right).$$

Then, by the previous lemma,

$$\begin{aligned}\mathscr{N}_E^{\varphi_E}\mathscr{N}_{LE/E}^{\infty}\left(\mathscr{I}_{LE}^{\varphi_{LE}}\right) &= \left(\mathscr{N}_{E/K}^{\infty}\right)^{-1}\left(\mathscr{N}_K^{\varphi_K}\mathscr{N}_{L/K}^{\infty}\left(\mathscr{I}_L^{\varphi_L}\right)\right)\\ &= \left(\mathscr{N}_{E/K}^{\infty}\right)^{-1}\left(\mathscr{N}_K^{\varphi_K}\mathscr{N}_{E/K}^{\infty}\left(\mathscr{I}_E^{\varphi_E}\right)\right)\\ &= \mathscr{I}_E^{\varphi_E},\end{aligned}$$

and the isomorphism

$$\mathsf{NR}_{LE/E}^{\varphi_E\ \mathrm{Weil}^*} : \mathscr{I}_E^{\varphi_E}/\mathscr{N}_E^{\varphi_E}\mathscr{N}_{LE/E}^{\infty}\left(\mathscr{I}_{LE}^{\varphi_{LE}}\right) \xrightarrow{\sim} \mathrm{Gal}(LE/E)$$

yields the equality $\mathrm{Gal}(LE/E) = \{\mathrm{id}_{LE}\}$. That is, $L \subseteq E$. Replacing the roles of L and E gives the reverse inclusion $E \subseteq L$, which completes the proof.

Next, in order to prove the surjectivity part of the non-Abelian global existence theorem, we shall prove the following lemma:

Lemma 8. *If S is an open normal subgroup of finite index in $\mathscr{I}_K^{\varphi_K}$ containing $\mathscr{N}_K^{\varphi_K}$, then there exists a finite Galois extension L_o of K inside K^{sep} such that*

$$S = \mathscr{N}_K^{\varphi_K}\mathscr{N}_{L_o/K}^{\infty}\left(\mathscr{I}_{L_o}^{\varphi_{L_o}}\right).$$

Proof. Let S be an open normal subgroup of finite index in $\mathscr{I}_K^{\varphi_K}$ containing $\mathscr{N}_K^{\varphi_K}$. Denote the open subgroup of finite index in $\mathscr{N}_K^{\varphi_K}\backslash\mathscr{I}_K^{\varphi_K}$ corresponding to S by $\overline{S}$. As Conjecture 1 is assumed to be true, the topological isomorphism

$$\mathsf{NR}_K^{\varphi_K\ \mathrm{Weil}^*} : \mathscr{N}_K^{\varphi_K}\backslash\mathscr{I}_K^{\varphi_K} \xrightarrow{\sim} W_K$$

defines an open normal subgroup $\mathsf{NR}_K^{\varphi_K\ \mathrm{Weil}^*}(\overline{S})$ of finite index in W_K. Now, choose a finite Galois extension L of K inside K^{sep} such that $W_L \subseteq \mathsf{NR}_K^{\varphi_K\ \mathrm{Weil}^*}(\overline{S})$. Thus, W_L is a normal subgroup of W_K, and $\mathsf{NR}_K^{\varphi_K\ \mathrm{Weil}^*}(\overline{S})/W_L$ is a normal subgroup of the finite quotient group $W_K/W_L \xrightarrow{\sim} \mathrm{Gal}(L/K)$. Hence, under the isomorphism $W_K/W_L \underset{\sim}{\xrightarrow{\beta_{L/K}}} G_K/G_L \underset{\sim}{\xrightarrow{\mathrm{res}_L^*}} \mathrm{Gal}(L/K)$, the normal subgroup $\mathsf{NR}_K^{\varphi_K\,\mathrm{Weil}^*}(\overline{S})/W_L$ of W_K/W_L is mapped isomorphically onto a normal subgroup $\mathrm{Gal}(L/L_o)$ of $\mathrm{Gal}(L/K)$ for some finite Galois subextension L_o/K of L/K. Now, we claim that

$$\mathsf{NR}_K^{\varphi_K\,\mathrm{Weil}^*}(\overline{S}) = W_{L_o} = \beta_K^{-1}(G_{L_o}).$$

In order to prove this claim, let $w \in \mathsf{NR}_K^{\varphi_K\,\mathrm{Weil}^*}(\overline{S})$. Then,

$$\mathrm{res}_L^* \circ \beta_{L/K} \circ \mathrm{red}_{W_L}(w) = \mathrm{res}_L^*(\beta_K(w) \pmod{G_L}) = \beta_K(w)\mid_L \in \mathrm{Gal}(L/L_o),$$

which shows that $\beta_K(w)\mid_{L_o} = \mathrm{id}_{L_o}$. Thus, $\beta_K(w) \in G_{L_o}$ and $w \in \beta_K^{-1}(G_{L_o}) = W_{L_o}$, proving that $\mathsf{NR}_K^{\varphi_K\,\mathrm{Weil}^*}(\overline{S})$ is a subgroup of W_{L_o}. Now, it follows that

$$[L : L_o] = \left(\mathsf{NR}_K^{\varphi_K\,\mathrm{Weil}^*}(\overline{S}) : W_L\right) \leq (W_{L_o} : W_L) = [L : L_o],$$

which proves the equality

$$\mathsf{NR}_K^{\varphi_K\,\mathrm{Weil}^*}(\overline{S}) = W_{L_o}.$$

Now, following the proof of Theorem 4,

$$S = \left(\mathsf{NR}_K^{\varphi_K\,\mathrm{Weil}}\right)^{-1}(W_{L_o}) = \mathscr{N}_K^{\varphi_K}\mathscr{N}_{L_o/K}^{\infty}\left(\mathscr{I}_{L_o}^{\varphi_{L_o}}\right),$$

which completes the proof. □

Now, the bijectivity part of Theorem 9 follows directly from Lemmas 7 and 8. Moreover, this bijective correspondence is inclusion-reversing. In fact, let L and E be any two finite Galois extensions of K inside K^{sep}. If $L \subseteq E$, then by Proposition 3,

$$\mathscr{N}_{E/K}^{\infty}\left(\mathscr{I}_E^{\varphi_E}\right) = \mathscr{N}_{L/K}^{\infty} \circ \mathscr{N}_{E/L}^{\infty}\left(\mathscr{I}_E^{\varphi_E}\right) \subseteq \mathscr{N}_{L/K}^{\infty}\left(\mathscr{I}_L^{\varphi_L}\right).$$

Therefore,

$$\mathscr{N}_K^{\varphi_K}\mathscr{N}_{E/K}^{\infty}\left(\mathscr{I}_E^{\varphi_E}\right) \subseteq \mathscr{N}_K^{\varphi_K}\mathscr{N}_{L/K}^{\infty}\left(\mathscr{I}_L^{\varphi_L}\right),$$

which proves that the bijective correspondence is inclusion-reversing. Hence, the proof of Theorem 9 is complete now.

Corollary 1. *If L is a finite separable extension of K inside K^{sep}, then*

$$S = \left(\mathsf{NR}_K^{\varphi_K \,\text{Weil}}\right)^{-1}(W_L)$$

is an open subgroup of $\mathcal{J}_K^{\varphi_K}$ containing $\mathcal{N}_K^{\varphi_K}$. Moreover,

$$S = \mathcal{N}_K^{\varphi_K}\mathcal{N}_{L_o/K}^{\infty}\left(\mathcal{J}_{L_o}^{\varphi_{L_o}}\right)$$

where L_o/K is the maximal Galois subextension of L/K.

Proof. Clearly S is an open subgroup of $\mathcal{J}_K^{\varphi_K}$ containing $\mathcal{N}_K^{\varphi_K}$, as W_L is an open subgroup of W_K. By Lemma 8, there exists a finite Galois extension L_o of K inside K^{sep} such that

$$S = \mathcal{N}_K^{\varphi_K}\mathcal{N}_{L_o/K}^{\infty}\left(\mathcal{J}_{L_o}^{\varphi_{L_o}}\right).$$

Now, we claim that L_o/K is the maximal Galois subextension of L/K. In fact, by Lemma 7 and Remark 4,

$$\begin{aligned}
\mathcal{N}_L^{\varphi_L}\mathcal{N}_{L_oL/L}^{\infty}\left(\mathcal{J}_{L_oL}^{\varphi_{L_oL}}\right) &= \left(\mathcal{N}_{L/K}^{\infty}\right)^{-1}\left(\mathcal{N}_K^{\varphi_K}\mathcal{N}_{L_o/K}^{\infty}\left(\mathcal{J}_{L_o}^{\varphi_{L_o}}\right)\right)\\
&= \left(\mathcal{N}_{L/K}^{\infty}\right)^{-1}(S)\\
&= \left(\mathcal{N}_{L/K}^{\infty}\right)^{-1}\left(\mathcal{N}_K^{\varphi_K}\mathcal{N}_{L/K}^{\infty}\left(\mathcal{J}_L^{\varphi_L}\right)\right)\\
&= \mathcal{J}_L^{\varphi_L}.
\end{aligned}$$

Therefore, the isomorphism

$$\mathsf{NR}_{L_oL/L}^{\varphi_L \,\text{Weil}*} : \mathcal{J}_L^{\varphi_L}/\mathcal{N}_L^{\varphi_L}\mathcal{N}_{L_oL/L}^{\infty}\left(\mathcal{J}_{L_oL}^{\varphi_{L_oL}}\right) \xrightarrow{\sim} \text{Gal}(L_oL/L)$$

yields the equality $\text{Gal}(L_oL/L) = \{\text{id}_{L_oL}\}$. That is, $L_o \subseteq L$. Now, let E/K be any Galois subextension of L/K. Then, as $E \subseteq L$,

$$\mathcal{N}_K^{\varphi_K}\mathcal{N}_{E/K}^{\infty}\left(\mathcal{J}_E^{\varphi_E}\right) \supseteq \mathcal{N}_K^{\varphi_K}\mathcal{N}_{L/K}^{\infty}\left(\mathcal{J}_L^{\varphi_L}\right) = S$$

by Proposition 3. Therefore,

$$\mathcal{N}_K^{\varphi_K}\mathcal{N}_{E/K}^{\infty}\left(\mathcal{J}_E^{\varphi_E}\right) \supseteq S = \mathcal{N}_K^{\varphi_K}\mathcal{N}_{L_o/K}^{\infty}\left(\mathcal{J}_{L_o}^{\varphi_{L_o}}\right),$$

which proves that $E \subseteq L_o$ by the existence theorem, completing the proof. □

4.9 Non-Abelian Ray Class Groups and Non-Abelian Ray Class Fields

Let $\mathfrak{m}$ be a cycle (=modulus) of the global field K. So, $\mathfrak{m}$ is a formal product of the form

$$\mathfrak{m} = \mathfrak{m}_{\mathbb{h}_K}\mathfrak{m}_{\mathbb{a}_K}.$$

Here, the "Henselian part" $\mathfrak{m}_{\mathbb{h}_K}$ is just a nonzero integral ideal in the ring of integers O_K of K viewed as a product $\prod_{\nu\in\mathbb{h}_K}\nu^{e_\nu}$ of Henselian primes ν in K. The "Archimedean part" $\mathfrak{m}_{\mathbb{a}_K}$ is any subset of the real Archimedean primes $\mathbb{a}_{K,\mathbb{R}}$ of K, if K is a number field. On the other hand, as $\mathbb{a}_K = \varnothing$ in case K is a function field, we just define $\mathfrak{m}_{\mathbb{a}_K} = \varnothing$ whenever K is a function field. In both cases, we shall identify $\mathfrak{m}_{\mathbb{a}_K}$ with a certain formal product $\prod_{\nu\in\mathbb{a}_K}\nu^{e_\nu}$ (so that $\mathfrak{m} = \mathfrak{m}_{\mathbb{h}_K}$ in the function field case, which is an effective divisor of the function field K if additive notation is used). Moreover, for any subset S of $\mathbb{h}_K \cup \mathbb{a}_K$, an S-cycle $\mathfrak{m}$ of K is by definition a cycle $\mathfrak{m}$ of K with support disjoint from S.

If K is assumed to be a number field, let S denote $\mathbb{a}_{K,\mathbb{C}}$. An S-cycle $\mathfrak{m} = \prod_{\nu\in\mathbb{h}_K\cup\mathbb{a}_K}\nu^{e_\nu}$ of K canonically defines a subgroup $\mathscr{U}_{S,\mathfrak{m}}^{\underline{\varphi}_K}\left(= \mathscr{U}_{\mathfrak{m}}^{\underline{\varphi}_K}\text{in short}\right)$ of $\mathscr{J}_K^{\underline{\varphi}_K}$ by the free product

$$\mathscr{U}_{\mathfrak{m}}^{\underline{\varphi}_K} = *_\nu\,\mathscr{U}_{\mathfrak{m},\nu}^{(\varphi_{K_\nu})},$$

where the local groups $\mathscr{U}_{\mathfrak{m},\nu}^{(\varphi_{K_\nu})}$ for $\nu \in \mathbb{h}_K \cup \mathbb{a}_K$ are defined as follows:

- $\nu \in \mathbb{h}_K$ and $e_\nu = 0$: $\mathscr{U}_{\mathfrak{m},\nu}^{(\varphi_{K_\nu})} = {}_1\nabla_{K_\nu}^{(\varphi_{K_\nu})^{\underline{0}}}$;
- $\nu \in \mathbb{h}_K$ and $e_\nu > 0$: $\mathscr{U}_{\mathfrak{m},\nu}^{(\varphi_{K_\nu})} = {}_1\nabla_{K_\nu}^{(\varphi_{K_\nu})^{\underline{e_\nu}}}$;
- $\nu \in \mathbb{a}_{K,\mathbb{R}}$ and $e_\nu = 0$: $\mathscr{U}_{\mathfrak{m},\nu}^{(\varphi_{K_\nu})} = W_{\mathbb{R}}$;
- $\nu \in \mathbb{a}_{K,\mathbb{R}}$ and $e_\nu = 1$: $\mathscr{U}_{\mathfrak{m},\nu}^{(\varphi_{K_\nu})} = W_{\mathbb{R},>0}$, where $W_{\mathbb{R},>0}$ is the subgroup of $W_{\mathbb{R}}$ which is defined as the pre-image of the subgroup $\mathbb{R}_{>0}$ of $\mathbb{R}^\times$ under the natural Abelianization homomorphism $W_{\mathbb{R}} \to W_{\mathbb{R}}^{ab} \xrightarrow{\sim} \mathbb{R}^\times$ of $W_{\mathbb{R}}$;
- $\nu \in \mathbb{a}_{K,\mathbb{C}}$: So $e_\nu = 0$ and we set $\mathscr{U}_{\mathfrak{m},\nu}^{(\varphi_{K_\nu})} = W_{\mathbb{C}}$.

In case K is a function field with full constant field $\mathbb{F}_q$, following [1], S denotes *any* fixed subset of places of K, and an S-cycle $\mathfrak{m} = \prod_{\nu\in\mathbb{h}_K}\nu^{e_\nu}$ of K is nothing but an effective divisor with support disjoint from S in additive notation. Moreover, assume that $S \neq \varnothing$ (look at Section 1 of [1]). Then, the S-cycle $\mathfrak{m}$ of K canonically defines a subgroup $\mathscr{U}_{S,\mathfrak{m}}^{\underline{\varphi}_K}\left(= \mathscr{U}_{\mathfrak{m}}^{\underline{\varphi}_K}\text{in short}\right)$ of $\mathscr{J}_K^{\underline{\varphi}_K}$ by

$$\mathscr{U}_{\mathfrak{m}}^{\underline{\varphi}_K} = \left(*_\nu\,\mathscr{U}_{\mathfrak{m},\nu}^{(\varphi_{K_\nu})}\right) \cap \mathscr{J}_K^{\underline{\varphi}_K},$$

where the local groups $\mathscr{U}_{\mathfrak{m},\nu}^{(\varphi_{K_\nu})}$ for $\nu \in \mathbb{h}_K$ are defined as follows:

- $\nu \in \mathbb{h}_K - S$ and $e_\nu = 0$: $\mathscr{U}_{\mathfrak{m},\nu}^{(\varphi_{K_\nu})} = {}_1\nabla_{K_\nu}^{(\varphi_{K_\nu})^{\underline{0}}}$;
- $\nu \in \mathbb{h}_K - S$ and $e_\nu > 0$: $\mathscr{U}_{\mathfrak{m},\nu}^{(\varphi_{K_\nu})} = {}_1\nabla_{K_\nu}^{(\varphi_{K_\nu})^{\underline{e_\nu}}}$;
- $\nu \in S$: $\mathscr{U}_{\mathfrak{m},\nu}^{(\varphi_{K_\nu})} = {}_{\mathbb{Z}}\nabla_{K_\nu}^{(\varphi_{K_\nu})}$.

For the definition and the basic properties of the subgroup ${}_1\nabla_{K_\nu}^{(\varphi_{K_\nu})^{\underline{i}}}$ of ${}_{\mathbb{Z}}\nabla_{K_\nu}^{(\varphi_{K_\nu})}$, where $\underline{i}$ is an "increasing net" in $\mathbb{R}_{\geq -1}$, look at [7].

Now, assume that Conjecture 1 holds in this subsection.

Theorem 10. *For any S-cycle $\mathfrak{m}$ of K,*

$$\mathscr{N}_K^{\underline{\varphi}_K} \mathscr{I}_K^{\underline{\varphi}_K^c} \mathsf{s}_K^{-1}\left(K^\times\right) \mathscr{U}_{\mathfrak{m}}^{\underline{\varphi}_K} = \mathscr{N}_K^{\underline{\varphi}_K} \mathsf{s}_K^{-1}\left(K^\times\right) \mathscr{U}_{\mathfrak{m}}^{\underline{\varphi}_K}$$

is an open normal subgroup of finite index in $\mathscr{I}_K^{\underline{\varphi}_K}$ containing $\mathscr{N}_K^{\underline{\varphi}_K}$, where

$$\mathsf{s}_K : \mathscr{I}_K^{\underline{\varphi}_K} \twoheadrightarrow \mathbb{J}_K$$

is the continuous surjective homomorphism defined by the Abelianization map of $\mathscr{I}_K^{\underline{\varphi}_K}$.

Proof. The inclusion[5] $\mathscr{I}_K^{\underline{\varphi}_K^c} \subseteq \mathsf{s}_K^{-1}(K^\times)$ is clear. Under the arrow $\mathsf{s}_K : \mathscr{I}_K^{\underline{\varphi}_K} \twoheadrightarrow \mathbb{J}_K$, the subgroup $\mathscr{I}_K^{\underline{\varphi}_K^c} \mathsf{s}_K^{-1}(K^\times)\mathscr{U}_{\mathfrak{m}}^{\underline{\varphi}_K}$ of $\mathscr{I}_K^{\underline{\varphi}_K}$ is mapped onto the subgroup $K^\times U_{\mathfrak{m}}$ of $\mathbb{J}_K$. Here, $U_{\mathfrak{m}}$ is the subgroup of $\mathbb{J}_K$ canonically defined by the cycle $\mathfrak{m}$. Moreover, $\mathscr{I}_K^{\underline{\varphi}_K^c} \mathsf{s}_K^{-1}(K^\times)\mathscr{U}_{\mathfrak{m}}^{\underline{\varphi}_K}$ is the pre-image of $K^\times U_{\mathfrak{m}}$ under the natural surjection $\mathsf{s}_K : \mathscr{I}_K^{\underline{\varphi}_K} \twoheadrightarrow \mathbb{J}_K$. As $K^\times U_{\mathfrak{m}}$ is a finite index open subgroup of $\mathbb{J}_K$ containing $K^\times$ (if K is assumed to be a function field, then we apply Proposition 1.1 of [1]), it follows that $\mathscr{I}_K^{\underline{\varphi}_K^c} \mathsf{s}_K^{-1}(K^\times)\mathscr{U}_{\mathfrak{m}}^{\underline{\varphi}_K}$ is a finite index open and normal subgroup of $\mathscr{I}_K^{\underline{\varphi}_K}$. Therefore, $\mathscr{N}_K^{\underline{\varphi}_K} \mathscr{I}_K^{\underline{\varphi}_K^c} \mathsf{s}_K^{-1}(K^\times)\mathscr{U}_{\mathfrak{m}}^{\underline{\varphi}_K}$ is an open normal subgroup of finite index in $\mathscr{I}_K^{\underline{\varphi}_K}$ containing $\mathscr{N}_K^{\underline{\varphi}_K}$, which completes the proof. □

Given any S-cycle, $\mathfrak{m} = \prod_{\nu \in \mathbb{h}_K \cup \mathbb{o}_K} \nu^{e_\nu}$ of K. By Theorem 9 of Sect. 4.8 on the *non-Abelian global existence*, there exists a finite Galois extension $R_{\mathfrak{m}}$ of K inside K^{sep}, called the *S-ray class field of* $\mathfrak{m}$, satisfying

$$\mathsf{NR}_{R_{\mathfrak{m}}/K}^{\underline{\varphi}_K\ \mathrm{Weil}^*} : \mathscr{I}_K^{\underline{\varphi}_K} / \mathscr{N}_K^{\underline{\varphi}_K} \mathsf{s}_K^{-1}\left(K^\times\right) \mathscr{U}_{\mathfrak{m}}^{\underline{\varphi}_K} \xrightarrow{\sim} \mathrm{Gal}(R_{\mathfrak{m}}/K).$$

The group $\mathscr{I}_K^{\underline{\varphi}_K} / \mathscr{N}_K^{\underline{\varphi}_K} \mathsf{s}_K^{-1}(K^\times)\mathscr{U}_{\mathfrak{m}}^{\underline{\varphi}_K}$ is called the *S-ray class group of* $\mathfrak{m}$.

[5] In fact, if $\mathsf{s}_K = \mathsf{a}_K$, then $\mathscr{I}_K^{\underline{\varphi}_K^c} = \mathsf{s}_K^{-1}(K^\times)$ by Theorem 2.

Moreover, we have the following two theorems about the Galois extension $R_{\mathfrak{m}}/K$, where $\mathfrak{m} = \prod_{\nu \in \mathfrak{h}_K \cup \mathfrak{a}_K} \nu^{e_\nu}$ is an S-cycle of K.

Theorem 11. *The Galois extension $R_{\mathfrak{m}}$ over K is unramified at all ν with $e_\nu = 0$.*

Proof. Let $\nu \in \mathfrak{h}_K$. The extension $R_{\mathfrak{m}}/K$ is unramified at ν if and only if the inertia group $I_\nu := \mathrm{I}_\nu(R_{\mathfrak{m}}/K)$ of ν in $\mathrm{Gal}(R_{\mathfrak{m}}/K)$ determined by the continuous homomorphism $e_\nu^{\mathrm{Weil}} : W_{K_\nu} \to W_K$ is trivial. Recall that the subgroup I_ν of $\mathrm{Gal}(R_{\mathfrak{m}}/K)$ is defined in Sect. 4.6 by

$$I_\nu = \mathrm{res}^*_{R_{\mathfrak{m}}} \circ \beta_{R_{\mathfrak{m}}/K} \circ \mathrm{red}_{W_{R_{\mathfrak{m}}}} \circ e_\nu^{\mathrm{Weil}} \left(W^0_{K_\nu}\right).$$

Now, assume furthermore that $e_\nu = 0$. By Theorem 6 combined with the fact that $q_\nu \left({}_1\nabla_{K_\nu}^{(\varphi_{K_\nu})^{\underline{0}}}\right) \subset \ker\left(\mathsf{NR}_{R_{\mathfrak{m}}/K}^{\underline{\varphi}_K \ \mathrm{Weil}}\right)$, it follows that

$$I_\nu = \mathsf{NR}_{R_{\mathfrak{m}}/K}^{\underline{\varphi}_K \ \mathrm{Weil}} \circ q_\nu \left({}_1\nabla_{K_\nu}^{(\varphi_{K_\nu})^{\underline{0}}}\right) = 1,$$

which proves that $R_{\mathfrak{m}}/K$ is unramified at such ν. Next, let $\nu \in \mathfrak{a}_K$ and assume that $e_\nu = 0$. Then $I_\nu = 1$ by Theorem 6 combined with the fact that $q_\nu\left(W^0_{K_\nu}\right) \subset \ker\left(\mathsf{NR}_{R_{\mathfrak{m}}/K}^{\underline{\varphi}_K \ \mathrm{Weil}}\right)$, where $W^0_{K_\nu} = W_{K_\nu}$. □

Let L be any finite Galois extension of K in K^{sep}. By the non-Abelian existence theorem (Theorem 9 of Sect. 4.8), the open normal subgroup of finite index in $\mathscr{J}_K^{\underline{\varphi}_K}$ containing $\mathscr{N}_K^{\underline{\varphi}_K}$ and corresponding to L is $\mathscr{N}_K^{\underline{\varphi}_K} \mathscr{N}_{L/K}^{\infty}\left(\mathscr{J}_L^{\underline{\varphi}_L}\right)$. Therefore, the subgroup $\mathsf{s}_K^{-1}\left(K^\times\right) \mathscr{N}_K^{\underline{\varphi}_K} \mathscr{N}_{L/K}^{\infty}\left(\mathscr{J}_L^{\underline{\varphi}_L}\right)$ of $\mathscr{J}_K^{\underline{\varphi}_K}$ is open and of finite index in $\mathscr{J}_K^{\underline{\varphi}_K}$ and corresponds to a subfield $L_!$ of L under the non-Abelian existence theorem.

Theorem 12. *Let L be any finite Galois extension of K in K^{sep}. Then, $L_!$ is a subfield of $R_{\mathfrak{m}}$ for some cycle $\mathfrak{m}$ of K.*

Proof. Let L be a finite Galois extension of K. The subgroup $\mathsf{s}_K\left(\mathscr{N}_K^{\underline{\varphi}_K} \mathscr{N}_{L/K}^{\infty}\left(\mathscr{J}_L^{\underline{\varphi}_L}\right)\right)$ of $\mathbb{J}_K$ is open and of finite index in $\mathbb{J}_K$. Thus, there exists a cycle $\mathfrak{m}$ of K such that

$$K^\times U_{\mathfrak{m}} \subseteq K^\times \mathsf{s}_K\left(\mathscr{N}_K^{\underline{\varphi}_K} \mathscr{N}_{L/K}^{\infty}\left(\mathscr{J}_L^{\underline{\varphi}_L}\right)\right),$$

where $U_{\mathfrak{m}}$ is the subgroup of $\mathbb{J}_K$ canonically defined by $\mathfrak{m}$. Hence, the following inclusions hold:

$$\mathsf{s}_K^{-1}\left(K^\times\right) \mathscr{U}_{\mathfrak{m}}^{\underline{\varphi}_K} \subseteq \mathscr{N}_K^{\underline{\varphi}_K} \mathsf{s}_K^{-1}\left(K^\times\right) \mathscr{U}_{\mathfrak{m}}^{\underline{\varphi}_K} \subseteq \mathsf{s}_K^{-1}\left(K^\times\right) \mathscr{N}_K^{\underline{\varphi}_K} \mathscr{N}_{L/K}^{\infty}\left(\mathscr{J}_L^{\underline{\varphi}_L}\right).$$

Thus, by the non-Abelian existence theorem and by Theorem 10,

$$L_! \subseteq R_{\mathfrak{m}},$$

which completes the proof. □

4.10 The Set of Primes in K that Split in a Finite Extension L/K

Let L be a finite Galois extension of a global field K. Denote the set of finite primes ν in K that split completely in L by $\mathrm{Spl}(L/K)$. The aim of this section is to characterize the set $\mathrm{Spl}(L/K)$ in terms of the base field K alone. We assume that Conjecture 1 holds in this section.

Recall that a finite prime ν in K splits completely in the Galois extension L over K if and only if the decomposition group $\mathrm{D}_\nu(L/K)$ of ν in $\mathrm{Gal}(L/K)$ is trivial.Thus, by Theorem 6 combined with Theorem 4, the following theorem follows immediately.

Theorem 13. *A prime ν in K splits completely in L/K if and only if* $q_\nu\left({}_{\mathbb{Z}}\nabla_{K_\nu}^{(\varphi_{K_\nu})}\right) \subset \ker\left(\mathsf{NR}_{L/K}^{\varphi_K\ \mathrm{Weil}}\right) = \mathscr{N}_K^{\varphi_K}\mathscr{N}_{L/K}^{\infty}\left(\mathscr{J}_L^{\varphi_L}\right)$.

Thus, by Theorem 13, the set $\mathrm{Spl}(L/K)$ is characterized in terms of the base field K by

$$\mathrm{Spl}(L/K) = \left\{\nu \in \mathfrak{h}_K \mid q_\nu\left({}_{\mathbb{Z}}\nabla_{K_\nu}^{(\varphi_{K_\nu})}\right) \subset \ker\left(\mathsf{NR}_{L/K}^{\varphi_K\ \mathrm{Weil}}\right)\right\}.$$

One final remark is in order.

Remark 6. In case L/K is a finite Abelian extension, then by Lemma 1 and Theorem 5 applied to the Abelian extension L/K,

$$\begin{aligned}\mathrm{Spl}(L/K) &= \left\{\nu \in \mathfrak{h}_K \mid q_\nu\left({}_{\mathbb{Z}}\nabla_{K_\nu}^{(\varphi_{K_\nu})}\right) \subset \ker\left(\mathsf{NR}_{L/K}^{\varphi_K\ \mathrm{Weil}}\right)\right\} \\ &= \{\nu \in \mathfrak{h}_K \mid \varepsilon_\nu\left(K_\nu^\times\right) \subset \ker(\bullet, L/K)\},\end{aligned}$$

which is nothing but the well-known formulation of the set of primes in K that split completely in the Abelian extension L/K.

References

1. Auer, R.: Ray class fields of global function fields with many rational places. Acta Arith. **95**, 97–122 (2000)
2. Fesenko, I.B.: Noncommutative local reciprocity maps. In: Miyake, K. (ed.) Class Field Theory–Its Centenary and Prospect, Tokyo 1998. Advanced Studies in Pure Mathematics, vol. 30, pp. 63–78. Mathematical Society of Japan, Tokyo (2001)
3. Hilton, P., Wu, Y.-C.: A Course in Modern Algebra. Wiley-Interscience, New York (1989)
4. Ikeda, K.I.: On the non-Abelian global class field theory. Annales mathématiques du Quebec **37**, 129–172 (2013)

5. Ikeda, K.I., Kazancıoğlu, S., Serbest, E.: On the relationship between the generalized Fesenko and the Laubie reciprocity maps (submitted)
6. Ikeda, K.I., Serbest, E.: Non-Abelian local reciprocity law. Manuscripta Math. **132**, 19–49 (2010)
7. Ikeda, K.I., Serbest, E.: Ramification theory in non-Abelian local class field theory. Acta Arith. **144**, 373–393 (2010)
8. Iwasawa, K.: On solvable extensions of algebraic number fields. Ann. Math. **58**, 548–572 (1953)
9. Koch, H.: Local class field theory for metabelian extensions. In: Behara M., Fritsch R., Lintz R.G. (eds.) Symposia Gaussiana, Conference A: Mathematics and Theoretical Physics, pp. 287–300. de Gruyter, Berlin/New York (1995)
10. Laubie, F.: Une théorie non abélienne du corps de classes local. Composit. Math. **143**, 339–362 (2007)
11. Manin Yu.I., Panchishkin, A.A.: Introduction to Modern Number Theory: Fundamental Problems, Ideas and Theories. Encyclopedia of Mathematical Sciences, vol. 49, 2nd edn. Springer, Berlin/Heidelberg (2005)
12. Miyake, K.: The arithmetic structure of the Galois group of the maximal nilpotent extension of an algebraic number field. Proc. Jpn. Acad. Ser. A **67**, 55–59 (1991)
13. Miyake, K.: Galois theoretic local-global relations in nilpotent extensions of algebraic number fields. In: David, S. (ed.) Séminaire de Théorie des Nombres, Paris 1989–90. Progress in Mathematics vol. 102, pp. 191–207. Birkhäuser, Boston (1992)
14. Morris, S.A.: Free products of topological groups. Bull. Aust. Math. Soc. **4**, 17–29 (1971)
15. Neukirch, J.: Algebraic Number Theory. Grundlehren Math. Wiss., vol. 322. Springer, Berlin/Heidelberg (1999)
16. Tate, J.: Number theoretic background. In: Borel, A., Casselman W. (eds.) Automorphic Forms, Representations, and L-Functions. Proceedings of the Symposium in Pure Mathematics, vol. 33, Part 2, pp. 3–26. American Mathematical Society, Providence (1979)

Cosmos and Its Furniture

Olav Arnfinn Laudal

1 Introduction

Vos calculs sont corrects, mais votre physique est abominable: (Albert Einstein 1927 à George Lemaître.)

If we want to study a natural phenomenon, called **P**, we must in the present scientific situation, describe **P** in some mathematical terms, say as a mathematical object, X, depending upon some parameters, in such a way that the changing aspects of **P** would correspond to altered parameter values for X. This object would be a *model for* **P** if, moreover, X with any choice of parameter values would correspond to some, possibly occurring, aspect of **P**.

Two mathematical objects X(1) and X(2), corresponding to the same aspect of **P**, would be called equivalent, and the set, $\mathcal{P}$, of equivalence classes of the objects **P** would correspond to (a quotient of) the *moduli space*, **M**, of the models, X. The study of the natural phenomena **P**, and its changing aspects, would then be equivalent to the study of the *structure* of $\mathcal{P}$ and therefore to the study of the dynamics of the moduli space **M**. In particular, the notion of *time* would, in agreement with Aristotle and St. Augustin, see [1] and [9], correspond to some *metric* on this space.

It turns out that to obtain a complete theoretical framework for studying the phenomenon **P**, or the model X, together with its *dynamics*, we should introduce the notion of *dynamical structure*, defined on the space, **M**. Assuming that **M** is an algebraic scheme of some sort, this is done via the construction of a universal

O.A. Laudal (✉)
Matematisk institutt, University of Oslo, Pb. 1053, Blindern, N-0316 Oslo, Norway
e-mail: arnfinnl@math.uio.no

P. Cartier et al. (eds.), *Mathematics in the 21st Century*, Springer Proceedings in Mathematics & Statistics 98, DOI 10.1007/978-3-0348-0859-0_6

noncommutative *phase space* functor, $\mathrm{Ph}(-) : \mathrm{Alg}_k \to \mathrm{Alg}_k$. It extends to the category of schemes, and its infinite iteration $\mathrm{Ph}^\infty(-)$ is outfitted with a universal *Dirac derivation*, $\delta \in \mathrm{Der}_k(\mathrm{Ph}^\infty(-), \mathrm{Ph}^\infty(-))$.

A dynamical structure defined on an associative k-algebra $A \in \mathrm{Alg}_k$ is now a δ-stable ideal $\sigma \subset \mathrm{Ph}^\infty(A)$, and the structure we are interested in is the *space* $\mathbf{U} := \mathrm{Ph}^\infty(\mathbf{M})/\sigma$, corresponding to an affine covering of $\mathbf{M}$ by algebras of the type $\mathrm{Ph}^\infty(A)/\sigma$; see [9, 10], and also [11].

But now we observe that there may be an action of a Lie algebra $\mathfrak{g}$, a *gauge group*, on $\mathbf{U}$, such that the dynamics of $\mathcal{P}$, really, corresponds to that of the quotient $\mathbf{U}/\mathfrak{g}$.

To any *open* subset V, of $\mathbf{U}$, there would be associated an, not necessarily commutative, affine k-algebra, $A := O_{\mathbf{U}}(V)$, with an action of the Lie algebra $\mathfrak{g}$, containing the available information about the structure of O. An element of this algebra would be called an *observable* and, wishing to measure the *values* of an observable, leads to the study of representations of this algebra. Finally, the Dirac derivation and the gauge group $\mathfrak{g}$ will act on the moduli space of representations of A, inducing the dynamical laws we are interested in.

With this philosophy in mind, and stimulated by the results in the deformation theory, obtained in [4] and [5], we embarked, in a series of papers, [6–8], on the study of moduli spaces of representations (modules) of associative algebras, in general, and on their quotients, modulo Lie algebra actions. Here is where the *invariant theory* and *noncommutative algebraic geometry* enter the play. The Dirac derivation translates into a vector field on these moduli spaces, together with a Hamiltonian operator on the universal family of representations, furnishing the equations of motions that we need.

In [9, 10], and also [11], we introduced a *toy model*, used to illustrate the general theory and to connect to present-day physics. It was shown to generalize both general relativity and quantum field theories. In particular, the definition of time made the space of velocities compact.

In this paper, this toy model has become the main figure. It is the *time-space* of an observer observing an observed in 3-space. In mathematical terms, this is the Hilbert scheme, $\mathbf{H}$, of unordered pairs of points in $\mathbf{A}^3$, the affine 3-space. It is easy to see that $\mathbf{H} = \underline{\tilde{H}}/\mathbf{Z}_2$, where $\underline{\tilde{H}}$ is the blowup of the space of ordered pairs of points, $\underline{H} := (\mathbf{A}^3 \times \mathbf{A}^3)$, along the diagonal $\underline{\Delta}$.

In this paper [9], we discussed the possibility of including a cosmological model in our toy models of *time-spaces*. The one-dimensional model we presented was created by the deformations of the trivial singularity, $O := k[x]/(x)^2$. Using elementary deformation theory for algebras, we obtained amusing results, depending upon some rather bold mathematical interpretations of the, more or less accepted, cosmological vernacular.

The main axiom of the leading branches of cosmology seems to be that the space-time of the existing universe can be described via a general relativistic model, somehow given by Einstein's equation with respect to some mass-stress tensor, mass and energy being homogeneously and isotropically distributed in space. This leads to the assumption that the universe is a four-dimensional space-time of a form commonly called a Friedman-Robertson-Walker model or, sometimes, the FLRW-model (where one includes Lemaître). In particular this space has an open-ended time coordinate, leaving out the 0 point, the *Big Bang*, but still assuming that this *point-like singularity* is in the closure of even the shortest complete history of the universe.

There are a lot of assumptions here. One is that the space-time is capable of containing something and that these things can be described as independent upon the space, even though they curve space, and otherwise intervene in the dynamical process. For example, even in the very start of the universe, spin is assumed to be present. So, in mathematical terms, the space-time must be outfitted with a $su(2)$, or a complexified $sl(2)$, tangent structure, obviously determined by the Big Bang event. Moreover, since the space-time of the model does not contain the prime event and the jump between that, supposedly singular point-like, event and the mathematically well-defined space-time is not part of the model, we do not have a model of the Big Bang itself, but rather of what may have happened in our usual space, a long time ago, with respect to a rather artificially chosen time parameter.

In this paper I propose to show how the *time-space*, **H**, can be thought of as an immediate product of a mathematical scenario incorporating a *Big Bang event*, making this event mathematically sound. This, however, should not be interpreted as if I actually propose this *mathematical model*, as a physical explanation of the universe we observe today, whatever that would mean. This paper is about mathematics, and should be considered as a continuation of my struggle to understand the language of physics, as I have explained in [11]. In this struggle I have been comforted by Dirac, in particular by his "lecture delivered on presentation of the James Scott prize, February 6, 1939"; see [2], where he talks about the *mathematical quality in Nature*. It is this quality that fascinates me, and it is the realization of a relationship between *numbers and nature* that goes back to the Pythagoreans that inspires me; see [3].

The more philosophical questions about the reality of physical notions, like the reality of time, or fantasies like time travel, will not be treated, not because I do not find such problems interesting but simply because I just work with the mathematical models. They may, or may not, be real, but that question does not seem to me to be of any importance.

2 Big Bang and Deformations of Associative Algebras

Starting with the pure notion of three-dimensionality, i.e., a k-scheme $\underline{U} = \mathrm{Spec}(U)$ with only one point and a three-dimensional tangent space, in algebraic terms, the singularity,

$$U := k[x_1, x_2, x_3]/(x_1, x_2, x_3)^2,$$

we shall see that we may construct, in a canonical way, a formal versal deformation base space, **M**, and a corresponding versal family $\mathbf{U}_\star$, containing all isomorphism classes of deformations of U, as associative algebra to artinian bases.

The technique for this general deformation theory can be found in [4]; see also [12]. In the last book we introduced the notion of *moduli suite*. This corresponds here to a partition of the space **M** into a series of *rooms*, containing an inner room; the *modular stratum*, composed of just one point $\star$, corresponding to the singularity we start with, U; and a very special component that turns out to be **H**, where the family **U**, the restriction of $\mathbf{U}_\star$ to **H**, is algebraic and where corresponding to a point $(o, p) \in \underline{H} - \underline{\Delta}$, the fiber is the four-dimensional associative algebra,

$$U(o,p) := k < x_1, x_2, x_3 > /(x_i x_j - o_i x_j - p_j x_i + o_i p_j),$$

where we have used the coordinates x_1, x_2, x_3 to express the two points, o and p, in 3-space $\mathbf{A}^3$, by coordinates, $\{o_i\}, \{p_j\}, i, j = 1, 2, 3$.

If $o = p$, $U(o, p)$ is isomorphic to U. But the family $U(o, p)$ has, never the less, a unique extension to all of $\underline{\tilde{H}}$, and the Z_2-action also extends, so we obtain a unique well-defined family of associative algebras, defined on **H**.

There is also a special *room* in the moduli suite, corresponding to the quaternions, **Q**. In fact, we have

$$\mathbf{Q} = k < x_1, x_2, x_3 > /(x_i x_j - \epsilon_{i,j,k} x_k + \delta_{i,j}),$$

where $\epsilon_{i,j,k}$ and $\delta_{i,j}$ are the usual notations for 0, 1.

2.1 The Construction of M and H

The tangent space of the formal moduli of the singularity

$$U := k[x_1, x_2, x_3]/(x_1, x_2, x_3)^2,$$

as an associative k-algebra is, by deformation theory (see [4] and [12])

$$T_\star := A^1(k, U; U) = \mathrm{Hom}_F(\ker\rho, U)/\mathrm{Der},$$

where $\rho : F \to U$ is any surjective homomorphism of a free k-algebra F, onto U, Hom_F means the F-bilinear maps and Der denotes the subset of the restrictions to $I := \ker\rho$ of the k-derivations from F to U. In our case we choose $F = k < x_1, x_2, x_3 >$ and the obvious surjection, making $\ker\rho = (\underline{x})^2$, generated as F bimodule by the family $\{x_{i,j} := x_i x_j\}$.

Any F-bilinear morphism $\phi : (\underline{x})^2 \to U$ must be of the form

$$\phi(x_{i,j}) = a^0_{i,j} + \sum_{l=1}^{3} a^l_{i,j} x_l$$

and the bilinearity is seen to imply that $a^0_{i,j} = 0$. Thus, the dimension of $\mathrm{Hom}_F(I, U)$ is 27.

Any derivation $\delta \in \mathrm{Der}$ must be given by

$$\delta(x_i) = b_i^0 + \sum_{l=1}^{3} b_i^l x_l$$

and the restriction of this map, to the generators of $I = (\underline{x})^2$, must have the form

$$\delta(x_{i,j}) = b_j^0 x_i + b_i^0 x_j,$$

therefore determined by the b_i^0, $i = 1, 2, 3$. It follows that the tangent space $T_\star$ is of dimension $27 - 3 = 24$.

Let $o, p \in \mathbf{A}^3$, with coordinates $o = (o_1, o_2, o_3)$, $p = (p_1, p_2, p_3)$, with respect to the coordinate system, $\underline{x}$, and put

$$\phi_{o,p}(x_{i,j}) = p_j x_i + o_i x_j,$$

then it is easy to see that the maps $\{\phi_{o,p}\}$ generate a six-dimensional sub vector space $\underline{H} = \mathbf{A}^3 \times \mathbf{A}^3 = \mathrm{Spec}(H)$, $H = k[o, p]$, of $T_\star$. Notice that if $o = p$ then $\phi_{o,p}$ is a derivation, thus 0 in $T_\star$.

Now, let us consider the Lie algebra of infinitesimal automorphisms of U, $\mathfrak{g}(U) := \mathrm{Lie}(\mathrm{Aut}(U)) = \mathfrak{gl}_3(k)$. By deformation theory, see, e.g., [12], $\mathfrak{g}(U)$ acts on the tangent space $T_\star$, and a simple calculation gives the result: $\alpha \in \mathfrak{g}(U)$, with $\alpha(x_i) = \sum_{l=1}^{3} \alpha_i^l x_l$, and acts on $\phi \in T_\star$, given in terms of its coordinates $\left\{a^l_{i,j}\right\}$ as

$$\alpha\left(a^l_{i,j}\right) = \sum_{p=1}^{3} \alpha_i^p a^l_{p,j} + \sum_{p=1}^{3} \alpha_j^p a^l_{i,p} + \sum_{p=1}^{3} \alpha^l_p a^p_{i,j}.$$

In particular,

$$\alpha\left(\phi_{o,p}\right) = \phi_{\alpha(o),\alpha(p)},$$

and the origin of the tangent space $T_\star$ is the only fix point of the Lie algebra $\mathfrak{g}(U)$. This implies, see [12], that the *modular stratum* of the versal base space $\mathbf{M}$ of U is reduced to the base point, $\star = [U]$.

Now, the rather unexpected happens. We may integrate the tangent subspace $\underline{H}$ and obtain a family of flat deformations of U. In fact, it is easy to see that for $(o, p) \in \underline{H}$,

$$U(o, p) := k < x_1, x_2, x_3 > / \left(x_i x_j - o_i x_j - p_j x_i + o_i p_j\right),$$

is an associative k-algebra of dimension 4 and a deformation of U, in a direction of $\underline{H}$. This defines a family of associative $H := \mathbf{R}[o, p]$ algebras,

$$\mathbf{U}' := H < x_1, x_2, x_3 > / \left(x_i x_j - o_i x_j - p_j x_i + o_i p_j\right),$$

Notice that if $o = p$ then $U(o, p)$ is isomorphic to U, as it should, and that $U(o, p) \simeq U(-o, -p)$. Moreover, for any nonzero element $\kappa \in k$, and any 3-vector $c \in \mathbf{A}^3$, we have

$$U(o, p) \simeq U(\kappa o, \kappa p), \; U(o, p) \simeq U(o - c, p - c).$$

Choosing $c = 1/2(p + o)$, we find $o' := o - c = -(p - c) =: -p'$, and it is easy to see that if $o' \neq 0$, the sub Lie algebra above generated by $\{x_1, x_2, x_3\}$ in $U(o', p')$ is isomorphic to the standard three-dimensional Lie algebra with relations $[y_1, y_2] = y_2, [y_1, y_3] = y_3, [y_2, y_3] = 0$. Moreover, choosing $c = (p + o)$, we find an isomorphism,

$$U(o, p) \simeq U(-p, -o) \simeq U(p, o),$$

which should be related to the action of Z_2 on $\underline{\tilde{H}}$; the blowup of $\underline{H}$ along the diagonal (see the Introduction); and thus, according to our philosophy, the mathematical reason for the CPT-equivalence in quantum theory; see [11], (4.9).

Notice also that the algebra,

$$\mathbf{Q} := k < x_1, x_2, x_3 > / \left(x_i x_j - \epsilon_{i,j,k} x_k + \delta_{i,j}\right),$$

where $\epsilon_{i,j,k}$ and $\delta_{i,j}$ are the usual indices, the first one nonzero only for $\{i, j, k\} = \{1, 2, 3\}$ and the last one the usual delta function, is isomorphic to the quaternions, which therefore is another nontrivial deformation of U.

Consider now the restriction to the subscheme $\underline{H} - \underline{\Delta}$, of the family $\mathbf{U}$, still denoted by

$$\nu' : \mathbf{U}' \to \underline{H} - \underline{\Delta}.$$

Since for all nonzero $\kappa \in \mathbf{R}$, we have $U(p, -p) \simeq U(\kappa p, \kappa(-p))$, this family extends uniquely, first to a family,

$$\nu : \tilde{\mathbf{U}} \to \underline{\tilde{H}},$$

then, since $\mathbf{Z}_2$ acts on this family, to the family

$$\nu : \mathbf{U} \to \mathbf{H},$$

2.2 The Metric Structure of H

This component of the versal base space of U, the Hilbert scheme of subschemes of length 2 in $\mathbf{A}^3$, is easily computed and has, as mentioned, the form $\mathbf{H} = \underline{\tilde{H}}/Z_2$, where $H = k[t_1, \ldots, t_6]$, $k = \mathbf{R}$, and $\underline{H} := \mathrm{Spec}(H)$ and is the space of all ordered pairs of points in $\mathbf{A}^3$, $\underline{\tilde{H}}$ is the blowup of the diagonal, and Z_2 is the obvious group action. The space $\mathbf{H}$, and by extension, $\underline{H}$ and $\underline{\tilde{H}}$, was, in [11], called the *time-space* of the model.

Measurable time, in this mathematical model, turned out to be a metric ρ on the time-space, measuring all possible infinitesimal changes of *the state* of the objects in the family we are studying. This implies that the notion of relative velocity may be interpreted as an oriented line in the tangent space of a point of $\underline{\tilde{H}}$. Thus, the space of velocities is compact.

This leads to a *physics* where there are no infinite velocities and where the principle of relativity comes for free. The Galilean group acts on $\mathbf{E}^3$ and therefore on $\underline{\tilde{H}}$. The Abelian Lie algebra of translations defines a three-dimensional distribution, $\tilde{\Delta}$ in the tangent bundle of $\underline{\tilde{H}}$, corresponding to 0-velocities. Given a metric on $\underline{\tilde{H}}$, we defined the distribution $\tilde{c}$, corresponding to light velocities, as the normal space of $\tilde{\Delta}$. We explained how the classical *space-time* can be thought of as a universal subspace, $\underline{\tilde{M}}(l)$, of $\underline{\tilde{H}}$, defined by a fixed line $l \subset \mathbf{E}^3$.

We also showed how the generator $\tau \in Z_2$, above, is linked to the operators C, P, T in classical physics, such that $\tau^2 = \tau PT = \mathrm{id}$. Moreover, we observed that the three fundamental gauge groups of current quantum theory $U(1)$, $SU(2)$, and $SU(3)$ are part of the structure of the fiber space,

$$\tilde{E} \longrightarrow \underline{\tilde{H}}.$$

In fact, for any point $\underline{t} = (o, x)$ in $\underline{H}$, outside the diagonal $\underline{\Delta}$, we may consider the line l in $\mathbf{E}^3$ defined by the pair of points $(o, x) \in \mathbf{E}^3 \times \mathbf{E}^3$. We may also consider the action of $U(1)$ on the normal plane $B_o(l)$, of this line, oriented by the normal (o, x), and on the same plane $B_x(l)$, oriented by the normal (x, o). Using parallel transport in $\mathbf{E}^3$, we find an isomorphism of bundles,

$$P_{o,x} : B_o \to B_x,\ P : B_o \oplus B_x \to B_o \oplus B_x,$$

the *partition isomorphism*. Using P we may write $(v, v) \in B_o \oplus B_x$ for $(v, P_{o,x}(v))$. We have also seen, loc.cit., that the line l defines a unique subscheme $\underline{H}(l) \subset \underline{H}$. The corresponding tangent space at (o, x) is called $A_{(o,x)}$. Together this defines a decomposition of the tangent bundle of $\underline{H}$,

$$T_{\underline{H}} = B_o \oplus B_x \oplus A_{(o,x)}.$$

If $\underline{t} = (o, o) \in \underline{\Delta}$, and if we consider a point $\underline{o}'$ in the exceptional fiber E_o of $\underline{\tilde{H}}$, we find that the tangent bundle, at this point, decomposes into

$$T_{\underline{\tilde{H}},o'} = C_{o'} \oplus A_{o'} \oplus \tilde{\Delta},$$

where $C_{o'}$ is the tangent space of E_o, at o'; $A_{o'}$ is the light velocity defining o'; and $\tilde{\Delta}$ is the 0-velocity. Both B_o and B_x as well as the bundle $C_{(o,x)} := \{(\psi, -\psi) \in B_o \oplus B_x\}$ become complex line bundles on $\underline{H} - \underline{\Delta}$. $C_{(o,x)}$ extends to all of $\underline{\tilde{H}}$, and its restriction to E_o coincides with the tangent bundle. Tensorising with $C_{(o,x)}$, we complexify all bundles. In particular we find complex 2-bundles $\mathbf{C}B_o$ and $\mathbf{C}B_x$, on $\underline{H}-\underline{\Delta}$, and we obtain a canonical decomposition of the complexified tangent bundle. Any real metric on $\underline{H}$ will decompose the tangent space into the light velocities $\tilde{\mathbf{c}}$ and the 0-velocities, $\tilde{\Delta}$, and obviously,

$$T_{\underline{H}} = \tilde{\mathbf{c}} \oplus \tilde{\Delta}, \ \mathbf{C}T_{\underline{H}} = \mathbf{C}\tilde{\mathbf{c}} \oplus \mathbf{C}\tilde{\Delta}.$$

This decomposition can also be extended to the complexified tangent bundle of $\underline{\tilde{H}}$. Clearly, $U(1)$ acts on $T_{\underline{H}}$, and $SU(2)$ and $SU(3)$ acts naturally on $\mathbf{C}B_o \oplus \mathbf{C}B_x$ and $\mathbf{C}\tilde{\Delta}$, respectively. Moreover $SU(2)$ acts on $\mathbf{C}C_{o'}$, in such a way that its action should be *physically* irrelevant. The groups, $U(1),\ SU(2), and\ SU(3)$, are our elementary *gauge groups*, and we shall consider the corresponding Lie algebra,

$$\mathfrak{g} := \mathfrak{u}(1) \oplus \mathfrak{u}(2) \oplus \mathfrak{u}(3)$$

as the *gauge group*, in the sense of the Introduction.

2.3 *Newton's and Kepler's Laws*

Let us study the geometry of $\mathbf{H}$. Recall that $\underline{\tilde{H}} \rightarrow \underline{H}$ is the (real) blowup of the diagonal $\underline{\Delta} \subset \underline{H}$, where $\underline{H}$ is the space of pairs of points in $\mathbf{E}^3$. Clearly any point $\underline{t} \in \underline{H}$ outside the diagonal determines a vector $\xi(o, x)$ and an oriented line $l(o, x) \subset \mathbf{E}^3$, on which both the observer o and the observed x sit. This line also determines a subscheme $\underline{H}(l) \subset \underline{H}$, see above and [9], and in $\underline{H}(l)$ there is a unique *light velocity curve* $\underline{l}(\underline{t})$, through $\underline{t}$, an integral curve of the distribution $\tilde{c}$, and this curve cuts the diagonal $\underline{\Delta}$ in a unique point $c(o, x)$, the *center of gravity of the observer and the observed*, and thus defines a unique point $\xi(\underline{t})$, of the blowup of the diagonal, in the fiber of $\underline{\tilde{H}} \rightarrow \underline{H}$, above $c(o, x)$.

Recall that the subspace $\underline{\tilde{M}}(l) \subset \underline{\tilde{H}}$, corresponding to a line $l \subset \mathbf{A}^3$, referred to above, consists of all points $(o, p) \in \underline{\tilde{H}}$ for which $c(o, x) \in l$.

There is a convenient parametrization of $\underline{\tilde{H}}$. Consider, as above, for each $\underline{t} \in \underline{\tilde{H}}$ the length ρ, in $\mathbf{E}^3$, the Euclidean space, of the vector $1/2(o, x)$. Given a point $\underline{\lambda} \in \underline{\Delta}$, and a point $\xi \in E(\underline{\lambda}) = \pi^{-1}(\underline{\lambda})$, the fiber of

$$\pi : \underline{\tilde{H}} \rightarrow \underline{H},$$

at the point $\underline{\lambda}$, for $o = x$. Since $E(\underline{\lambda})$ is isomorphic to S^2, parametrized by $\underline{\phi}$, any element of $\underline{\tilde{H}}$ is now uniquely determined in terms of the triple $\underline{t} = \left(\underline{\lambda}, \underline{\phi}, \rho\right)$, such that $c\ (\underline{t}) = c(o, x) = \underline{\lambda}$ and such that ξ is defined by the line $\underline{ox}$; here $\rho \geq 0$. Notice also that, at the exceptional fiber, i.e., for $\rho = 0$, the momentum corresponding to $d\rho$ is not defined.

Consider any metric on $\underline{\tilde{H}}$, of the form

$$g = h_\rho\left(\underline{\lambda}, \underline{\phi}, \rho\right) d\rho^2 + h_{\underline{\phi}}\left(\underline{\lambda}, \underline{\phi}, \rho\right) d\underline{\phi}^2 + h_{\underline{\lambda}}\left(\underline{\lambda}, \underline{\phi}, \rho\right) d\underline{\lambda}^2,$$

where $d\underline{\phi}^2$ is the natural metric in $S^2 = E\ (\underline{\lambda})$.

It is reasonable to believe that the geometry of $(\underline{\tilde{H}}, g)$ might explain the notions like *energy*, mass, *charge*, etc. In fact, we tentatively propose that the source of mass, charge, etc., is located in the *black holes* $E\ (\underline{\lambda})$. This would imply that mass, charge, etc., are properties of the five-dimensional superstructure of our usual three-dimensional Euclidean space, essentially given by a *density*, $h\ (\underline{\lambda}, \phi, \theta)$. This might bring to mind Kaluza-Klein-theory. However, it seems to me that there are important differences, making comparison very difficult.

Let us first treat the following simple case,

$$h_\rho = \left(\frac{\rho - h}{\rho}\right)^2,\ h_\phi = (\rho - h)^2,\ h_\lambda = 1,$$

where h is a positive real number. This metric is everywhere defined in the subspace of $\underline{M}(l)$, where we have reduced the spherical coordinates $\underline{\phi}$, to just ϕ. Notice that for $\rho = 0$, there are no tangent vectors in $d\rho$ direction.

It clearly reduces to the Euclidean metric far away from $\underline{\Delta}$, and it is singular on the *horizon* of the black hole, given by $\rho = h$, which in $\underline{H}$ is simply a sphere in the *light-space*, of radius h. Moreover it is clear that h is also the *radius of the exceptional fiber*, since the length of the circumference of $\rho = 0$ is $2\pi h$. Clearly, the exceptional fiber, the *black hole* itself, is not visible and does not bound anything. However, the horizon bounds a piece of space. Moreover, if we reduce the horizon to a point in $\underline{H}$, then the circumference, or area of the exceptional fiber, as measured using the above metric, reduces to zero, and the metric becomes the usual Euclidean metric. We shall reduce to a plane in the light directions, i.e., we shall just assume that $S^2 = E\ (\underline{\lambda})$, which is reduced to a circle, with coordinate ϕ. There is actually no restriction made, as is easily seen.

The corresponding equations for the geodesics in $\underline{\tilde{H}}$ are, see [11],

$$\frac{d^2\rho}{dt^2} = -\left(\frac{h}{\rho(\rho - h)}\right)\left(\frac{d\rho}{dt}\right)^2 + \left(\frac{\rho^2}{(\rho - h)}\right)\left(\frac{d\phi}{dt}\right)^2,$$

$$\frac{d^2\phi}{dt^2} = -2/(\rho - h)\frac{d\rho}{dt}\frac{d\phi}{dt}$$

$$\frac{d^2\lambda}{dt^2} = 0.$$

where t is time. But time is, by definition, the distance function in $\underline{\tilde{H}}$, so we must have

$$\left(\frac{\rho-h}{\rho}\right)^2\left(\frac{d\rho}{dt}\right)^2+(\rho-h)^2\left(\frac{d\phi}{dt}\right)^2+\left(\frac{d\lambda}{dt}\right)^2=1,$$

from which we find

$$\left(\frac{d\rho}{dt}\right)^2=\rho^2(\rho-h)^{-2}\left(1-\left(\frac{d\lambda}{dt}\right)^2\right)-\rho^2\left(\frac{d\phi}{dt}\right)^2.$$

From the third equation, we find that $\frac{d\lambda}{dt}$, the *rest-mass* of the system, is constant. Put $K^2=\left(1-|\frac{d\lambda}{dt}|^2\right)$, then K is the *kinetic energy of the system*. The definition of time therefore gives us

$$\rho^{-2}\left(\frac{d\rho}{dt}\right)^2=(\rho-h)^{-2}K^2-\left(\frac{d\phi}{dt}\right)^2.$$

Put this into the first equation above, and obtain

$$\frac{d^2\rho}{dt^2}=-hK^2\left(\frac{\rho}{\rho-h}\right)\frac{1}{(\rho-h)^2}+\left(\frac{\rho+h}{\rho-h}\right)\rho\left(\frac{d\phi}{dt}\right)^2.$$

Assume now $r:=\rho-h\approx\rho$, we find

$$\frac{d^2r}{dt^2}=-\frac{hK^2}{r^2}+r\left(\frac{d\phi}{dt}\right)^2,$$

i.e., Kepler's first law. The constant h, i.e., the radius of the exceptional fiber, is thus related to mass. Recall that the Schwarzschild radius, the Einstein equivalent to h, is assumed to be

$$r_s=2GM/c^2,$$

where G = Newton's gravitational constant, M = mass, and c = the speed of light, which here, of course, is put equal to 1. As we have hinted at above, this suggests that *mass* is a property of the space $\underline{\tilde{H}}$. In this case it is a function of the surface of the exceptional fiber, i.e., the *black hole*, associated with the point $\underline{\lambda}$ in the ordinary 3-space $\underline{\Delta}$. In the same way, the second equation above gives us Kepler's second law,

$$r\left(\frac{d^2\phi}{dt^2}\right)+2\left(\frac{dr}{dt}\right)\left(\frac{d\phi}{dt}\right)=0.$$

Notice that with the chosen metric, time, in light velocity direction, is *standing still* on the *horizon* $\rho=h$, of the *black hole* at $\underline{\lambda}\in\underline{\Delta}$. Therefore no light can escape from the black hole. In fact, no geodesics can pass through $\rho = h$. Notice also that, for a photon with light velocity, we have $K = 1$, so we may measure h, by measuring the trajectories of photons in the neighborhood of the *black hole*. Finally, see that if the distance between the two interacting points is close to constant, i.e., if we have a circular movement, the left side of the time equation becomes zero, and we therefore have the following equation:

$$(\rho - h)d\phi = Kdt,$$

which may be related to the perihelion precession and also to the *Thomas Precession*; see [14] and [15].

If we, in our metric, permit the radius of the black hole, h, to depend on λ, i.e., be given by a function of the form $h(\lambda)$ so that the metric looks like

$$\left(\frac{\rho - h(\lambda)}{\rho}\right)^2 d\rho^2 + (\rho - h(\lambda))^2 d\phi^2 + \kappa(\lambda) d\lambda^2,$$

then the force law formulas above become more involved:

$$\frac{d^2\rho}{dt^2} = -\left(\frac{h(\lambda)}{\rho(\rho - h(\lambda))}\right)\left(\frac{d\rho}{dt}\right)^2 + \left(\frac{2}{(\rho - h(\lambda))}\right)\left(\frac{dh}{d\lambda}\right)\left(\frac{d\rho}{dt}\right)\left(\frac{d\lambda}{dt}\right)$$
$$+\left(\frac{\rho^2}{(\rho - h(\lambda))}\right)\left(\frac{d\phi}{dt}\right)^2,$$

$$\frac{d^2\phi}{dt^2} = -2/(\rho - h(\lambda))\frac{d\rho}{dt}\frac{d\phi}{dt} + 2/(\rho - h(\lambda))\left(\frac{dh}{d\lambda}\right)\left(\frac{d\phi}{dt}\right)\left(\frac{d\lambda}{dt}\right)$$

$$\frac{d^2\lambda}{dt^2} = \left(\frac{(\rho - h(\lambda)}{\rho}\right)\left(\frac{1}{\kappa(\lambda)}\right)\left(\frac{dh}{d\lambda}\right)\left(\frac{d\rho}{dt}\right)^2 + (\rho - h(\lambda))\left(\frac{1}{\kappa(\lambda)}\right)\left(\frac{dh}{d\lambda}\right)\left(\frac{d\phi}{dt}\right)^2$$
$$+ 1/2\left(\frac{d\ln(\kappa)}{d\lambda}\right)\left(\frac{d\lambda}{dt}\right)^2$$

where t, as above, is time.

We find that if $(\frac{dh}{d\lambda})$ is negative, then for $\rho \leq h(\lambda)$ the acceleration of ρ is positive, and unlimited for ρ close to $h(\lambda)$, and the acceleration of λ is negative. We shall, at the end of this paper, come back to this in relation to the Big Bang and Inflation.

2.4 Thermodynamics, the Heat Equation and Navier-Stokes

Let us now go back and consider any metric on $\underline{\tilde{H}}$, of the form

$$g = h_1\left(\underline{\lambda}, \underline{\phi}, \rho\right) d\rho^2 + h_2\left(\underline{\lambda}, \underline{\phi}, \rho\right) d\underline{\phi}^2 + h_3\left(\underline{\lambda}, \underline{\phi}, \rho\right) d\underline{\lambda}^2,$$

where $d\underline{\phi}^2$ is the natural metric in $S^2 = E\left(\underline{\lambda}\right)$.

Recall for $C = H = k[t_1, \dots, t_6]$ and for a nonsingular Riemannian metric $g = 1/2 \sum_{i,j=1,\dots,r} g_{i,j} dt_i dt_j \in \mathrm{Ph}(C)$; (see [11], the notations),

$$\Gamma_p^{j,i} = \sum_k g^{j,k} \Gamma_{k,p}^i, \; R_{i,j} := \left[dt_i, dt_j\right], \; F_{i,j} := R_{i,j} - \sum_p \left(\Gamma_p^{j,i} - \Gamma_p^{i,j}\right) dt_p,$$

and we have the general force law in $\mathrm{Ph}(C)$,

$$d^2 t_i = -\sum_{p,q} \Gamma_{p,q}^i dt_p dt_q - 1/2 \sum_{p,q} g_{p,q} \left(F_{i,p} dt_q + dt_p F_{i,q}\right)$$
$$+ 1/2 \sum_{l,p,q} g_{p,q} \left[dt_p, \left(\Gamma_l^{i,q} - \Gamma_l^{q,i}\right)\right] dt_l + [dt_i, T],$$

generating the dynamical structure $\mathfrak{c} := \mathfrak{c}(g)$.

Remark 1. In principle, according to our philosophy, the natural common quantization of classical general relativity and Yang-Mills theory would be based on the dynamical properties of $\mathrm{Simp}_{\leq \infty}\left(\underline{\tilde{H}}(\mathfrak{c})\right)$, with respect to the versal family, $\tilde{\rho} : O_{\underline{\tilde{H}}(\mathfrak{c})} \to \mathrm{End}_{\underline{\tilde{H}}(\mathfrak{c})}\left(\tilde{V}\right)$, where we have to consider $O_{\underline{\tilde{H}}(\mathfrak{c})}$ as a presheaf of associative k-algebras, defined in $\underline{\tilde{H}}$. As a first try, we shall concentrate on singular situations and, in particular, on the structure of the Levi-Civita representation.

Recall that at a point $\underline{t} = (o, x) \in \underline{H} - \underline{\Delta}$, the tangent space, $\Theta_{\underline{\tilde{H}}}(\underline{t})$, is represented by the space of all pairs of 3-vectors, $\xi(\underline{t}) = (\xi_o, \xi_x)$, ξ_o, fixed at o, and ξ_x, fixed at the point x in $\mathbf{E}^3$. Moreover, any such tangent vector may, depending only upon the choice of metric, be decomposed into the sum $\xi = \underline{p} + \underline{m}$, with $\underline{m} \in \tilde{\Delta}$, and $\underline{p} \in \tilde{c}$.

Now, consider the Levi-Civita connection,

$$D : \Theta_H \to \mathrm{End}_R\left(\Theta_{\underline{\tilde{H}}}\right),$$

see [11], corresponding to the representation,

$$\rho_\Theta : \tilde{H}\left(\sigma_g\right) \to \mathrm{End}_R\left(\Theta_{\underline{\tilde{H}}}\right),$$

together with the Hamiltonian, i.e., the Laplace-Beltrami operator,

$$Q \in \mathrm{End}_{\underline{\tilde{H}}}\left(\Theta_{\underline{\tilde{H}}}\right).$$

Any *state* $\xi \in \Theta_{\underline{\tilde{H}}}$ may be interpreted as a (relative) momentum (ξ_o, ξ_x) of the pair of points $(o, x) \in \mathbf{E}^3$, defined for all pairs of points in the domain of definition for ξ. Write, as above,

$$\underline{\xi} = \underline{p} + \underline{m}, \; \underline{p} = (p_1, p_2, p_3, 0, 0, 0) \in \tilde{c}, \; \underline{m} = (0, 0, 0, m_1, m_2, m_3) \in \tilde{\Delta},$$

where we have introduced local coordinates, $(\lambda_1, \lambda_2, \lambda_3, x_1, x_2, x_3)$, such that

$$\tilde{\Delta} = (\delta_{\lambda_1}, \delta_{\lambda_2}, \delta_{\lambda_3}), \text{ and } \tilde{c} = (\delta x_1, \delta x_2, \delta x_3)$$

Put

$$D_{\lambda_i} := D_{\delta_{\lambda_i}}, \; D_{x_i} := D_{\delta_{x_i}}.$$

The norms,

$$\mu := |\xi|, \; \nu := |\underline{m}|, \; \kappa := |\underline{p}|$$

defined by ξ, are the *energy density*, the *density of mass*, and the *density of kinetic momentum*, respectively. We find that (p_1, p_2, p_3) is a classical relative momentum-vector and $v = (v_1, v_2, v_3) := \mu^{-1}(p_1, p_2, p_3)$ is a classical velocity vector.

Consider now the corresponding Schrødinger equation,

$$\frac{d}{dt}(\xi) = Q(\xi).$$

Time, here, is the notion used in quantum theory, and Q is the Laplace-Beltrami operator. We have, however, introduced another notion of time, the metric in general relativity theory, and these two notions of time should be equal. Therefore, we must have

$$\frac{d}{dt} = \mu^{-1} D_\xi,$$

as operators on $\Theta_{\underline{\tilde{H}}}$.

Let us compute the left-hand side of the Schrødinger equation. It is clear that

$$\begin{aligned}\frac{d}{dt}(\xi) &= \frac{d}{dt}(\underline{m}) + \frac{d}{dt}\left(\underline{p}\right)\\ &= \sum_1^3 \mu^{-1} m_j D_{\lambda_j}(\underline{m}) + \sum_1^3 \mu^{-1} m_j D_{\lambda_j}\left(\underline{p}\right)\\ &+ \sum_1^3 v_j D_{x_j}(\underline{m}) + \sum_1^3 v_j D_{x_j}\left(\underline{p}\right).\end{aligned}$$

Reduce to the subscheme $\underline{\tilde{M}}(l) \subset \underline{\tilde{H}}$, corresponding to a chosen line $l \subset \mathbf{A}^3$; see (2.2), above, and consult [11]. Then the three-dimensional vector $\underline{m} = \sum_{i=1}^3 m_i \frac{\partial}{\partial \lambda_i}$ reduces to $m\frac{\partial}{\partial \lambda}$, where $m = \nu$, and the term $\mu^{-1}\nu\frac{\partial}{\partial \lambda}$ may be compared to $\frac{\partial}{\partial \tau}$, where τ is the relativistic *proper time*.

The outcome of this is that, reduced to the subscheme $\underline{M}(l)$, the Schrødinger equation is, in a realistic classical Euclidean situation, the coupled equation, containing the general relativistic heat equation,

$$\frac{d\nu}{dt} = \frac{\partial \nu}{\partial \tau} + \sum_{j=1}^{3} v_j \frac{\partial \nu}{\partial x_j} = \Delta(\nu)$$

and the "relativistic" Navier-Stokes equation (NSE),

$$\frac{d p_i}{dt} = \frac{\partial p_i}{\partial \tau} + \sum_{j=1}^{3} v_j \frac{\partial p_i}{\partial x_j} = \Delta(p_i),\ i = 1,2,3.$$

where $\Delta = Q = \frac{\partial^2}{\partial \lambda^2} + \sum_{j=1}^{3} \frac{\partial^2}{\partial x_j^2}$.

Computing, we find for NSE,

$$\frac{\partial p_i}{\partial \tau} = \mu \frac{\partial v_i}{\partial \tau} + \frac{\partial \mu}{\partial \tau} v_i,\ \sum_{j=1}^{3} v_j \frac{\partial p_i}{\partial x_j} = \mu(v \nabla v_i) + (v \nabla \mu) v_i$$

where ∇ is the dimension 3 del-operator, with respect to the parameters $\underline{x} = (x_1, x_2, x_3)$. Let $\overline{\nabla}$ be the dimension 4 del-operator, with respect to the parameters (λ, x_1, x_2, x_3); put $\overline{v} = (\nu/\mu, v_1, v_2, v_3)$; and compute

$$\Delta(p_i) = \mu \Delta(v_i) + 2\overline{\nabla}\mu\overline{\nabla}v_i + \Delta(\mu)v_i,$$

from which we deduce an equation, close to the classical Navier-Stokes equation,

$$\mu\left(\frac{\partial v_i}{\partial \tau} + v\nabla v_i\right) = \mu\Delta(v_i) + 2\overline{\nabla}\mu\overline{\nabla}v_i + \Delta(\mu)v_i - \left(\overline{v}\overline{\nabla}\mu\right)v_i,\ i = 1,2,3.$$

Remark 2. Any vector field $\xi \in \Theta_{\underline{\tilde{H}}}$ may be interpreted as a description of the *relative state* of the space, everywhere, a kind of mass-stress-tensor, describing the situation of all pairs of points in our three-dimensional space, together with any corresponding pair of momenta. This is the *Furniture* of our *Cosmos*, referred to in the title of this paper.

In general, we might hope that knowing ξ, i.e., the six functions defined in $\underline{\tilde{H}}$, locally defining the vector ξ, the Schrødinger equation would determine the metric, g, i.e., the six functions $h_\rho, h_\phi, h_{\underline{\lambda}}$. This would presumably lead to time developments $\xi(T)$ and $g(T)$, determined by any given ground state, $\xi_\star$, and clocked by some parameter T. This would again have as a consequence that any cyclic behavior of the phenomenon modeled by $\xi(T)$ would lead to a gravitational wave defined by $g(T)$.

In particular, the collapsing of a star and the Big Bang event, both usually modeled as a fluid depending on pressure, temperature, energy density, viscosity etc., would in the above scenario define a generalized gravitational wave, $g(T)$.

We would therefore be tempted to consider the Schrødinger equation,

$$\frac{d}{dt}(\xi) = Q(\xi),$$

as our field equation, replacing the Einstein field equation. A solution would be a metric g determining the dynamics of the *past and the future* of our space. To make this reasonably understandable, we need a mathematical model of the beginning of it all, of the Big Bang. This is, however, the subject of the last section in this paper.

As a first example of the usefulness of the Schrødinger equation, in studying the geometry of our universe, consider the very special case of the metric of the last section, defined by

$$h_\rho = \left(\frac{\rho - h}{\rho}\right)^2, \; h_\phi = (\rho - h)^2, \; h_3 = 1,$$

where h is a positive real number.

Put $\rho = t_1, \phi = t_2, \lambda = t_3$, then we found the following formulas:

$$\begin{aligned}
\Gamma^1_{1,1} &= h/\rho(\rho - h), \; \Gamma^1_{2,2} = -\rho^2/(\rho - h) \\
\Gamma^2_{1,2} &= 1/(\rho - h), \; \Gamma^2_{2,1} = 1/(\rho - h) \\
\Gamma^3_{i,j} &= 0
\end{aligned}$$

All other components vanish. From this we find the following formulas:

$$\begin{aligned}
D_\rho &:= \nabla_{\delta_1} = \frac{\partial}{\partial \rho} + \nabla_\rho, \\
D_\phi &:= \nabla_{\delta_2} = \frac{\partial}{\partial \phi} + \nabla_\phi, \\
D_\lambda &:= \nabla_{\delta_3} = \frac{\partial}{\partial \lambda} + \nabla_\lambda
\end{aligned}$$

$$Q = \sum_{i=1}^{3} 1/h_i \nabla^2_{\delta_i}$$

$$\rho(\delta^2(t_i)) = [Q, \rho(dt_i)] = 1/h_i \,[Q, \nabla_{\delta_i}].$$

Here, the h_i is the function defined above, i.e., $g_{i,i}$ in our metric, and

$$\nabla_\rho = \begin{pmatrix} h/\rho(\rho-h) & 0 & 0 \\ 0 & 1/(\rho-h) & 0 \\ 0 & 0 & 0 \end{pmatrix}$$

$$\nabla_\phi = \begin{pmatrix} 0 & -\rho^2/(\rho-h) & 0 \\ 1/(\rho-h) & 0 & 0 \\ 0 & 0 & 0 \end{pmatrix}$$

$$\nabla_\lambda = 0$$

$$[\nabla_\rho, \nabla_\phi] = \begin{pmatrix} 0 & -1/\rho(\rho-h) & 0 \\ -\rho/(\rho-h) & 0 & 0 \\ 0 & 0 & 0 \end{pmatrix}$$

$$\frac{\partial}{\partial\rho}\nabla_\rho = \rho^2(\rho-h)^{-2}\begin{pmatrix} -h\rho^{-2}(\rho-h)^{-1} - h\rho^{-1}(\rho-h)^{-2} & 0 & 0 \\ 0 & -(\rho-h)^{-2} & 0 \\ 0 & 0 & 0 \end{pmatrix}$$

$$\frac{\partial}{\partial\rho}\nabla_\phi = (\rho-h)^{-2}\begin{pmatrix} 0 & 2h\rho & 0 \\ -1 & 0 & 0 \\ 0 & 0 & 0 \end{pmatrix}$$

The left-hand side of the Schrødinger equation,

$$\frac{\partial\xi}{\partial t} = Q(\xi),$$

for a general vector field, $\xi = (f_1, f_2, f_3)$, takes the form

$$\begin{aligned} D_\xi(\xi) = &\left(f_1\frac{\partial f_1}{\partial\rho}, f_1\frac{\partial f_2}{\partial\rho}, f_1\frac{\partial f_3}{\partial\rho}\right) + f_1\left(f_1h(\rho-h)^{-1}, f_2h(\rho-h)^{-1}, 0\right) \\ &+ \left(f_2\frac{\partial f_1}{\partial\phi}, f_2\frac{\partial f_2}{\partial\phi}, f_2\frac{\partial f_3}{\partial\phi}\right) + f_2\left(-f_2\rho^2(\rho-h)^{-1}, f_1h(\rho-h)^{-1}, 0\right) \\ &+ \left(f_3\frac{\partial f_1}{\partial\lambda}, f_3\frac{\partial f_2}{\partial\lambda}, f_3\frac{\partial f_3}{\partial\lambda}\right), \end{aligned}$$

and the right-hand side becomes, with the obvious simplified notations,

$$\begin{aligned} Q(\xi) = &-(\rho-h)^{-4}\left(\rho^4 f_{1:\rho\rho} + h^2 f_1 - \rho^2 f_1 + \rho^3 f_{2:\rho}\right. \\ &+ \rho^4 f_{1:\rho,\rho} + h^4 f_{1:\lambda,\lambda} - 2\rho^3 f_{2:\phi} \\ &+ \rho^2 f_{1:\phi,\phi} + h^2 f1 : \phi, \phi - 4h\rho^3 f_{1:\lambda,\lambda} + 6h^2\rho^2 f_{1:\lambda,\lambda} - 2h\rho f_1 - h^2\rho f_{1:\rho} \end{aligned}$$

$$
\begin{aligned}
&- 2h\rho^3 f_{1:\rho,\rho} + h^2\rho^2 f_{1:\rho,\rho} - 4h^3\rho f_{1:\lambda,\lambda} + 2h\rho^2 f_{2:\phi} - 2h\rho f_{1:\phi,\phi}\big) D_\rho \\
&- (\rho - h)^{-4}\big(3\rho^3 f_{3:\rho} + 2\rho f_{1:\phi} - 2h f_{1:\phi} + \rho^2 f 2 : \phi, \phi + \rho^4 f_{2:\lambda,\lambda} + h^4 f_{2:\lambda,\lambda} \\
&+ \rho^4 f_{2:\rho,\rho} h^2 f_{2:\phi,\phi} - 4h\rho^2 f_{2:\rho} - 2h\rho f_{2:\phi,\phi} + 6h^2\rho^2 f_{2:\lambda,\lambda} - 2h\rho f_2 + h^2\rho f_{2:\rho} \\
&+ h^2\rho^2 f_{2:\rho,\rho} - 2h\rho^3 f_{2:\rho,\rho} - 4h^3\rho f_{2:\lambda,\lambda} - 4h\rho^3 f_{2:\lambda,\lambda}\big) D_\phi \\
&+ \rho_{-1}(\rho - h)^{-1} \big(\rho^2 f_{3:\rho} + \rho^3 f_{3:\rho,\rho} + \rho f_{3:\phi,\phi} + \rho^3 f_{3:\lambda,\lambda} \\
&- 2h\rho^2 f_{3:\lambda,\lambda} + h^2\rho f_{3:\lambda,\lambda}\big) D_\lambda
\end{aligned}
$$

Put

$$\xi = (0, f(\rho), 0),$$

then $D_\xi(\xi) = (0,0,0)$, and the Schrødinger equation, *the furniture equation*, reduces to the second-order differential equation,

$$\rho^2(\rho - h)^2 \frac{d^2 f}{d\rho^2} + \rho(\rho - h)(3\rho - h)\frac{df}{d\rho} - 2h\rho f = 0,$$

with the easy solution,

$$f = (\rho - h)^{-2},$$

which means that the *fluid*, the content of the space, rotates about the Black Hole $\rho = 0$ with speed $(\rho - h)^{-1}$, so with infinite speed close to the horizon, almost standing still at great distances, and therefore with lots of *shear*.

Notice that

$$\xi = \left(0, (\rho - h)^{-2}(\rho - h)^{-2}((1/2\rho - 2h)\rho + h^2 ln(\rho)), p\right),$$

with p as constant, is also a solution.

3 Spin, Isospin, and SUSY

Consider the Lie algebra,

$$\mathfrak{g} := \mathrm{Der}_H(\mathbf{U})$$

as a principal Lie algebra bundle on the space, $\underline{H}$.

3.1 Spin Structure of H

Any element $\delta \in \mathrm{Der}_H(\mathbf{U})$ must be given by its values on the coordinates, i.e., by the values

$$\delta(x_i) = \delta_i^0 + \delta_i^1 x_1 + \delta_i^2 x_2 + \delta_i^3 x_3,\ \delta_i^j \in H.$$

Put

$$\bar{o} = (1, o_1, o_2, o_3), \bar{p} = (1, p_1, p_2, p_3),$$

and consider the 4-vectors

$$\delta_i = \left(\delta_i^0, \delta_i^1, \delta_i^2, \delta_i^3\right), i = 1, 2, 3.$$

Computing, we find the formula, in $\mathbf{U}$,

$$\delta(x_i x_j - o_i x_j - p_j x_i + o_i p_j) = (\delta_i \cdot \bar{o})x_j - (\delta_i \cdot \bar{o})p_j + x_i(\delta_j \cdot \bar{p}) - o_i(\delta_j \cdot \bar{p})$$

which leads to

$$\delta \in \mathrm{Der}_H(\mathbf{U})$$

if and only if

$$\delta_i \cdot \bar{o} = \delta_i \cdot \bar{p} = 0, i = 1, 2, 3;$$

(see the proof of the next theorem). Put, from now on,

$$x_{i,j} = x_i x_j - o_i x_j - p_j x_i + o_i p_j \in H < x_1, x_2, x_3 >=: \tilde{F}$$

We find that if $\delta \in \mathrm{Der}_H(\mathbf{U})$ then, in the free H-algebra $\tilde{F}$, we have

$$\delta\left(x_{i,j}\right) = \sum_{p=1}^{3} \delta_i^p \left(x_{p,j}\right) + \sum_{p=1}^{3} \delta_j^p \left(x_{i,p}\right).$$

Let $\xi \in A^1(H, \mathbf{U}; \mathbf{U})$ be represented by $\xi\left(x_{i,j}\right) = \xi_{i,j}^0 + \sum_{p=1}^3 \xi_{i,j}^p x_p$. Recall that ξ is zero if it is of the form

$$\xi\left(x_{i,j}\right) = -(\eta_i.\bar{o})p_j - o_i(\eta_j.\bar{p}) + (\eta_i.\bar{o})x_j + (\eta_j.\bar{o})x_i,$$

for some derivation $\eta \in \mathrm{Der}_H(\tilde{F}, \mathbf{U})$.

Let $\underline{t} = (o, p) \in \underline{H} - \underline{\Delta}$, and consider the relation of $U(\underline{t})$,

$$x_{i,j} := x_i x_j - o_i x_j - p_j x_i + o_i p_j,$$

Let us compute the Lie algebra $\mathfrak{g}(\underline{t}) := \mathrm{Der}_k(U(\underline{t}))$. Any element $\delta \in \mathrm{Der}_k(U(\underline{t}))$ must have the form

$$\delta(x_i) = \delta_i^0 + \delta_i^1 x_1 + \delta_i^2 x_2 + \delta_i^3 x_3.$$

Put, as above, $\bar{o} = (1, o_1, o_2, o_3)$, $\bar{p} = (1, p_1, p_2, p_3)$, and consider the 4-vectors $\delta_i = \left(\delta_i^0, \delta_i^1, \delta_i^2, \delta_i^3\right), i = 1, 2, 3.$

As above, we find that

$$\delta \in \mathrm{Der}_k(U(\underline{t}), U(\underline{t}))$$

if and only if

$$\delta_i \cdot \bar{o} = \delta_i \cdot \bar{p} = 0, i = 1, 2, 3.$$

Moreover we find that if $\xi \in A^1(k, U(\underline{t}), U(\underline{t}))$ is represented by $\xi\left(x_{i,j}\right) = \xi_{i,j}^0 + \sum_{p=1}^3 \xi_{i,j}^p x_p$, then it is easy to see that the action of any $\delta \in \mathrm{Der}_k(U(\underline{t}))$ on the tangent space, $A^1(k, U(\underline{t}), U(\underline{t}))$, of the versal deformation space, of $U(\underline{t})$, is given as follows:

$$\delta(\xi)\left(x_{i,j}\right) := \xi\left(\delta\left(x_{i,j}\right)\right) - \delta\left(\xi\left(x_{i,j}\right)\right) = \sum_{p=1}^{3} \delta_i^p\left(\xi_{p,j}\right) + \sum_{p=1}^{3} \delta_j^p\left(\xi_{i,p}\right) - \delta\left(\xi_{i,j}\right)$$

$$= \sum_{p=1}^{3} \delta_i^p \xi_{p,j}^0 + \sum_{p=1}^{3} \delta_j^p \xi_{i,p}^0 - \sum_{p=1}^{3} \delta_p^0 \xi_{i,j}^p$$

$$+ \sum_{r=1}^{3}\left(\sum_{p=1}^{3} \delta_i^p \xi_{p,j}^q + \sum_{p=1}^{3} \delta_j^p \xi_{i,p}^q - \sum_{p=1}^{3} \delta_p^q \xi_{i,j}^p\right) x_q.$$

If we consider the particular interesting part of the tangent space, given by the $\xi \in A^1(k, U(\underline{t}), U(\underline{t}))$, of $U(\underline{t})$ represented by

$$\xi\left(x_{i,j}\right) = \xi_{i,j}^0 + \xi_i x_j + \nu_j x_i,$$

then we find that the action above simplifies to

$$\delta(\xi)\left(x_{i,j}\right) = \sum_{p=1}^{3} \delta_i^p \xi_{p,j}^0 + \sum_{p=1}^{3} \delta_j^p \xi_{i,p}^0 - \delta_i^0 \xi_j - \delta_j^0 \nu_i + \sum_{p=1}^{3} \delta_i^p \xi_p x_j + \sum_{p=1}^{3} \delta_j^p \nu_p x_i.$$

If $o \neq p$, it follows that $\bar{o}$ and $\bar{p}$ are linearly independent, in a four-dimensional vector space; therefore, each vector $\delta_i, i = 1, 2, 3$ is confined to a two-dimensional vector space. Consequently, $\mathfrak{g}(\underline{t}) := \mathrm{Der}_k(U(\underline{t}))$ is of dimension 6. Using the isomorphism, $U(o, p) \simeq U(o - c, p - c)$, mentioned above, we may choose coordinates such that $o = (0, 0, 0), p = (1, 0, 0)$. In fact we may, first, put $c = o$, and reduce to the situation where $o = 0$ and where p is a nonzero 3-vector. Any $\delta \in \mathrm{Der}_k(U(o, p))$ will then be represented by a matrix of the form

$$M := \begin{pmatrix} \delta_1^1 & \delta_1^2 & \delta_1^3 \\ \delta_2^1 & \delta_2^2 & \delta_2^3 \\ \delta_3^1 & \delta_3^2 & \delta_3^3 \end{pmatrix},$$

where $M(p) = 0$, and we know that the Lie structure is the ordinary matrix Lie products. Now, clearly we may find a nonsingular matrix N such that $N(p) = (1, 0, 0)$ and the Lie algebra of matrices M will be isomorphic to the Lie algebra of the matrices, NMN^{-1}, which are those corresponding to $p = e_1 := (1, 0, 0)$, and we are working with $U(0, e_1)$. Notice that in this picture, the fundamental vector $\overline{op} = (1, 0, 0)$. With this it is easy to see that $\delta \in \mathfrak{g}(\underline{t})$ imply

$$\delta_i^0 = \delta_i^1 = 0, i = 1, 2, 3.$$

It follows that the Lie algebra $\mathfrak{g}(\underline{t})$ is isomorphic to the Lie algebra of matrices of the form

$$\begin{pmatrix} 0 & \delta_1^2 & \delta_1^3 \\ 0 & \delta_2^2 & \delta_2^3 \\ 0 & \delta_3^2 & \delta_3^3 \end{pmatrix}$$

The radical $\mathfrak{r}$ is generated by three elements, $\{u, r_1, r_2\}$, with

$$u = \begin{pmatrix} 0 & 0 & 0 \\ 0 & 1 & 0 \\ 0 & 0 & 1 \end{pmatrix}, r_1 = \begin{pmatrix} 0 & 1 & 0 \\ 0 & 0 & 0 \\ 0 & 0 & 0 \end{pmatrix}, r_2 = \begin{pmatrix} 0 & 0 & 1 \\ 0 & 0 & 0 \\ 0 & 0 & 0 \end{pmatrix}.$$

where $u \notin [\mathfrak{g}, \mathfrak{g}]$, $[u, r_i] = -r_i, [r_1, r_2] = 0$ and the quotient

$$\mathfrak{g}(\underline{t})/\mathfrak{r} = \mathfrak{sl}(2).$$

with the usual generators u_0, u_1, u_2,

$$u_0 = \begin{pmatrix} 0 & 0 & 0 \\ 0 & 1 & 0 \\ 0 & 0 & -1 \end{pmatrix}, u_1 = \begin{pmatrix} 0 & 0 & 0 \\ 0 & 0 & 1 \\ 0 & 0 & 0 \end{pmatrix}, u_2 = \begin{pmatrix} 0 & 0 & 0 \\ 0 & 0 & 0 \\ 0 & -1 & 0 \end{pmatrix}.$$

In particular, we find that $\mathfrak{sl}(2) \subset \mathfrak{g}(\underline{t})$. Now, put

$$\sigma^0 := u,\ \sigma^1 := u_1 - u_2,\ \sigma^2 := -iu_1 - iu_2,\ \sigma^3 := u_1 + u_2.$$

These are the Dirac matrices, a basis for $su(2)$, which we shall see operate on the each of the complexified sub-bundles,

$$B_o \subset \Theta_{\underline{\tilde{H}}},\ B_p \subset \Theta_{\underline{\tilde{H}}}$$

corresponding to *left and right* Weyl spinors.

Given a point $\underline{t} \in \underline{H}$, the tangent space at this point is, of course, nicely represented by the space of all pairs of 3-vectors, (ξ, μ) and, as we have seen, it is easy to compute the action of $\mathfrak{g}(\underline{t})$ on this six-dimensional vector space. Just as in the case of the action of $\mathrm{Der}_k(U)$ on $\underline{H}$, any $\delta \in \mathfrak{g}(\underline{t})$ with $\delta_i^0 = 0,\ i = 1, 2, 3$, acts as $\delta(\xi, \mu) = (\delta(\xi), \delta(\mu))$, where in each coordinate, the action is that of the matrix algebra above.

The Lie algebra $\mathfrak{sl}_2(\underline{t})$ therefore acts as follows. There are natural three-dimensional sub-bundles Θ_o, Θ_p of the tangent bundle of $\underline{H}' := \underline{H} - \underline{\Delta}$, such that $\Theta_{\underline{H}'} = \Theta_o \oplus \Theta_p$. We may find a natural basis for both components, for Θ_o and for Θ_p, $\{\mathfrak{l}, \nu_1, \nu_2\}$, where $\mathfrak{l}$ is the special tangent vector given by $\overline{po}$, i.e., the tangent direction in our affine (or Euclidean) 3-space, *in which we are looking*. It is obvious from the above matrix bases of $\mathfrak{g}(\underline{t})$ that $\mathfrak{g}(\underline{t})$ kills $\mathfrak{l}$. Therefore, assuming some metric given, in this basis, $\mathfrak{s}l(2)$ acts on the planes normal to $\mathfrak{l} = \overline{po}$. As a consequence, if we pick any line $\mathbf{l} \subset \mathbf{A}^3$, then the tangent space of $\underline{H}(\mathbf{l})$, the subspace of $\underline{H}$ corresponding to the points $\{\underline{t} = (o, p), o, p \in \mathbf{l}\}$, is killed by $\mathfrak{g}(\underline{t})$.

We have therefore seen that for any point $\underline{t} \in \underline{\tilde{H}}$, the $\mathfrak{s}l(2)$ component of the Lie algebra of infinitesimal automorphisms of the universal algebra $U(\underline{t})$ acts on $\Theta_{\underline{\tilde{H}}}$ in a particular nice way. The generators u_0, u_1, u_2 act on sections of the sub-bundle $B_o \oplus B_p$ of the tangent bundle $\Theta_{\underline{\tilde{H}}}$, just as we described, geometrically, in [9] and [11]. In particular, $sl(2)$ acts on $\tilde{\Delta}$, fixing the special tangent $(\mathfrak{l}, \mathfrak{l})$; see [11], (4.14), p. 119.

Having found natural representations for the Dirac matrices $\sigma^i,\ i = 0, 1, 2, 3$, we may also treat the γ-matrices, in this language. They are now simply acting on the sum, $\Theta_o \oplus \Theta_p$, and are reasonably given as

$$\gamma^0 := \begin{pmatrix} \sigma^0 & 0 \\ 0 & -\sigma^0 \end{pmatrix},\ \gamma^1 := \begin{pmatrix} 0 & \sigma^1 \\ -\sigma^1 & 0 \end{pmatrix},\ \gamma^2 := \begin{pmatrix} 0 & \sigma^2 \\ -\sigma^2 & 0 \end{pmatrix},\ \gamma^3 := \begin{pmatrix} 0 & \sigma^3 \\ -\sigma^3 & 0 \end{pmatrix}.$$

Now, as in [9], consider the Kodaira-Spencer map of the family,

$$\nu : \mathbf{U} \longrightarrow \underline{\tilde{H}},$$

It is the linear map,

$$\eta : \Theta_H = \mathrm{Der}_k\left(\tilde{H}, \tilde{H}\right) \to A^1(H, \mathbf{U}, \mathbf{U}),$$

defined by,

$$\eta\left(\frac{\partial}{\partial o_i}\right) = \left\{r_{i,j} = \left(x_i x_j - o_i x_j - p_j x_i + o_i p_j\right) \mapsto \frac{\partial}{\partial o_i}(r_{i,j}) = -x_j + p_j\right\}$$

$$\eta\left(\frac{\partial}{\partial p_j}\right) = \left\{r_{i,j} = \left(x_i x_j - o_i x_j - p_j x_i + o_i p_j\right) \mapsto \frac{\partial}{\partial p_i}(r_{i,j}) = -x_i + o_i\right\}.$$

Recall that

$$A^1(H,\mathbf{U},\mathbf{U}) = \mathrm{Hom}_{H,H}(\tilde{I},\mathbf{U})/\mathrm{Der}$$

where we have picked a surjection ρ of a free H-algebra $\tilde{F}$ onto $\mathbf{U}$ and $\tilde{I} = \ker(\rho)$. Der is then the sub module of $\mathrm{Hom}_{H,H}\left(\tilde{I},\mathbf{U}\right)$ generated by the elements of $\mathrm{Der}_H\left(\tilde{F},\mathbf{U}\right)$, restricted to $\tilde{I}$. In our case $\tilde{I}$ is generated by $\left\{r_{i,j} := x_i x_j - o_i x_j - p_j x_i + o_i p_j, i, j = 1,2,3\right\}$. Therefore, since any derivation $\delta \in \mathrm{Der}_H\left(\tilde{F},\mathbf{U}\right)$ has the form

$$\eta(x_i) = \eta_i^0 + \eta_i^1 x_1 + \eta_i^2 x_2 + \eta_i^3 x_3,\ \eta_i^l \in H, i = 1,2,3, l = 0,1,2,3,$$

we find that

$$\begin{aligned}\eta\left(r_{i,j}\right) &= \eta\left(x_i\right)x_j + x_i\eta\left(x_j\right) - o_i\eta\left(x_j\right) - \eta\left(x_i\right)p_j \\ &= \left(\eta_j\cdot\bar{o}\right)\left(x_i - o_i\right) + \left(\eta_i\cdot\bar{o}\right)\left(x_j - p_j\right).\end{aligned}$$

Since we have seen that all $\mathfrak{g}(\underline{t})$, $\underline{t} \in \underline{\tilde{H}}$ are isomorphic, it is reasonable that the Kodaira-Spencer map η vanishes on $\underline{\tilde{H}}$, implying:

Lemma 1. *The kernel of the Kodaira-Spencer map, also called the Gauss-Manin kernel of the family* ν, *is*

$$\tilde{\mathbf{G}} := \ker(\eta) = \left\{\xi = \sum_{i=1}^{3}(\eta_i\cdot\bar{o})\frac{\partial}{\partial o_i} + \sum_{j=1}^{3}(\eta_i\cdot\bar{p})\frac{\partial}{\partial p_j} \mid \eta \in \mathrm{Der}_H(\tilde{F},\mathbf{U})\right\} \simeq \Theta_{\underline{\tilde{H}}}.$$

3.2 SUSY-Like Structure of H

Notice that, so far, we have not been forced to choose an origin in our affine space $\mathbf{A}^3$. All results regarding the vectors $o, p, \overline{o}, \overline{p}$, have been independent upon this choice. However, for the next Theorem we shall have to fix an origin, $* \in \mathbf{A}^3$. This may seem strange, but as it will be explained in the next section, this is a natural consequence of the introduction of a Big Bang event, the singularity U, and its versal base space.

Theorem 1. *There is a well-defined $\tilde{H}$-linear map,*

$$\kappa : \mathfrak{g} \to \tilde{\Delta} \subset \Theta_{\underline{\tilde{H}}},$$

defined by

$$\kappa(\delta) = \sum_{i=1}^{3} \left((\delta_i \cdot o) \frac{\partial}{\partial o_i} + (\delta_i \cdot p) \frac{\partial}{\partial p_i} \right) = - \sum_{i=1}^{3} \delta_i^0 \left(\frac{\partial}{\partial o_i} + \frac{\partial}{\partial p_i} \right).$$

Assume the vectors $o := (o_1, o_2, o_3)$, $p := (p_1, p_2, p_3)$ are linearly independent, and let $\alpha := (\alpha_1, \alpha_2, \alpha_3) = o \times p$. Then,

$$\mathfrak{k} := \ker(\kappa).$$

is a rank 3 $\tilde{H}$-sub Lie algebra of $\mathfrak{g}$ generated by elements $\{u, v, w\}$, with Lie structure,

$$[u, v] = -\alpha_2 u + \alpha_1 v, \ [u, w] = -\alpha_3 u + \alpha_1 w, \ [v, w] = -\alpha_3 v + \alpha_2 w,$$

which, at each point $\underline{t} \in \underline{\tilde{H}}$, is isomorphic to $rad(\mathfrak{g}(\underline{t})) \subset \mathfrak{g}(\underline{t})$.

If we for any $\delta \in \mathfrak{g}$ define

$$\tilde{\delta} := \delta + \kappa(\delta),$$

then, applying the adjoint (regular) representation of $\mathfrak{g}$, we have the following:

$$\begin{aligned} &\tilde{\delta} \in \mathrm{End}_k(\mathfrak{g}) \\ &\kappa\left(\tilde{\delta}\right) = \kappa(\delta) \\ &\kappa\left(\left[\tilde{\delta}, \tilde{\eta}\right] - \widetilde{[\delta, \eta]}\right) = [\kappa(\delta), \kappa(\eta)]. \end{aligned}$$

Finally κ defines an isomorphism of $\tilde{H}$-modules,

$$\tilde{\kappa} : \mathfrak{g}/\mathfrak{k} \to \tilde{\Delta}.$$

where $\mathfrak{g}/\mathfrak{k} \simeq \mathfrak{sl}(2)$

Proof. Put

$$\begin{aligned} \bar{\delta}_i &: = \left(\delta_i^0, \delta_i^1, \delta_i^2, \delta 3_i\right) \\ \bar{o} &: = (1, o_1, o_2, o_3) \\ \bar{p} &: = (1, p_1, p_2, p_3). \end{aligned}$$

Let $\delta \in$ Der, and compute, as above,

$$\begin{aligned}\delta\left(r_{i,j}\right) &= \delta(x_i)x_j + x_i\delta\left(x_j\right) - o_i\delta\left(x_j\right) - \delta(x_i)p_j \\ &= \left(\bar{o}\cdot\bar{\delta}_i\right)x_j + \left(\bar{p}\cdot\bar{\delta}_j\right)x_i - \left(\bar{o}\cdot\bar{\delta}_i\right)p_j - \left(\bar{p}\cdot\bar{\delta}_j\right)o_i.\end{aligned}$$

From this we learn two things. First, we see that $\delta \in \mathrm{Der}_H(\mathbf{U},\mathbf{U})$, if and only if $\bar{o}\cdot\bar{\delta}_i = 0,\ \bar{p}\cdot\bar{\delta}_i = 0$, for all $i = 1,2,3$, which we already knew; second, we see that the definition of κ is well defined. The computation of $\left[\tilde{\delta},\tilde{\eta}\right]$, and its image by κ is left as an exercise.

But then we see that $\mathfrak{k} := \ker(\kappa) = \left\{\delta \in \mathrm{Der}_H(\mathbf{U},\mathbf{U}) | \delta_i^0 = 0,\ i = 1,2,3\right\}$ is a sub Lie algebra, though not necessarily a Lie ideal. By definition of $\mathfrak{k}$, if $\delta \in \mathfrak{k}$, we must have $\delta_i^0 = 0$, for all $i = 1,2,3$, and so also

$$o\cdot\delta_i = 0,\ p\cdot\delta_i = 0, \forall i = 1,2,3.$$

It follows that

$$\left(\delta_i^1,\delta_i^2,\delta_i^3\right) = c_i(\alpha_1,\alpha_2,\alpha_3),\ c_i \in H.$$

If δ,δ' corresponds to the two vectors $c := (c_1,c_2,c_3)$ and $c' := (c_1',c_2',c_3')$, then $[\delta,\delta']$ is seen to correspond to the vector, $c \times c' \times \alpha$, from where the structural constants may be read off. Moreover, since the determinant of the matrix corresponding to the Lie structure of $\mathfrak{k}$

$$\begin{pmatrix} -\alpha_2 & \alpha_1 & 0 \\ -\alpha_3 & 0 & \alpha_1 \\ 0 & -\alpha_3 & \alpha_2 \end{pmatrix}$$

is zero, it is clear that $[\mathfrak{k},\mathfrak{k}]$ is of dimension 2 and Abelian. Since $\mathfrak{k}$ is solvable, and obviously maximal, the contention follows.

Notice that the map κ is $\tilde{H}$-linear but that it is not a $\tilde{H}$-Lie algebra morphism. Nevertheless the kernel is an $\tilde{H}$-sub Lie algebra, identified, at every point $\underline{t} \in \underline{\tilde{H}}$, with the radical of $\mathfrak{g}(\underline{t})$. Therefore we may, for every $\delta \in \mathfrak{g}$, talk about its class, modulo $\mathfrak{k}$, i.e., in $\mathfrak{sl}(2)$. The last contention therefore follows from the surjectivity of κ. □

This result is going to have several important consequences, in particular for the derivation of a general analogy of the Dirac equation and for the construction, in our case, of a kind of supersymmetry.

But first, let us consider the singular scheme of the morphism κ above, i.e., the subspace $M(B)$ of $\underline{\tilde{H}}$, where o and p are not linear independent. It turns out to be a four-dimensional subspace of $\underline{\tilde{H}}$, fibered over the exceptional fiber $E\ (\underline{0})$, associated to the Big Bang, $\underline{0} \in \underline{\Delta}$. We have the following diagram:

$$\begin{array}{ccccc} M(B) \xrightarrow{\mu} E(\underline{0}) \hookrightarrow & & \underline{\tilde{\Delta}} \hookrightarrow & & \underline{\tilde{H}} \\ \downarrow & & \downarrow & & \downarrow \\ \{\underline{0}\} \hookrightarrow & & \underline{\Delta} \hookrightarrow & & \underline{H} \end{array}$$

where μ is a map that for all directed lines $l \subset \mathbf{A}^3$, through the Big Bang, i.e., such that $\underline{0} \in l$ maps the subspace $\underline{\tilde{H}}(l) \subset \underline{\tilde{H}}$ into the point of $E(\underline{0})$ corresponding to l.

M(B) looks like the product $\underline{c}(\underline{0}) \times \mathbf{R}$, where $\underline{c}(\underline{0})$ is the light-space of the Big Bang and where the $\mathbf{R}$-coordinate is the unique 0-velocity coordinate of $\underline{\tilde{H}}(l)$, which we have called λ_0. If we make a picture of this, with λ pointing upward, we have, as we shall see better at the end of the next section, the usual cosmological four-dimensional space-time picture with, the not so unimportant difference that, the time coordinate pointing upward in the classical picture, now replaced by the *proper time*.

Recall now, for later use, that if we start with $U = U(0,0)$, the Big Bang, we observe that the *Dirac derivation* $\delta = \mathrm{id}$ acts on $\underline{\tilde{H}}$, like the vector field,

$$\delta := \sum_{i=1,2,3} o_i \frac{\partial}{\partial o_i} + \sum_{j=1,2,3} p_j \frac{\partial}{\partial p_j}.$$

Evaluated at a point $\underline{t} = (o, p)$, we see that

$$\delta(o,p) = (-1/2(\overline{op}), 1/2(\overline{op})) + (1/2(o+p), 1/2(o+p).$$

The last tangent is in $\tilde{\Delta}$, i.e., it is a 0-velocity, and the only one contained in $M(B)$. One might say that *standing still* in $M(B)$ means to be carried along by the *cosmic stream* defined by the Dirac derivation of U. The first part of δ is a light direction and represents the extension of the visible universe.

Now, having shown that the *gauge Lie algebra* $u(1) \times su(2) \times su(3)$, of the standard model (SM), operates naturally on the noncommutative space $\mathrm{Ph}(\mathbf{H})$, see [11], and that there is defined on $\underline{\tilde{H}}$ a canonical principal bundle,

$$\mathfrak{g} := \mathrm{Der}_H(\mathbf{U}),$$

we now have a good measure of the ingredients of the SM. In fact, we see that the choice of a metric g defines a complex structure on $\Theta_{\mathbf{H}}$. Moreover we have seen that $su(2)$, and also complexified $sl(2)$, act naturally on complexified B_o and B_p, the *Weyl spinors*, and that $su(3)$ acts on complexified $\tilde{\Delta}$. We have also available most of the ingredients of a *canonical Yang-Mills Theory*, defined on $\mathbf{H}$; see [11].

It is therefore tempting to propose that the SM, itself, is concerned with the geometry of the noncommutative quotient scheme, $\underline{\tilde{H}}(\sigma_g)/\mathbf{G}$, with $\mathbf{G} := \mathfrak{g} \oplus \mathfrak{su}(\tilde{\Delta} \otimes \mathbf{C})$; see again [11]. Consider the following definition:

Definition 1. This noncommutative algebraic quotient space will be denoted

$$\mathbf{GQR} := \mathrm{Simp}\left(\tilde{H}(\sigma_g)\right), \mathbf{G}).$$

Taking this as a model for a combined GR and SM, let us make sure that the main ingredients of SM are available here.

First of all, we have the necessary *Gauge Fields*, since we have the principal bundles

$$\mathfrak{u}(1) \simeq B_o \simeq B_p \tag{1}$$

$$\mathfrak{g} \simeq \mathrm{Der}_{\tilde{H}}(\mathbf{U}) \tag{2}$$

$$\mathfrak{su}(3) := \mathfrak{su}_{\tilde{H}}(\tilde{\Delta} \otimes \mathbf{C}). \tag{3}$$

The canonical action of these principal bundles on the complexified tangent bundle of **H** gives us a lot of possible *Matter Fields*. Moreover, we have the $\tilde{H}$-linear isomorphisms,

$$\kappa : \mathfrak{sl}(2) \simeq \mathfrak{g}/\mathfrak{k} \simeq \tilde{\Delta}, \tag{4}$$

which defines the two obvious $\tilde{H}$-endomorphisms,

$$Q_i \in \mathrm{End}_{\tilde{H}}(\mathfrak{sl}(2) \oplus \tilde{\Delta}),\ i = 1, 2, \tag{5}$$

corresponding to $(\kappa, 0)$ and $\left(0, \kappa^{-1}\right)$, respectively, such that

$$\{Q_1, Q_2\} = \mathrm{id},\ \left[Q_i, P_\mu\right] = 0, \tag{6}$$

where P_μ is *the infinitesimal translation operator*, in the physicists notation.

We have got a graded Lie algebra,

$$\mathfrak{G} \subset \mathrm{End}_{\tilde{H}}(\mathfrak{sl}(2) \oplus \tilde{\Delta})$$

generated by the complexified adjoint operation of $\mathfrak{sl}(2)$ and the operation of $\mathfrak{su}(3)$ on the complexified $\tilde{\Delta}$, together with $Q_i,\ i = 1, 2$. An element is *even*, or *odd*, according to whether it contains an even or odd number of factors of the type Q_i. Even operators take "Bosonic states," $\mathfrak{s}l(2)$, into bosonic states and also "Fermionic states," $\tilde{\Delta}$, into fermionic states. And, obviously, odd operators take bosons into fermions and vice versa. With this interpretation, our model has aquired an $N = 1$, a SUSY-like structure, defined in $\underline{\tilde{H}}$ outside of $M(B)$. On this singular subset, the symmetry is "broken." The tangent bundle is no longer isomorphic to the Lie bundle $\mathfrak{g}$, but only to the sub-bundle generated by the $\left\{\sigma^i, i = 0, 1, 2, 3\right\}$. This is, however, the point of departure for the classical treatment of SUSY; see, e.g., [13].

But, of course, in quantum field theory, bosons and fermions are observables of type a^+ or a, having the correct "statistics," i.e., being eigen-operators of $ad(Q)$, where Q is our Hamiltonian, the Laplace-Beltrami operator on $\Theta_{\underline{\tilde{H}}}$, the least eigenvalue of which is the Planck's constant. See [11], (4.6), p.70.

From this point of view, bosons are observables corresponding to even elements in $\mathfrak{G}$, and Fermions are observables corresponding to odd elements.

Notice also that to any element $\delta_0 \in \mathfrak{g}(o,p)$ there exists a unique section $\delta \in \mathfrak{g}$, and so an element

$$\tilde{\delta} := \delta + \kappa(\delta) \in \mathrm{End}_k(\tilde{\mathfrak{g}}),$$

where δ acts on $\mathfrak{g}$ via the adjoint (regular) representation, i.e., as $ad(\delta)$. Picking a basis $\delta_{0,i}$ (say, of the $sl(2)$ part) of $\mathfrak{g}_0$, we might ask about the solutions of the Dirac type,

$$\sum_i \kappa(\delta_i)\delta_i(\psi) = m\psi, \psi \in \Theta_{\mathbf{H}}.$$

We have in the book [11] discussed the notions of chirality; the PST invariance, stemming from the Hilbert scheme structure of $\mathbf{H} = \underline{\tilde{H}}/Z_2$; and spinors, with the action on the tangent bundle, of two copies of $\mathfrak{sl}(2)$, together with $\mathfrak{su}(3)$. We saw how the charges of the up and down quarks were defined by the split form of the Cartan sub-algebras $\mathfrak{h}_1 \times \mathfrak{h}_2$ of the gauge groups, $\mathfrak{su}(2) \times \mathfrak{u}(1) \subset \mathfrak{su}(3)$, canonically defined at any point in $\underline{\tilde{H}}$.

Here we have made all this a unique consequence of the Big Bang event, mathematically played by the versal family of the associative noncommutative four-dimensional k-algebras, deforming the innocent-looking algebra $U := k[x_1, x_2, x_3]/(x_1, x_2, x_3)^2$. As we have seen, U (and thus also $\mathbf{M}$) contains a lot of information, in its nine-dimensional Lie algebra of derivations, although it is just modeling a single point, together with a three-dimensional tangent space.

We are tempted to express the content of this subsection, by saying that the SM with its spin structure and with a canonical SUSY structure turns out to be an immediate consequence of a Big Bang scenario.

Nothing less.

4 The Universe as a Versal Base Space

So, where was the Big Bang, in relation to our time-space, and what on earth is the meaning of the terms cosmological time, expansion of the universe, and read-shift? How can one fill into this geometric picture the more down-to-earth notions like matter, stress, pressure, charge, and forces like gravitation, electromagnetism, and weak and strong forces, acting on elementary particles, quarks, and their multiple combinations?

We should not have to goose-feed the Big Bang-created geometric picture, with this additional structure. It should all be part of the creation! Otherwise it must be difficult to believe in the existence of this prime event.

Going back to the constructed family, the universal family of the Hilbert scheme of subschemes of length 2 in $\mathbf{A}^3$,

$$\pi : \mathbf{E} \longrightarrow \mathbf{H},$$

we have just proved that this family may be complemented with another family, no longer a universal one, but just a versal family,

$$\nu : \mathbf{U} \longrightarrow \mathbf{H},$$

of four-dimensional associative algebras. The three-dimensional space $\underline{\Delta}$ is not a subspace of $\mathbf{H}$; in fact, any point of this ghost space corresponds to the same four-dimensional algebra, namely, to U, the Big Bang (BB) itself. A metric defined on $\underline{\Delta}$ therefore measures time at BB, before the creation of the universe, when "God did nothing"; see St. Augustin [1]!

So let us fix a point $* \in \underline{\Delta}$, the origin of the coordinate system (x_1, x_2, x_3), used to define U, thereby fixing the whereabouts of BB, clearly outside of our universe, even though time is already there, as the metric in $\underline{\Delta}$, measuring 0-velocities of U.

Now, as we have seen, $\mathfrak{g}(*) = \mathrm{Der}_k(U) = \mathfrak{gl}_3(k)$ acts on the tangent space of the versal base space and in fact on the subspace identified with $\underline{H}$. And we know that there is a very special derivation, the Dirac derivation, in this situation,

$$\delta \in \mathfrak{g}(*), \delta = \begin{pmatrix} 1 & 0 & 0 \\ 0 & 1 & 0 \\ 0 & 0 & 1 \end{pmatrix},$$

the unit element. We have seen above that δ acts on $\underline{H}$ as

$$\delta := \sum_{i=1,2,3} o_i \frac{\partial}{\partial o_i} + \sum_{j=1,2,3} p_j \frac{\partial}{\partial p_j}.$$

The value of δ at a point (o, p) is the tangent,

$$\overline{(o, p)} = (1/2(\overline{op}), -1/2(\overline{op})) + (1/2((o + p), 1/2((o + p)).$$

The last tangent is in $\tilde{\Delta}$; i.e., it is a 0-velocity.

Now, as a start, assume we concentrate on the first question of this subsection, and assume the metric is the trivial one so that mass, stress, charge, etc., can be neglected.

Then we find that the velocity associated to the direction of the tangent vector $(\rho o, \rho p)$ at $\underline{t}$ is given as $v = \sin(\theta)$, where

$$tg(\theta) = |1/2(\rho \cdot \overline{op}), -1/2(\rho \cdot \overline{op}))|/|1/2(\rho \cdot (o+p), 1/2(\rho \cdot (o+p))|.$$

From this we deduce two versions of the Hubble formula:

$$v = |\overline{op}|/\sqrt{|\overline{op}|^2 + |(o+p)|^2} = r/t,$$

and,

$$v/\sqrt{1-v^2} = |\overline{op}|/|(o+p)| = r/T,$$

where T is cosmological time and t is real time since the BB. The term r/T in the last formula is, in an obvious sense, the speed of the expansion of the universe, with respect to cosmological time. It is seen to become infinite when the real speed of the expansion v comes close to maximum, 1. This might lead us to think about the Inflation scenario more or less accepted in cosmology.

We have, in this section up to now, completely neglected the geometry of Cosmos. To be able to say something non-nonsensical about the metric, or the corresponding Laplace-Beltrami operator Q, defining the gravitation of Cosmos, one should have to guess about a content of the universe, about the *furniture*, call it ξ, and deduce the metric from

$$\frac{d\xi}{dt} = Q(\xi).$$

This seems to be what cosmologists are trying out, and I shall, hopefully, be able to return to the question at a later time.

Recall that we have nice coordinates of $\tilde{\underline{H}}$. For any element in $\underline{\lambda} \in \underline{\Delta} \simeq \mathbf{A}^3$, pick a point $\omega \in E(\underline{\lambda})$, in the spherical coordinates of the blowup of $\underline{H}$ in $\underline{\Delta}$, at the point $\underline{\lambda}$. This ω corresponds to an oriented real line through $\underline{\lambda}$ (perpendicular in some metric) to $\tilde{\Delta}$, and we may measure a (positive) length ρ along this oriented line. The triple $(\underline{\lambda}, \omega, \rho)$ is a good coordinate system for $\tilde{\underline{H}}$.

At this point we shall, as an example, try out the metric

$$\left(\frac{\rho - h(\lambda)}{\rho}\right)^2 d\rho^2 + (\rho - h(\lambda))^2 d\phi^2 + \kappa(\lambda) d\lambda^2,$$

introduced in [11], for the simplified space, in which ω is reduced to the angle ϕ and the coordinates $\underline{\lambda}$ reduced to one parameter $\lambda = |\underline{\lambda}|$. This corresponds to considering the sub-universe of $M(B)$, parametrized by (λ, ϕ, ρ). We may compute the force laws, and they look like

$$\frac{d^2\rho}{dt^2} = -\left(\frac{h(\lambda)}{\rho(\rho-h(\lambda))}\right)\left(\frac{d\rho}{dt}\right)^2 + \left(\frac{2}{(\rho-h(\lambda))}\right)\left(\frac{dh}{d\lambda}\right)\left(\frac{d\rho}{dt}\right)\left(\frac{d\lambda}{dt}\right)$$
$$+\left(\frac{\rho^2}{(\rho-h(\lambda))}\right)\left(\frac{d\phi}{dt}\right)^2,$$
$$\frac{d^2\phi}{dt^2} = -2/(\rho-h(\lambda))\frac{d\rho}{dt}\frac{d\phi}{dt} + 2/(\rho-h(\lambda))\left(\frac{dh}{d\lambda}\right)\left(\frac{d\phi}{dt}\right)\left(\frac{d\lambda}{dt}\right)$$
$$\frac{d^2\lambda}{dt^2} = -\left(\frac{(\rho-h(\lambda))}{\rho}\right)\left(\frac{1}{\kappa(\lambda)}\right)\left(\frac{dh}{d\lambda}\right)\left(\frac{d\rho}{dt}\right)^2$$
$$-(\rho-h(\lambda))\left(\frac{1}{\kappa(\lambda)}\right)\left(\frac{dh}{d\lambda}\right)\left(\frac{d\phi}{dt}\right)^2 + 1/2\left(\frac{dln(\kappa)}{d\lambda}\right)\left(\frac{d\lambda}{dt}\right)^2,$$

where t, as above, is time.

In the general case, we put $h(\underline{\lambda}) = h_0/\lambda$, with h_0 positive, then the area of the exceptional fiber, $E(\underline{\lambda})$, is $4\pi h(\lambda)^2$. The integral of this area over any sphere $S(\lambda) \subset \underline{\Delta}$ with center at BB comes out as $16\pi^2 h_0^2$. So considering λ as the *cosmological time*, and the area of the Black hole $E(\underline{\lambda})$ as the mass density, this corresponds to the conservation of mass in our universe, with respect to cosmological time.

In the simplified "universe" above, the corresponding constant mass is, of course, $2\pi h_0$. We observe that $(\frac{dh}{d\lambda})$ is always negative. This means that for $\rho \leq h(\lambda)$, the acceleration of ρ is positive, and unlimited close to the horizon, i.e., for ρ close to $h(\lambda)$. In the same region, assuming that $\kappa(\lambda)$ is constant, the acceleration of λ is negative, vanishing on the horizon. Moreover $h(0)$ is infinite, and we have found a startling analogy to the present-day assumption of Inflation. In particular we see that the *gravitation* is expanding inside the horizon and contracting outside.

We shall leave the situation here. The rest of the story depends on the usefulness of the *furniture equation*, to which I shall return.

References

1. Augustin, St.: Les Confessions de Saint Augustin. par Paul Janet. Charpentier, Libraire-Éditeur, Paris (1861)
2. Dirac, P.A.M.: The relation between mathematics and physics. Proc. R. Soc. (Edinburgh) **59**(Part II), 122–129 (1938–1939)
3. Ferguson, K.: Pythagoras. His Lives and the Legacy of a Rational Universe. Icon Books, London (2010)
4. Laudal, O.A.: Formal Moduli of Algebraic Structures. Lecture Notes in Mathematics, vol. 754, Springer, Berlin/New York (1979)
5. Laudal, O.A.: Matric Massey products and formal moduli I. In: Roos, J.E. (ed.) Algebra, Algebraic Topology and Their Interactions. Lecture Notes in Mathematics, vol. 1183, pp. 218–240. Springer, Berlin/New York (1986)
6. Laudal, O.A.: Noncommutative Algebraic Geometry. Max-Planck-Institut für Mathematik, Bonn (2000) (115)

7. Laudal, O.A.: Noncommutative deformations of modules. In: Inassaridze, H. (ed.) Special Issue in Honor of Jan-Erik Roos, Homology, Homotopy, and Applications. International Press, (2002). See also: Homol. Homotopy Appl. **4**, 357–396 (2002)
8. Laudal, O.A.: Noncommutative algebraic geometry. In: Proceedings of the International Conference in Honor of Prof. José Luis Vicente Cordoba, Sevilla, 2001. Revista Matematica Iberoamericana, vol. 19, pp. 1–72 (2003)
9. Laudal, O.A.: Time-space and space-times. In: Fuchs, J. et al. (eds.) Conference on Non-commutative Geometry and Representatioon Theory in Mathematical Physics. Karlstad, 5–10 July 2004. Contemporary Mathematics, vol. 391. American Mathematical Society (2005). ISSN:0271-4132
10. Laudal O.A.: Phase spaces and deformation theory (2007). Preprint, Institut Mittag-Leffler, 2006–07. See also the part of the paper published in: Acta Appl. Math. 25 January 2008
11. Laudal, O.A.: Geometry of Time Spaces. World Scientific, Singapore (2011)
12. Laudal, O.A., Pfister, G.: Local Moduli and Singularities. Lecture Notes in Mathematics, vol. 1310. Springer, Berlin/New York (1988)
13. Quevedo, F.: Cambridge lectures on supersymmetry and extra dimensions (2010). arXiv:1011.1491v1 [hep-th] 5 Nov 2010
14. Sachs, R.K., Wu, H.: General Relativity for Mathematicians. Springer, New York (1977)
15. Weinberg, S.: The Quantum Theory of Fields, vols. I, II, III. Cambridge University Press, Cambridge/New York (1995)

About Phase Transition and Zero Temperature

Renaud Leplaideur

1 Introduction

The goal of this communication is to present some open questions in smooth ergodic theory. These questions are inspired from statistical mechanics. The two phenomena are a priori independent, but it turns out that at least due to the way they arise in dynamical systems theory, they get some connections.

The first question deals with the notion of *phase transition*. What is a good way to define and model this phenomenon, which has so many different expressions: from the simple boiling water to the Big Bang, following new developments in astrophysics?

The second phenomenon is the zero temperature one. It is known that the physical properties of materials change with respect to temperature. It also appears that materials have a strong tendency to be highly ordered at low temperature. They reach the so-called ground states which usually are crystal or quasicrystal configurations. Supraconductivity is may be the most famous example of such phenomenon.

These questions have been studied a lot with the physicist viewpoint. Some have also been studied from the probability viewpoint (see, e.g., [13, 14]). Our goal

I would like to thank the organizers of the 6th World Conference on 21st Century Mathematics 2013 for having invited me. I also would like to thank S. Arshad for having introduced me to Abdus Salam School of Mathematical Sciences. And finally, I would like to thank Prof. A.D.R. Choudary and F. Watbled for reading the present communication and suggestions and corrections.

R. Leplaideur (✉)
Laboratoire de Mathématiques de Bretagne Atlantique, UMR 6205 Université de Brest. 6, rue V. Le Gorgeu, 29238 Brest, France
e-mail: renaud.leplaideur@univ-brest.fr

P. Cartier et al. (eds.), *Mathematics in the 21st Century*, Springer Proceedings in Mathematics & Statistics 98, DOI 10.1007/978-3-0348-0859-0_7

here is to present the point of view from dynamical systems. They are related to the *thermodynamic formalism*. This formalism was introduced during the 1970s, and the program was essentially achieved by Bowen, Ruelle, and Sinai (see [5, 23, 24]) for the uniformly hyperbolic case. Then, people working in dynamical systems were more interested in extending this formalism for the nonuniformly hyperbolic case. These systems do not have structural stability and are more difficult to study. In particular, there is still no general theory of thermodynamic formalism for nonuniformly hyperbolic dynamical systems.

Since the early 2000s, some people started to study maximizing measures. At that moment, they somehow rediscovered some phenomena studied by physicists, as, e.g., the *ground states*. These studies also naturally led people to study phase transition.

The rest of the paper proceeds as follows. In Sect. 2, we recall some basic notions of dynamical systems. In Sect. 3, we present the thermodynamic formalism and introduce the notion of pressure associated to a potential. In Sect. 4, we talk about phase transition and state our main question concerning this topic. In Sect. 5, we present the phenomenon of zero temperature and state and explain two open questions related to ground states and selection.

2 Basic Notions on Dynamical Systems

2.1 *Subshift of Finite Type*

Here, we shall refer to a *dynamical system* as a pair (X, T) where X is a compact metric space and $T : X \to X$ is globally continuous and locally a homeomorphism. We are interested in understanding the *orbits* of points under the iteration of T: x, $T(x), T^2(x) = T \circ T(x), \ldots$.

In other words, we have a $\mathbb{N}$- or $\mathbb{Z}$-action, which is usually seen as the natural action of time.

The results or notions stated in this note mainly hold for general *uniformly hyperbolic* dynamical systems, but sometimes we shall only consider the case of a subshift of finite type and, more precisely, the full shift with two symbols. In this case, X will be the set of infinite words in 0 and 1, $\{0, 1\}^{\mathbb{N}}$. The map T will be the shift map:

$$T(x_0x_1x_2\ldots) = x_1x_2x_3\ldots.$$

The distance is given by

$$d(x, y) = 2^{-\min\{i,\ x_i \neq y_i\}}$$

and is usually represented by the graphic

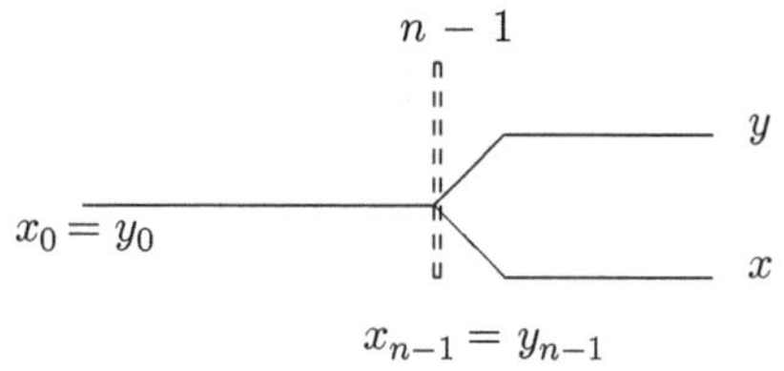

2.2 Necessity to Get Invariant Measures

Unless the system is very simple, it is usually impossible to describe all the orbits. Moreover, dynamical systems are examples of chaotic situations: two points may be very close but behave in very different ways after some time.

The main theorem in ergodic theory allows to recover some uniform behavior.

Theorem 1 (Birkhoff). *If μ is a T-invariant and ergodic probability, then, for every continuous $\phi : X \to \mathbb{R}$ and for μ-a.e. x,*

$$\lim_{n\to+\infty} \frac{1}{n}\sum_{k=0}^{n-1} \phi \circ T^k(x) = \int \phi \, d\mu.$$

To understand this theorem, before explaining what ergodic and T-invariant means, we give a very practical illustration.

The sentence "The average cost of car ownership rises to $8,946 per year" can be understood in two different ways:

- It is a time average, which means that the average is done after counting the cost of, say, one single person, for his car and for several years.
- It is a spatial average, which means that we consider for one single year the mean value over some region.

Theorem 1 says that under some assumptions, both mean values are equal.

Let us now explain these assumptions.

We denote by $\mathcal{B}$ the set of Borel sets in X. A Borel set A satisfying $T^{-1}(A) \supset A$ is said to be T-invariant.

Definition 1. A probability measure μ on X is said to be T-invariant if it satisfies

$$\forall A \in \mathcal{B},\ \mu(T^{-1}(A)) = \mu(A).$$

Roughly speaking, this definition means that we are studying a closed set along time evolution: there is neither creation nor disappearance of mass. We emphasize that a T-invariant Borel subset A also satisfies $\mu(T^{-1}(A) \setminus A) = 0$, thus $T^{-1}(A) = A$ in the sense of μ.

We recall that a probability measure can also be seen as a normalized linear form on $\mathcal{C}^0(X)$. Consequently, the set of probabilities is compact and convex for the weak* topology.

Lemma 1. *A probability measure μ is T-invariant, if and only if, for every $f : X \to \mathbb{R}$ continuous,*

$$\int f \circ T \, d\mu = \int f \, d\mu.$$

From Lemma 1, we easily get that the set of T-invariant (probability) measures is also compact and convex. Then we get:

Definition 2. The T-invariant probability μ is said to be ergodic if it satisfies one of the next following equivalent properties:

(1) Any T-invariant Borel A has μ-measure either equal to 1 or to 0.
(2) μ is an extremal element in the compact convex set of T-invariant probabilities.

Usually, a dynamical system admits a lot of T-invariant probabilities. For instance, in the case of $X = \{0,1\}^{\mathbb{N}}$, the n-cylinder $[x_0 \dots x_{n-1}]$ is the set of points in X starting as $x_0, \dots, x_{n-1}$. These sets generate the σ-field $\mathcal{B}$. Then, we let the reader check that, for $p \in (0,1)$, the measure ν_p defined by

$$\nu_p \leq ([x_0 \dots x_{n-1}]) = p^{\text{number of 0s}}(1-p)^{\text{number of 1s}}$$

is T-invariant and ergodic. This shows that there coexist uncountably many different ergodic measures.

3 Basic Notions on Thermodynamic Formalism

Due to the large number of invariant probabilities, it is normal to try to singularize some "good" T-invariant measures. The *thermodynamic formalism* is one way to do so. Let $\phi : X \to \mathbb{R}$ be a *potential*; we want to study the existence and uniqueness of measure maximizing their free energy:

$$h_\mu(T) + \int \phi \, d\mu = \sup_\nu \left\{ h_\nu(T) + \int \phi \, d\nu \right\}. \tag{1}$$

Such a measure is called an *equilibrium state*. The maximal value is called the *pressure*. The quantity $h_\nu(T)$ is the Kolmogorov entropy. The present communication does not intend to explain so briefly what the entropy means. We refer the reader to [20, 22]. We just mention that the entropy is a nonnegative real number, lower (in our case) than $\log 2$. It measures the disorder seen by ν. In the case of ergodic measure, the entropy means that for μ-a.e. $x = x_0x_1\ldots$,

$$\mu([x_0,\ldots x_{n-1}]) \approx_{n\to+\infty} e^{-nh_\mu(T)}.$$

The main property to know is that the entropy is upper semicontinuous: if $\mu_n \to \mu$ for the weak* topology, then

$$h_{\mu(T)} \geq \limsup_{n\to+\infty} h_{\mu_n}(T).$$

Following statistical mechanics, we want to study $\mathcal{P}(\beta)$, the pressure for $\beta.\phi$, with $\beta \in \mathbb{R}_+$. In statistical mechanics, β is the inverse of temperature. By definition, we immediately get that $\mathcal{P}(\beta)$ is convex.

Moreover, the upper semicontinuity of the entropy yields the existence of an equilibrium state for every ϕ continuous. With some more work, one can prove that $\mathcal{P}(\beta)$ admits an asymptote at $+\infty$.

But the main result concerns uniqueness.

Theorem 2 ([23]). *If the system is uniformly hyperbolic and the potential is Hölder continuous, then there exists a unique equilibrium state μ_β for $\beta.\phi$ and the pressure $\beta \mapsto \mathcal{P}(\beta)$ is analytic.*

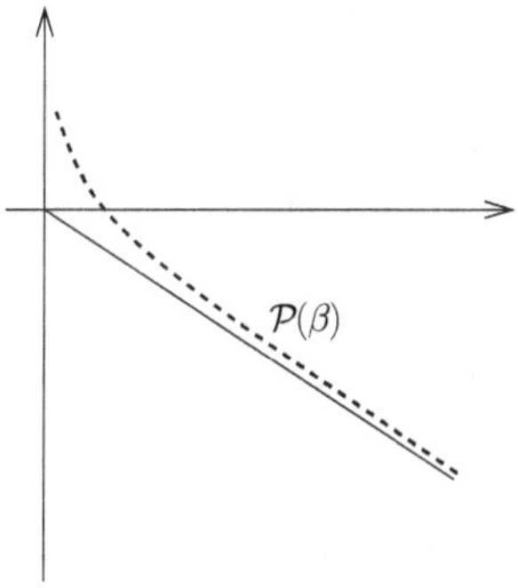

Our dynamical system $(\{0,1\}^{\mathbb{N}}, T)$ satisfies these assumptions. The heuristic way to explain this theorem is that hyperbolicity and Hölder continuity combine themselves to yield robust properties for an operator called the Ruelle-Perron-Frobenius operator (see [5]). It turns out that all the thermodynamic quantities (pressure, unique equilibrium state, conformal measure, etc.) are related to the spectrum of this operator. More precisely, this operator has a simple and unique dominating eigenvalue,[1] which is equal to $e^{\mathcal{P}(\beta)}$. The unique equilibrium state is

[1]This operator is actually quasi-compact.

constructed from the unique associated eigenfunction and the eigenmeasure of the dual operator.

In the case when A depends only on two coordinates, that is,

$$A(x) = A(x_0, x_1, \ldots, x_n, \ldots) = A(x_0, x_1),$$

we can introduce the matrix $\begin{pmatrix} e^{\beta.A(0,0)} & e^{\beta.A(0,1)} \\ e^{\beta.A(1,0)} & e^{\beta.A(1,1)} \end{pmatrix}$. In that case, for each β, the operator is finally reduced to this matrix, and the unique equilibrium state μ_β can be obtained via a Markov chain related to this matrix (see [12] for notions on Markov chains).

4 Phase Transitions

The first open question we want to mention deals with *phase transition.*

Definition 3. We say that there is a phase transition at β_0 if the pressure function fails to be analytic at β_0.

Our main question is:

Question 1. *Is it possible to exhibit a "good" class of continuous potentials $\phi : X \to \mathbb{R}$ which have a phase transition?*

Even if analyticity is a very rigid property, and thus rare, it is curiously not so easy to prove lack of analyticity for $\mathcal{P}(\beta)$. This is related to the fact that the Ruelle-Perron-Frobenius operator is quasi-compact. To get examples of phase transition, and due to Theorem 2, there are at least two strategies. The first is to weaken the hyperbolicity of the system, and the second is to weaken the regularity of the potential. This has to be done sufficiently to deteriorate the spectral properties of the operator, but it has also to be done sufficiently smoothly to not forbid any study of the operator.

Moreover, these two strategies may also sometimes coincide. The two well-known and historical examples in dynamical systems (but not in physics or probability theory) are the so-called Hofbauer potential in the shift (see [15])

$$\phi(x) = \log\left(1 + \frac{1}{n}\right) \text{ if } x = \underbrace{0\ldots0}_{n \text{ digits}}1\ldots,$$

or the Manneville-Pomeau map in the interval (see [21]) $f_{\mathrm{MP}} : [0,1] \to [0,1]$. This map is not uniformly hyperbolic due to the presence of some parabolic fixed point. In that case, the natural potential to study is $-\log f'_{\mathrm{MP}}$.

However, it is noteworthy that these two examples are related by some *renormalization* principle (see [2]). Actually, some recent works (see [2, 7, 8]) study the relations between renormalization substitution and phase transition, the goal being to get a machinery to exhibit examples of potentials in $\{0,1\}^{\mathbb{N}}$ with phase transitions.

We also address another question, which needs the next section to be understood.

Question 2. *What kind of phase transition on the pressure can we have, and for which potentials?*

In that direction, we mention [19] where it is proven that analyticity of the pressure does not imply uniqueness of the equilibrium state. The same example also shows other types of phase transitions than the previous ones. We also refer to [16] for other kinds of phase transitions.

5 Zero Temperature

In dynamical systems, most of the known phase transitions are *freezing phase transitions*. This means that the pressure is analytic and then reaches its asymptote:

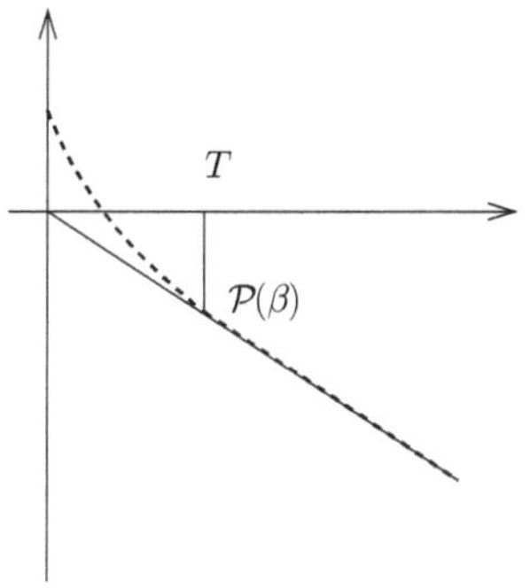

Then, for convexity reason, after the contact, it is equal to the asymptote, and there is a loss of analyticity. In statistical mechanics, one says that the equilibrium reaches a ground state.

Definition 4. A T-invariant probability measure μ is said to be ϕ-maximizing if it satisfies

$$\int \phi \, d\mu = \max_{\nu} \left\{ \int \phi \, d\nu \right\}.$$

A ground state is an accumulation point for μ_β as $\beta \to +\infty$.

It is easy to see that any ground state is a ϕ-maximizing measure (see [11]). In the case of a freezing phase transition, there exists some ϕ-maximizing measure μ such that, for every $\beta > \beta_c$, $\mu_\beta = \mu$. In other words, one reaches the ground state at positive temperature.

The two main questions we address concerning ground states and maximizing measures are the following:

Question 3. *For which potential do we have convergence of μ_β as $\beta \to +\infty$?*

Question 4. *If there is convergence, how does μ_β choose the limit?*

Question 3 is motivated by the fact that if ϕ depends only on finitely many coordinates (in the shift), then μ_β converges as $\beta \to +\infty$ (see [6, 10, 17]). On the other hand, there exists one example of non-convergence of μ_β for ϕ Hölder continuous (see [9]).

A nice theorem (see [4]) states that generically[2] (for the continuous norm, but this also holds for any "natural" Banach space), there exists a unique maximizing measure. In that case, the answer to Question 3 is easy because there is a unique possible accumulation point.

Nevertheless, it is extremely easy to construct some potential with at least two maximizing measures: consider two periodic orbits, say, $\mathcal{O}_1$ and $\mathcal{O}_2$, and set

$$\phi(x) = -d(x, \mathcal{O}_1 \cup \mathcal{O}_2).$$

Clearly, the two invariant measures with support on each periodic orbit are ϕ-maximizing. Consequently, any convex combination of them is also ϕ-maximizing. This simple example explains Question 4. The set of maximizing measures is a simplex, because any convex combination of two T-invariant maximizing probabilities is again a maximizing T-invariant probability. Therefore, it makes sense to study if there is convergence of μ_β as $\beta \to +\infty$ what singularizes this ground state among all the maximizing measures. Some works in that direction tend to show that flatness is a criteria for this *selection* of maximizing measure (see [1, 18]). We refer the reader to [3] for a lecture notes concerning selection, ground states, and other tools related to maximizing measures.

References

1. Baraviera, A.T., Leplaideur, R., Lopes, A.O.: Selection of ground states in the zero temperature limit for a one-parameter family of potentials. SIAM J. Appl. Dyn. Syst. **11**(1), 243–260 (2012)
2. Baraviera, A., Leplaideur, R., Lopes, A.O.: The potential point of view for renormalization. Stoch. Dyn. **12**(4), 1250005, 34 (2012)
3. Baraviera, A.T., Leplaideur, R., Lopes, A.O.: Ergodic optimization, zero temperature limits and the max-plus algebra. Publicações Matemáticas do IMPA. [IMPA Mathematical Publications], pp. ii+108. Instituto Nacional de Matemática Pura e Aplicada (IMPA), Rio de Janeiro. ISBN 978-85-244-0356-9 (2013). ArXiv e-prints.
4. Bousch, T.: La condition de Walters. Ann. Sci. École Norm. Sup. (4) **34**(2), 287–311 (2001)

[2]We mean for a G_δ set of potentials with respect to the considered norm.

5. Bowen, R.: Equilibrium States and the Ergodic Theory of Anosov Diffeomorphisms. Lecture Notes in Mathematics, vol. 470, Rev. edn. Springer, Berlin (2008). With a preface by David Ruelle, Edited by Jean-René Chazottes
6. Brémont, J.: Gibbs measures at temperature zero. Nonlinearity **16**(2), 419–426 (2003)
7. Bruin, H., Leplaideur, R.: Renormalization, freezing phase transitions and quasicrystals. a step to the general case: Fibonacci case. Annales de L'ENS (2013, in preparation)
8. Bruin, H., Leplaideur, R.: Renormalization, thermodynamic formalism and quasi-crystals in subshifts. Commun. Math. Phys. **321**, 209–247 (2013)
9. Chazottes, J.-R., Hochman, M.: On the zero-temperature limit of Gibbs states. Commun. Math. Phys. **297**(1), 265–281 (2010)
10. Chazottes, J.-R., Gambaudo, J.-M., Ugalde, E.: Zero-temperature limit of one-dimensional Gibbs states via renormalization: the case of locally constant potentials. Ergodic Theory Dyn. Syst. **31**(4), 1109–1161 (2011)
11. Contreras, G., Lopes, A.O., Thieullen, Ph.: Lyapunov minimizing measures for expanding maps of the circle. Ergodic Theory Dyn. Syst. **21**(5), 1379–1409 (2001)
12. Feller, W.: An Introduction to Probability Theory and Its Applications, vol. I, 3rd edn. Wiley, New York (1968)
13. Georgii, H.-O.: Gibbs Measures and Phase Transitions. de Gruyter Studies in Mathematics, vol. 9, 2nd edn. Walter de Gruyter, Berlin (2011)
14. Grimmett, G.: Probability on Graphs: Processes on Graphs and Lattices. Institute of Mathematical Statistics Textbooks, vol. 1. Cambridge University Press, Cambridge (2010). Random
15. Hofbauer, F.: Examples for the nonuniqueness of the equilibrium state. Trans. Am. Math. Soc. **228**, 223–241 (1977)
16. Iommi, G., Todd, M.: Transience in dynamical systems. Ergodic Theory Dynam. Systems **33**(5), 1450–1476 (2010). ArXiv e-prints
17. Leplaideur, R.: A dynamical proof for the convergence of Gibbs measures at temperature zero. Nonlinearity **18**(6), 2847–2880 (2005)
18. Leplaideur, R.: Flatness is a criterion for selection of maximizing measures. J. Stat. Phys. **147**(4), 728–757 (2012)
19. Leplaideur, R.: Chaos: Butterflies also generate phase transitions and parallel universes (2013)
20. Petersen, K.: Ergodic Theory. Cambridge Studies in Advanced Mathematics, vol. 2. Cambridge University Press, Cambridge (1989). Corrected reprint of the 1983 original
21. Pomeau, Y., Manneville, P.: Intermittent transition to turbulence in dissipative dynamical systems. Commun. Math. Phys. **74**(2), 189–197 (1980)
22. Rudolph, D.J.: Fundamentals of Measurable Dynamics: Ergodic Theory on Lebesgue Spaces. Oxford Science Publications/Clarendon Press/Oxford University Press, New York (1990)
23. Ruelle, D.: Thermodynamic Formalism: The Mathematical Structures of Equilibrium Statistical Mechanics. Cambridge Mathematical Library, 2nd edn. Cambridge University Press, Cambridge (2004)
24. Sinai, Y.G.: Introduction to Ergodic Theory. Princeton University Press, Princeton (1976). Translated by V. Scheffer, Mathematical Notes, 18

Hamiltonian Connectedness of Toeplitz Graphs

Muhammad Faisal Nadeem, Ayesha Shabbir, and Tudor Zamfirescu

1 Introduction

A simple undirected graph T with vertices $1, 2, \ldots, n$ is called a *Toeplitz graph* if its adjacency matrix $A(T)$ is Toeplitz. A *Toeplitz matrix* is an $(n \times n)$ symmetric matrix which has constant values along all diagonals parallel to the main diagonal. Therefore, a Toeplitz graph T is uniquely defined by the first row of $A(T)$, a (0–1) sequence. If the 1's in that sequence are placed at positions $1 + t_1, 1 + t_2, \ldots, 1 + t_k$ with $1 \leq t_1 < t_2 < \cdots < t_k < n$, we may simply write $T = T_n \langle t_1, t_2, \ldots, t_k \rangle$, two vertices x, y being connected by an edge iff $|x - y| \in \{t_1, t_2, \ldots, t_k\}$.

Let G be a graph of order n. It is called *Hamiltonian* if it contains a cycle of order n. It is called *traceable*, if it contains a path of order n; that path is then called a *Hamiltonian path* of G. The graph G is said to be *Hamiltonian connected* if for any pair of distinct vertices u and v of G, there exists a Hamiltonian path from u to v. The property of being Hamiltonian connected is stronger than being Hamiltonian.

The third author's work was supported by a grant of the Romanian National Authority for Scientific Research, CNCS – UEFISCDI, project number PN-II-ID-PCE-2011-3-0533.

M.F. Nadeem • A. Shabbir
Abdus Salam School of Mathematical Sciences, GC University, 68-B, New Muslim Town, Lahore, Pakistan
e-mail: mfaisalnadeem@ymail.com; ashinori@hotmail.com

T. Zamfirescu (✉)
Faculty of Mathematics, University of Dortmund, 44221 Dortmund, Germany

Institute of Mathematics "Simion Stoïlow" Roumanian Academy, Bucharest, Roumania

Abdus Salam School of Mathematical Sciences, GC University, 68-B, New Muslim Town, Lahore, Pakistan
e-mail: tuzamfirescu@googlemail.com.

P. Cartier et al. (eds.), *Mathematics in the 21st Century*, Springer Proceedings in Mathematics & Statistics 98, DOI 10.1007/978-3-0348-0859-0_8

References [1–4] contain results about connectivity, bipartiteness, planarity, and colorability of Toeplitz graphs. Some Hamiltonian properties of undirected Toeplitz graphs have been investigated in [1] and [5], while the directed case was studied in [6–8]. In [9], S. Malik and T. Zamfirescu started the investigation of the Hamiltonian connectedness of directed Toeplitz graphs. For the indirected case, in [9] it is proven that $T_n\langle 1,2\rangle$ is Hamiltonian connected only for $n = 3$, while $T_n\langle 1,2,s\rangle$ is Hamiltonian connected for all values of n and s. It will become clear that, concerning k, the first relevant case is $k = 3$. In this paper, we are completing the picture of Hamiltonian connectedness of Toeplitz graphs, more precisely of $T_n\langle t_1,t_2\rangle$, $T_n\langle 1,3,s\rangle$ and $T_n\langle 1,4,s\rangle$.

Let T be a Toeplitz graph and p,q be two vertices of T, such that $p < q$. By $P_{p,q}$ we mean a path from p to $p+1$ containing all vertices in $\{p, p+1, p+2, \dots, q-2, q-1, q\}$, and by $P_{q,p}$ we mean a path from q to $q-1$, containing the same vertices. The existence of $P_{p,q}$ or $P_{q,p}$ is not guaranteed.

We start with a few simple results.

Theorem 1. *For $n \neq 3$, $T_n\langle t_1,t_2\rangle$ is not Hamiltonian connected.*

Proof. Assume $T = T_n\langle t_1,t_2\rangle$ for $n \neq 3$ is Hamiltonian connected. Then there exists a Hamiltonian path from t_1+1 to t_2+1. But the path from t_1+1 to t_2+1 containing 1 is unique and is of length 2. This leads to a contradiction. Hence, T is not Hamiltonian connected. □

Theorem 2. *The Toeplitz graph $T_n\langle t_1,t_2,t_3,\dots, t_k\rangle$ is not Hamiltonian connected if $t_1, t_2, t_3, \dots, t_k$ are all odd.*

Proof. A bipartite graph is not Hamiltonian connected, and if $t_1, t_2, t_3, \dots, t_k$ are all odd, then the graph $T_n\langle t_1,t_2,t_3,\dots, t_k\rangle$ is bipartite. □

Corollary. *$T_n\langle 1,3,s\rangle$ is not Hamiltonian connected, when s is odd.*

Theorem 3. *If both n and t are odd, then $T_n\langle 1,t,n-1\rangle$ is not Hamiltonian connected.*

Proof. Let, for t odd, $T = T_n\langle 1,t,n-1\rangle$, where $n \geq t+2$ is an odd integer. Assume that T is Hamiltonian connected, then there exists a Hamiltonian path H between two even vertices x and y of T. The path H either contains the edge $(1,n)$ or not.

If H contains the edge $(1,n)$, we can contract it to a single vertex, because both vertices of the edge have the same parity (both are odd). After contraction, the resulting path H' is of even order, and the number of even vertices is equal to the number of odd vertices. But the end vertices of H' are even, which leads to a contradiction.

Next, we assume that H does not contain the edge $(1,n)$. But T without the edge $(1,n)$ becomes $T_n\langle 1,t\rangle$, which is a bipartite graph. Again, H cannot be a Hamiltonian path of T, and this completes the proof. □

Lemma 1. *If n is even, then $T_n\langle 1,3\rangle$ admits a Hamiltonian path from 1 to 2 and, by symmetry, another one from n to $n-1$.*

Fig. 1

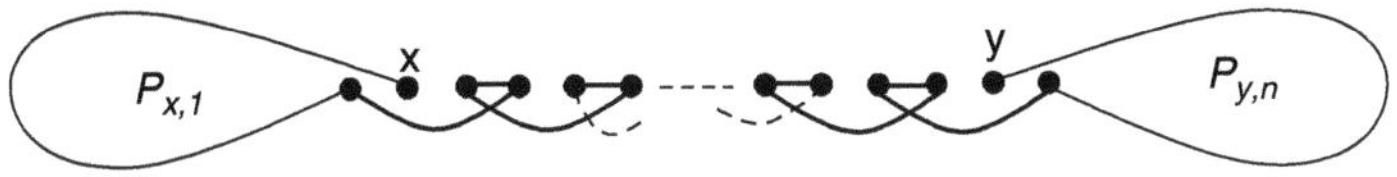

Fig. 2

Fig. 3

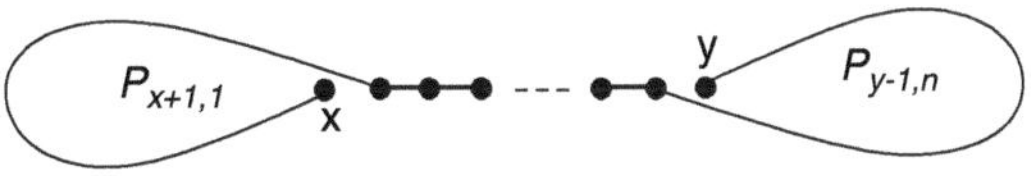

Fig. 4

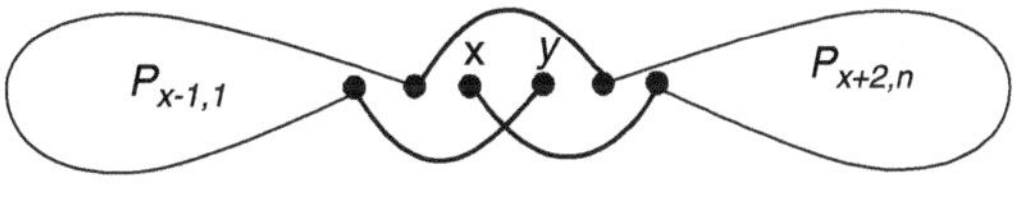

Proof. See Fig. 1, for a Hamiltonian path in $T = T_n\langle 1,3\rangle$, from vertex 1 to vertex 2, for even $n \geq 4$. A similar Hamiltonian path from vertex n to vertex $n-1$ exists in T, due to the symmetry of Toeplitz graphs. This completes the proof. □

Lemma 2. *Let p, q be two distinct vertices of $T_n\langle 1,3\rangle$. If $q-p$ is odd then paths $P_{p,q}$ and $P_{q,p}$ exist in $T_n\langle 1,3\rangle$.*

Proof. Apply Lemma 1 to the subgraph of $T_n\langle 1,3\rangle$ spanned by $p, p+1, \ldots q$. □

Theorem 4. *$T_n\langle 1,3,s\rangle$ is Hamiltonian connected for all $n \geq s+2$, if s is an even integer.*

Proof. Let $T = T_n\langle 1,3,s\rangle$ be the Toeplitz graph, where $n \geq s+2$. Then, there exist paths $P_{p,q}$ and $P_{q,p}$ in T, whenever $q-p$ is odd for $p < q$, by Lemma 2. Now, by using such paths of T, we prove that for any two distinct vertices x and y of T, there exists a Hamiltonian path from x to y. Take $x < y$. We split our proof into two main cases:

Case 1. n is even.

The following four subcases arise:

(*i*) *x is even, y is odd.*
In this case, $P_{x,1}$ and $P_{y,n}$ exist in T, and, with the help of these two paths, we obtain a Hamiltonian path $\left(P_{x,1}, x+2, x+1, \ldots, y-1, y-2, P_{y,n}\right)$ in T from x to y; see Fig. 2.

(*ii*) *x is odd, y is even.*
If $y \neq x+1$, then a possible Hamiltonian path of T from x to y, $\left(P_{x+1,1}, x+2, \ldots, P_{y-1,n}\right)$, is shown in Fig. 3.

When $y = x+1$, then for $x = 1$ or $x = n-1$, we use the path of Lemma 1, and for other values of x, we consider the path $(x, P_{x+2,n}, P_{x-1,1}, y)$; see Fig. 4.

Fig. 5

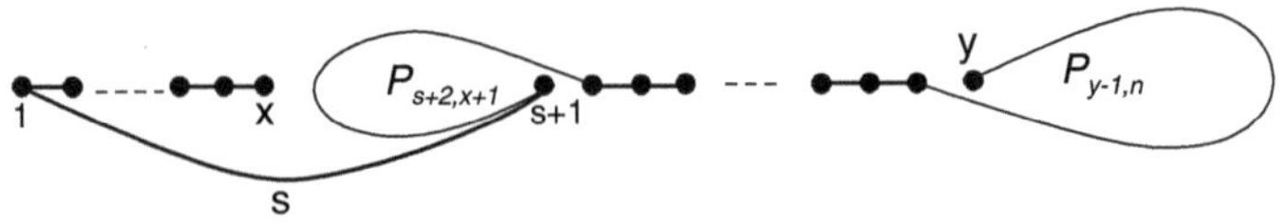

Fig. 6

Fig. 7

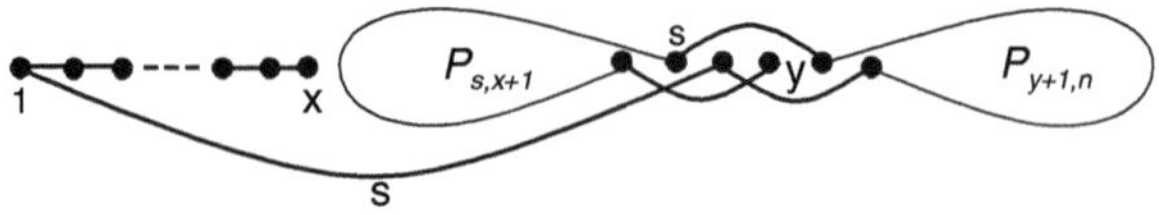

Fig. 8

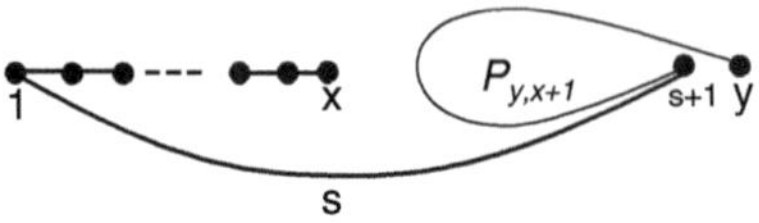

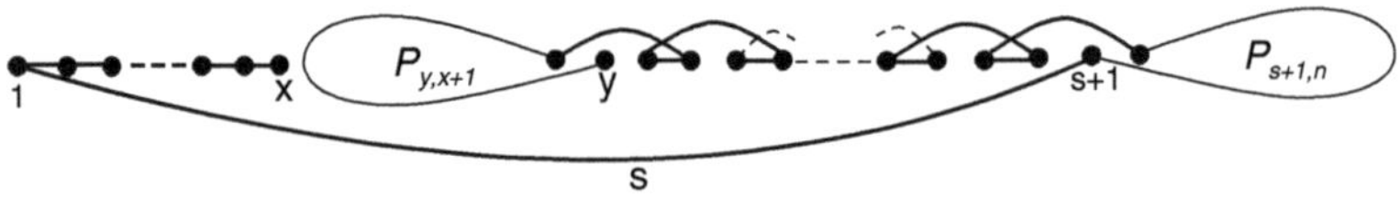

Fig. 9

(*iii*) *x and y are even.*
If $x > s$, then a Hamiltonian path from x to y is $(x,\ x-1,\ P_{x-s,1},\ x-s+1,\ \ldots,\ x-2,\ x+1,\ x+2,\ \ldots,\ y-2,\ P_{y-1,n})$ (see Fig. 5).
If $x \leq s$, then we have four subcases to discuss:

(*a*) For $y > s+2$, we consider a Hamiltonian path $(x,\ x-1,\ x-2,\ \ldots,\ 2,\ 1,\ P_{s+2,x+1},\ s+3,\ \ldots,\ y-2,\ P_{y-1,n})$ between x and y; see Fig. 6.
(*b*) When $y = s+2 \neq n$, then a possible Hamiltonian path joining x and y is $(x,\ x-1,\ x-2,\ \ldots,\ 2,\ 1,\ s+1,\ P_{y+1,n}, P_{s,x+1},\ s+2 = y)$; see Fig. 7.
(*c*) If $y = s+2 = n$, then a Hamiltonian path from x to y is $(x,\ x-1,\ x-2,\ \ldots,\ 3,\ 2,\ 1, P_{n,x+1})$; see Fig. 8.
(*d*) Finally, for $y \leq s$. A Hamiltonian path joining x and y is $(x,\ x-1,\ x-2,\ \ldots,\ 2,\ 1,\ P_{s+1,n},\ s-1,\ s,\ s-3,\ s-2,\ \ldots,\ y+1,\ y+2, P_{y,x+1})$ (see Fig. 9).

(*iv*) *x is odd, y is odd.*

This case is symmetric to case (*iii*). (Denote vertex i by $n+1-i$.)

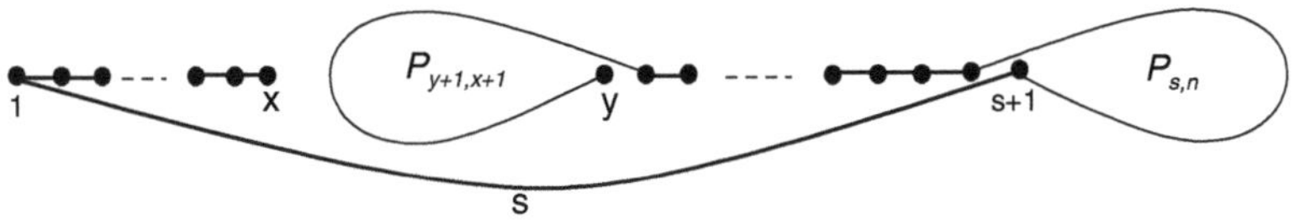

Fig. 10

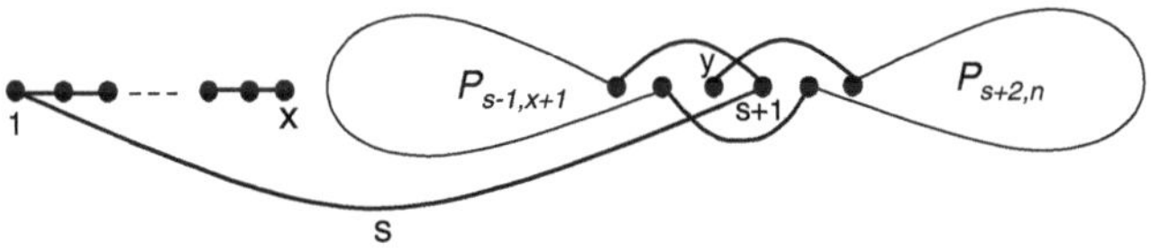

Fig. 11

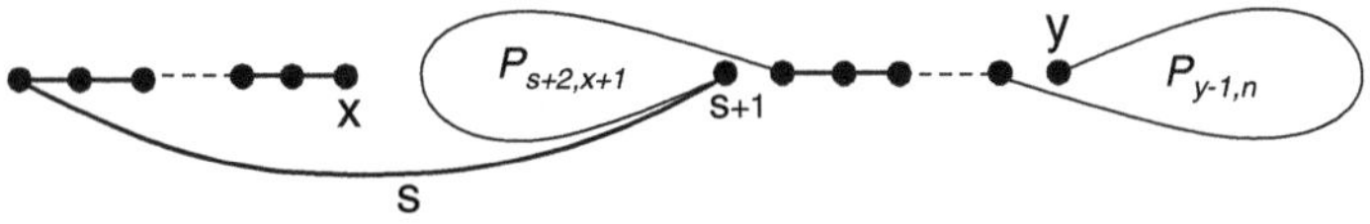

Fig. 12

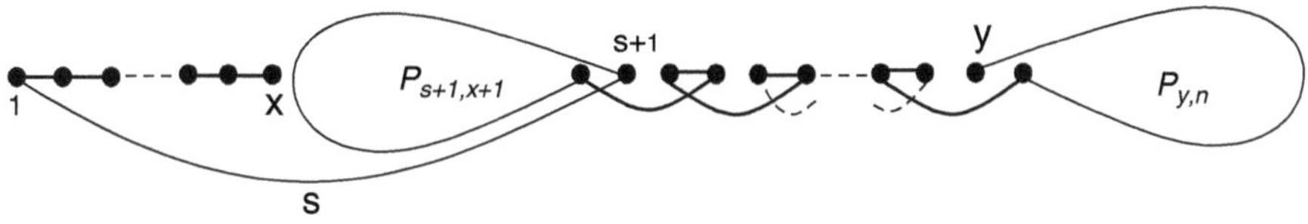

Fig. 13

Case 2. n is odd.

Again, we consider the following subcases:

(i) *x and y are of different parity.*
First, we assume that $y < s$. Then a Hamiltonian path joining x and y is $(x, x-1, x-2, \ldots, 2, 1, P_{s,n}, s-1, \ldots, y+2, P_{y+1,x+1})$ (see Fig. 10).
If $y = s$, then a Hamiltonian path joining x and y is $(x, x-1, \ldots, 2, 1, s+1, P_{s-1,x+1}, P_{s+2,n}, y)$ (see Fig. 11).
Next suppose that $x \leq s$ and $y > s+1$.

(a) If x is even, then a Hamiltonian path joining x and y is $(x, x-1, x-2, \ldots, 2, 1, P_{s+2,x+1}, s+3, \ldots, y-2, P_{y,n})$; see Fig. 12.
(b) If x is odd, then a Hamiltonian path $(x, x-1, \ldots, 1, P_{s+1,x+1}, s+3, \ldots, y-1, y-2, P_{y,n})$, joining x and y, is shown in Fig. 13.

When $2 < x < s$ and $y = s+1$, we consider a Hamiltonian path $(x, x-1, \ldots, 4, 1, 2, 3, P_{s+2,n}, P_{s+2,x+1})$ from x to y (see Fig. 14).

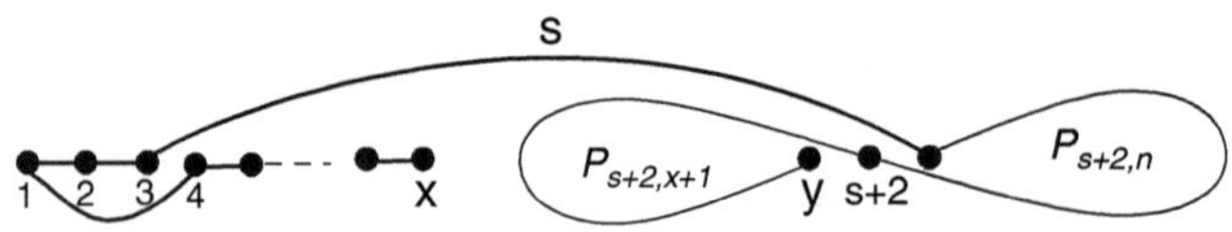

Fig. 14

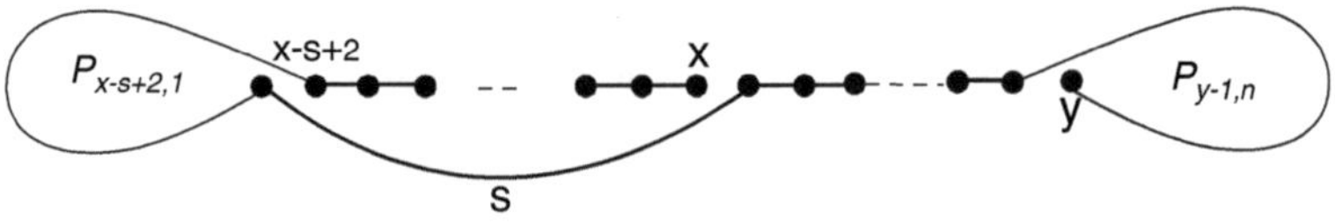

Fig. 15

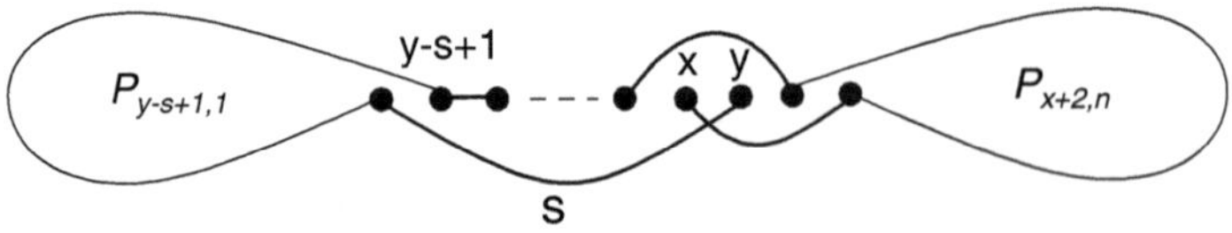

Fig. 16

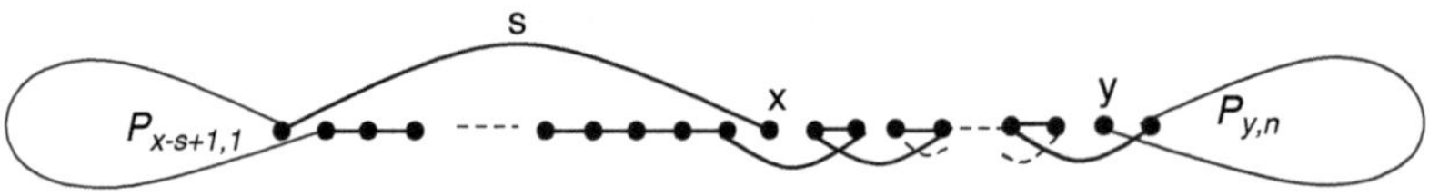

Fig. 17

If $x = 2$ and $y = s + 1$, then a Hamiltonian path joining x and y is $(x = 2, 1, P_{3,y-1}, P_{s+2,n}, y)$.
Finally, here we consider the case $x > s$.

(a) If x is even and $y \neq x + 1$, then a Hamiltonian path from x to y is $(x, x-1, \ldots, x-s+1, P_{x-s+2,1}, x+1, x+2, \ldots, y-2, P_{y-1,n})$; see Fig. 15.

If $y = x + 1$, then a Hamiltonian path from x to y is $(x, P_{x+2,n}, x-1, x-2, \ldots, y-s+2, P_{y-s+1,1}, y)$ (see Fig. 16).

(b) If x is odd, then a Hamiltonian path from x to y is $(x, P_{x-s+1,1}, x-s+2, \ldots, x-1, x+2, x+1, \ldots, y-1, y-2, P_{y,n})$; see Fig. 17.

(ii) *x and y are even.*
The following subcases arise:

If $x = 2$ and $y \geq s + 2$, we use $(2, 1, s+1, s, \ldots, 4, 3, s+3, s+2, s+5, \ldots, y-1, y-2, y+1, P_{y,n})$, the Hamiltonian path between x and y (see Fig. 18).

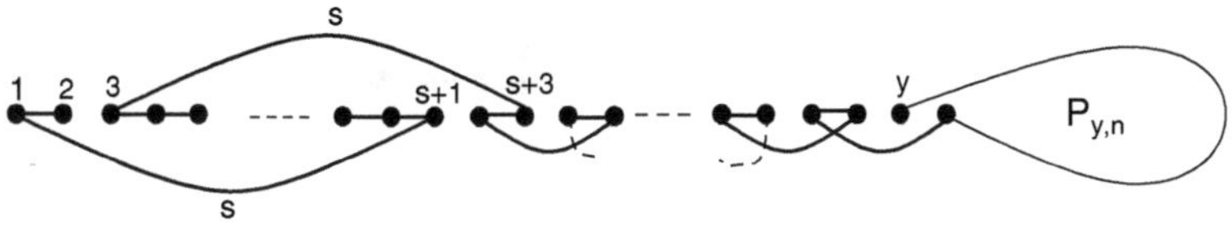

Fig. 18

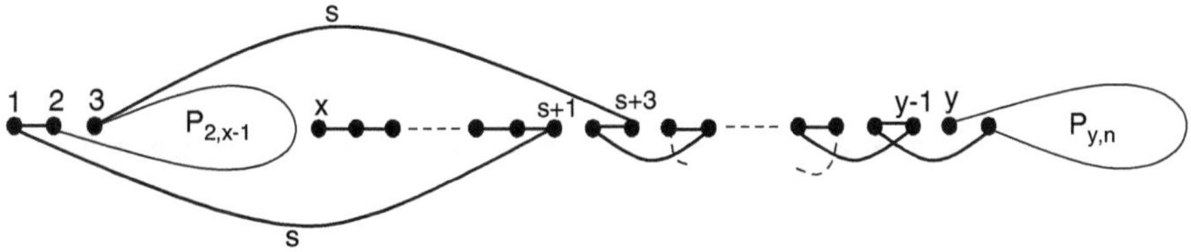

Fig. 19

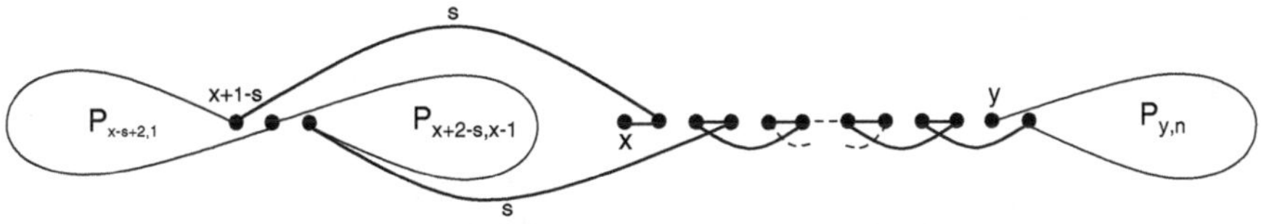

Fig. 20

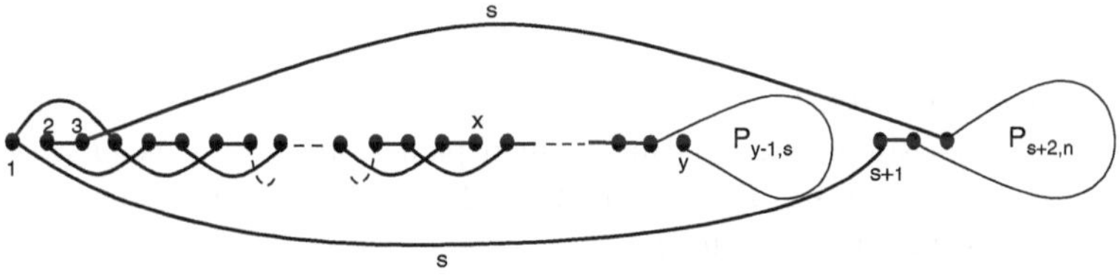

Fig. 21

When $4 \leq x \leq s$ and $y \geq s+2$, then a Hamiltonian path joining x and y is $(x, x+1, \ldots, s+1, 1, P_{2,x-1}, s+3, s+2, s+5, \ldots, y-1, y-2, P_{y,n})$, shown in Fig. 19.

If $x > s$, we have $(x, x+1, P_{x-s+2,1}, P_{x-s+2,x-1}, x+3, x+2, \ldots, y-1, y-2, P_{y,n})$, the Hamiltonian path from x to y (see Fig. 20).

If $y < s$, the path from x to y as desired is $(x, x-1, x-4, x-5, \ldots, 4, 1, s+1, P_{s+2,n}, 3, 2, 5, 6, \ldots, x-3, x-2, x+1, x+2, \ldots, y-2, P_{y-1,s})$, when $x \equiv 0 (\bmod 4)$, and $(x, x-1, x-4, x-5, \ldots, 6, 5, 2, 3, P_{s+2,n}, s+1, 1, 4, 7, 8, \ldots, x-3, x-2, x+1, x+2, \ldots, y-2, P_{y-1,s})$, when $x \equiv 2 (\bmod 4)$; see also Fig. 21.

Fig. 22

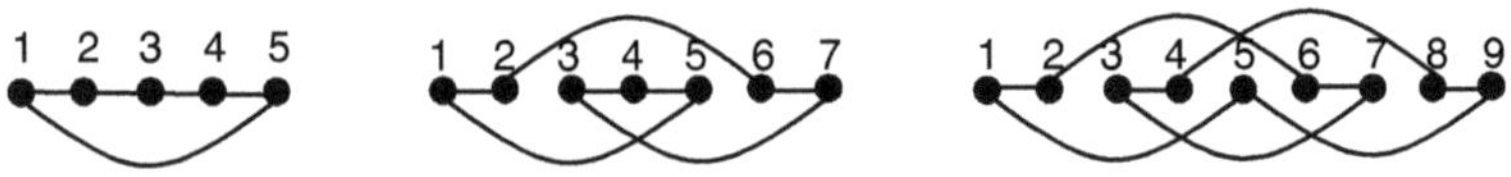

Fig. 23

Fig. 24

Fig. 25

Fig. 26

(*iii*) *x and y are odd.*

In this simple case a Hamiltonian path from x to y is $(P_{x+1,1}, x + 2, \ldots, y-3, y-2, P_{y-1,n})$ (see Fig. 22).

Now the proof is complete.

□

Lemma 3. *For $n = 5$ and all $n \geq 7$, $T_n\langle 1, 4\rangle$ admits a Hamiltonian path from* 1 *to* 2 *and, by symmetry, another one from n to $n - 1$.*

Proof. $T_n\langle 1, 4\rangle$ is Hamiltonian for all values of n except 6. See Fig. 23 for a Hamiltonian cycle in $T_n\langle 1, 4\rangle$, when $n \in \{5, 7, 9\}$. These cycles are unique and we use them to find a Hamiltonian path from 1 to 2 in $T_n\langle 1, 4\rangle$.

For any $n \equiv 0 (\mathrm{mod}\, 3)$, a suitable path is obtained by using the Hamiltonian cycle in $T_9\langle 1, 4\rangle$; see Fig. 24.

To obtain such a path when $n \equiv 1 (\mathrm{mod}\, 3)$, we use the Hamiltonian cycle found in $T_7\langle 1, 4\rangle$; see Fig. 25.

For $n \equiv 2 (\mathrm{mod}\, 3)$, the cycle $T_5\langle 1, 4\rangle$ is employed; see Fig. 26.

Now, because of the symmetry of the Toeplitz graph, we also have a Hamiltonian path from n to $n - 1$. □

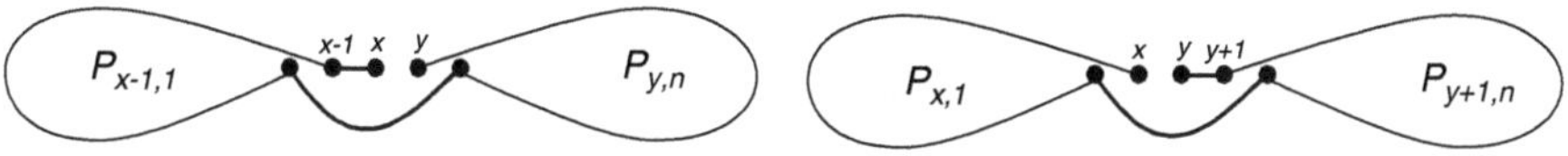

Fig. 27

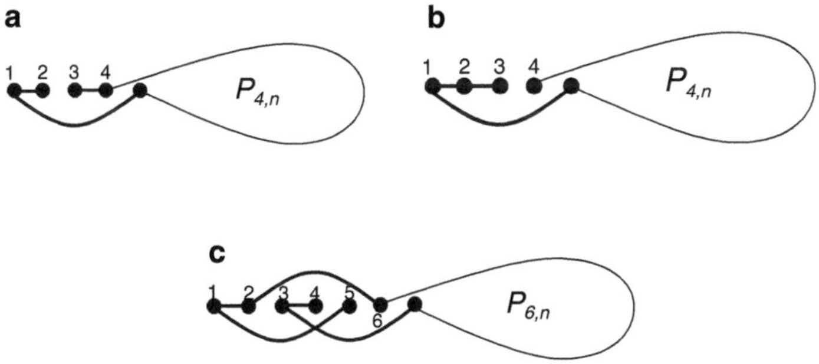

Fig. 28

Lemma 4. *Let p, q be two distinct vertices of $T_n\langle 1,4\rangle$. If $q - p \neq 2,3,5$, then there exist paths $P_{p,q}$ and $P_{q,p}$ in $T_n\langle 1,4\rangle$.*

Proof. See Lemma 3. □

Theorem 5. *$T_n\langle 1,4,s\rangle$ is Hamiltonian connected for all s and $n \geq 15$.*

Proof. For $n \geq 15$, let x and y be distinct vertices of the Toeplitz graph $T = T_n\langle 1,4,s\rangle$. Assume that $x < y$. To prove the result we show that there exists a Hamiltonian path between x and y.

Case 1. $y = x + 1$.

If $x = 1$ or $n - 1$, we have a desired path due to Lemma 3.

When $5 \leq x \leq n - 5$, then a Hamiltonian path between x and y is either $(x, P_{x-1,1}, P_{y,n})$ or $(P_{x,1}, P_{y+1,n}, y)$ (see Fig. 27).

When $2 \leq x \leq 4$, see Fig. 28 for a Hamiltonian path between x and y.

For $n - 4 \leq x \leq n - 2$, the desired Hamiltonian paths are symmetric to the paths for $x \in \{2,3,4\}$.

Case 2. If $y \neq x + 1$, the following three subcases arise:

(*i*) $x \leq n - 5$ and $y \in \{3,4,\ldots,n-5,n-3\}$.
(*ii*) $x \leq n - 5$ and $y \in \{n-4,n-2,n-1,n\}$.
(*iii*) $x \geq n - 4$.

Subcase (*i*). Let $y \in \{3,4,\ldots,n-5,n-3\}$.

(*a*) First, we assume the case when $x \in \{4,6,7,\ldots,n-5\}$. Now $(P_{x+1,1}, x+2, x+3, \ldots, y-2, P_{y-1,n})$ is a required Hamiltonian path between x and y (see Fig. 29).

(*b*) If $x = 1$, then a desired path between 1 and y is shown in Fig. 30.

Fig. 29

Fig. 30

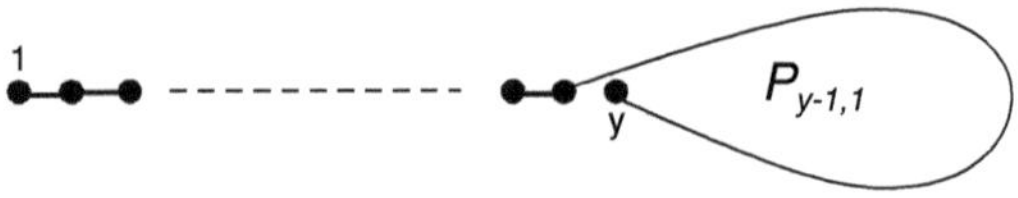

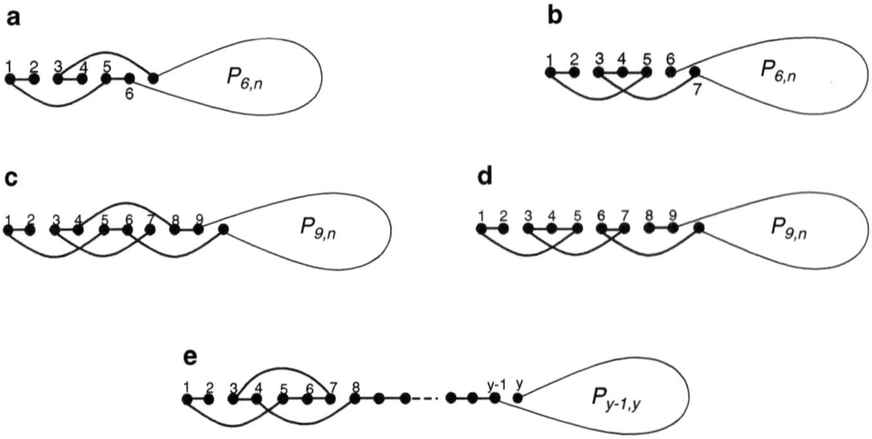

Fig. 31 (**a**) A Hamiltonian path between 2 and 4. (**b**) A Hamiltonian path between 2 and 6. (**c**) A Hamiltonian path between 2 and 7. (**d**) A Hamiltonian path between 2 and 8. (**e**) A Hamiltonian path between 2 and y, where $y \geq 9$

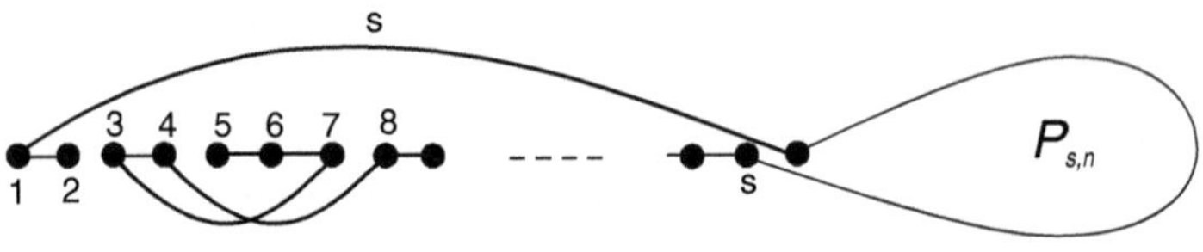

Fig. 32

(*c*) If $x = 2$ and $y \neq 5$, then Hamiltonian paths between 2 and different values of y are shown in Fig. 31.

When $x = 2$ and $y = 5$, to get a desired path, we use the difference s along with differences 1 and 4. See Fig. 32, for such a path when $s \in \{8, 9, 10, \ldots, n-6, n-4\}$.

When $s = 5, 6, 7$, see Fig. 33.

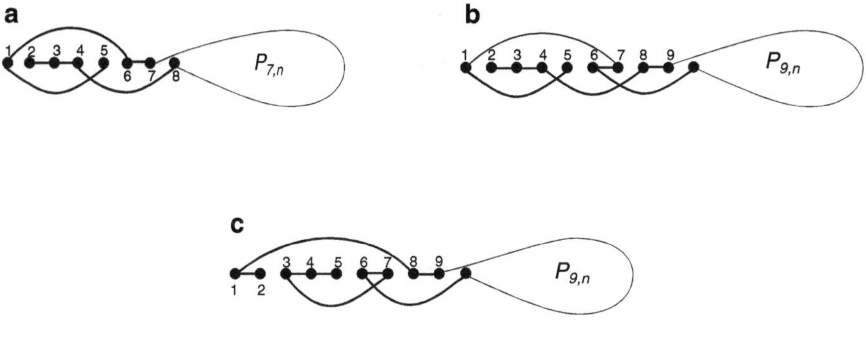

Fig. 33 (**a**) $s = 5$. (**b**) $s = 6$. (**c**) $s = 7$

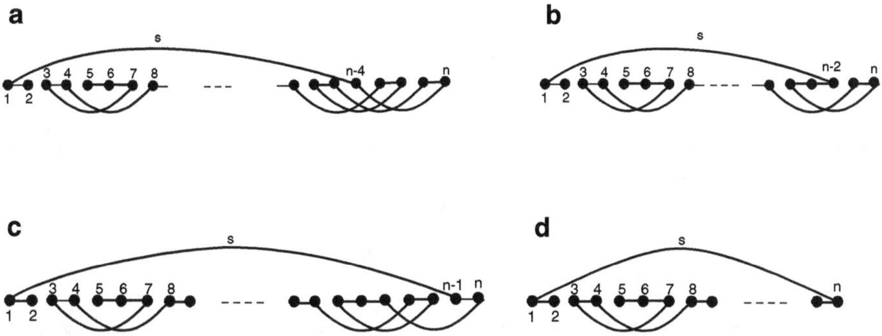

Fig. 34 (**a**) $s = n - 5$. (**b**) $s = n - 3$. (**c**) $s = n - 2$. (**d**) $s = n - 1$

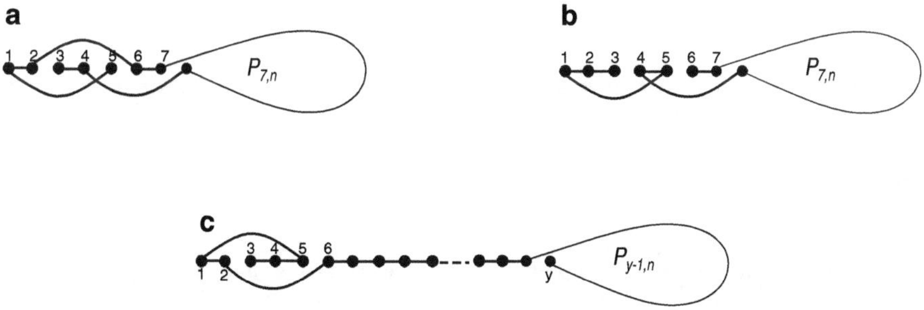

Fig. 35 (**a**) A Hamiltonian path between 3 and 5. (**b**) A Hamiltonian path between 3 and 6. (**c**) A Hamiltonian path between 3 and $y \geq 7$

And, for $s = n - 5, n - 3, n - 2, n - 1$, see Fig. 34

(d) If $x = 3$, then for a Hamiltonian path between 3 and y, see Fig. 35.

(e) If $x = 5$ and $y \neq 8$, a desired Hamiltonian path is shown in Fig. 36. When $y = 8$ and $n \neq 15, 17$, we use the path shown in Fig. 37. For $n = 15$ and $n = 17$, see Figs. 38 and 39, respectively.

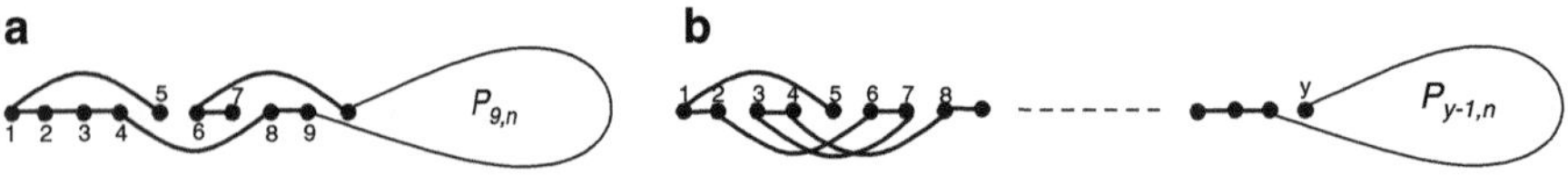

Fig. 36 (**a**) A Hamiltonian path between 5 and 7. (**b**) A Hamiltonian path between 5 and y, where $y \geq 9$

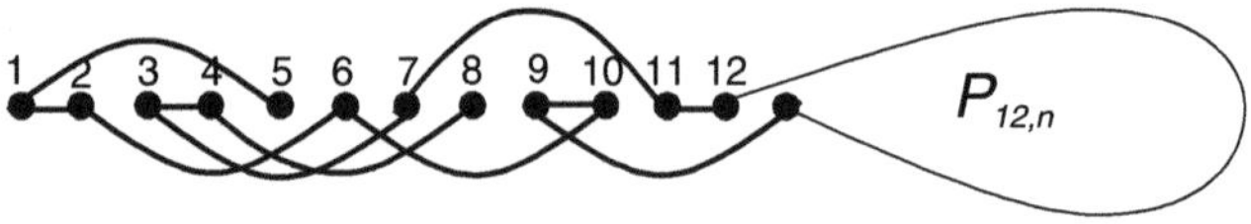

Fig. 37

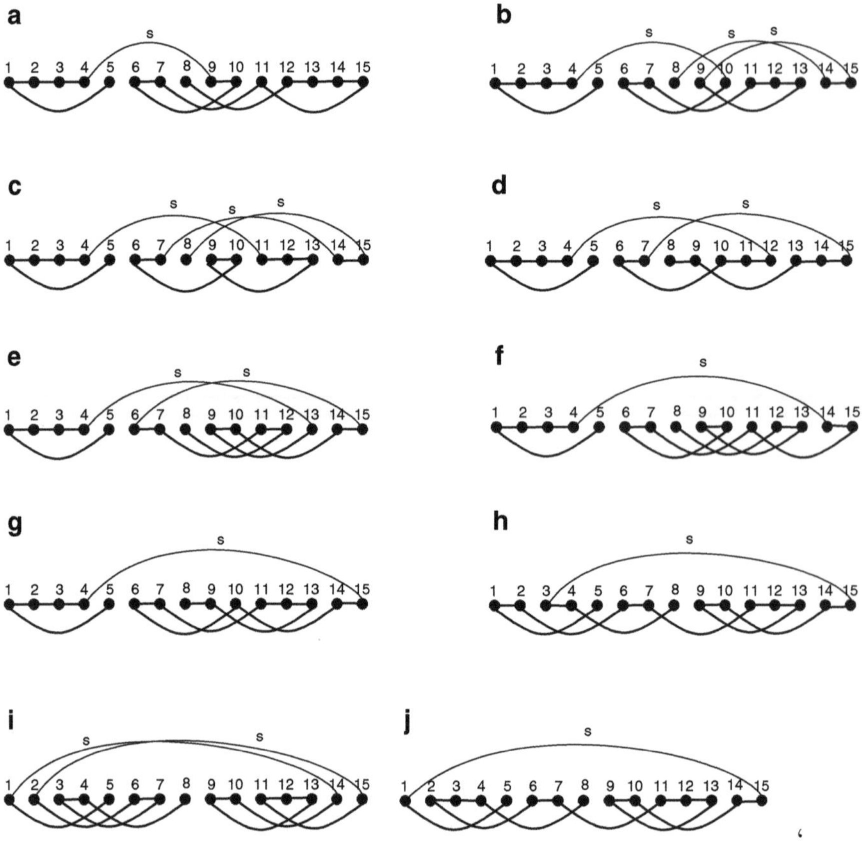

Fig. 38 Hamiltonian paths between 5 and 8 for different values of s, when $n = 15$. (**a**) $s = 5$. (**b**) $s = 6$. (**c**) $s = 7$. (**d**) $s = 8$. (**e**) $s = 9$. (**f**) $s = 10$. (**g**) $s = 11$. (**h**) $s = 12$. (**i**) $s = 13$. (**j**) $s = 14$

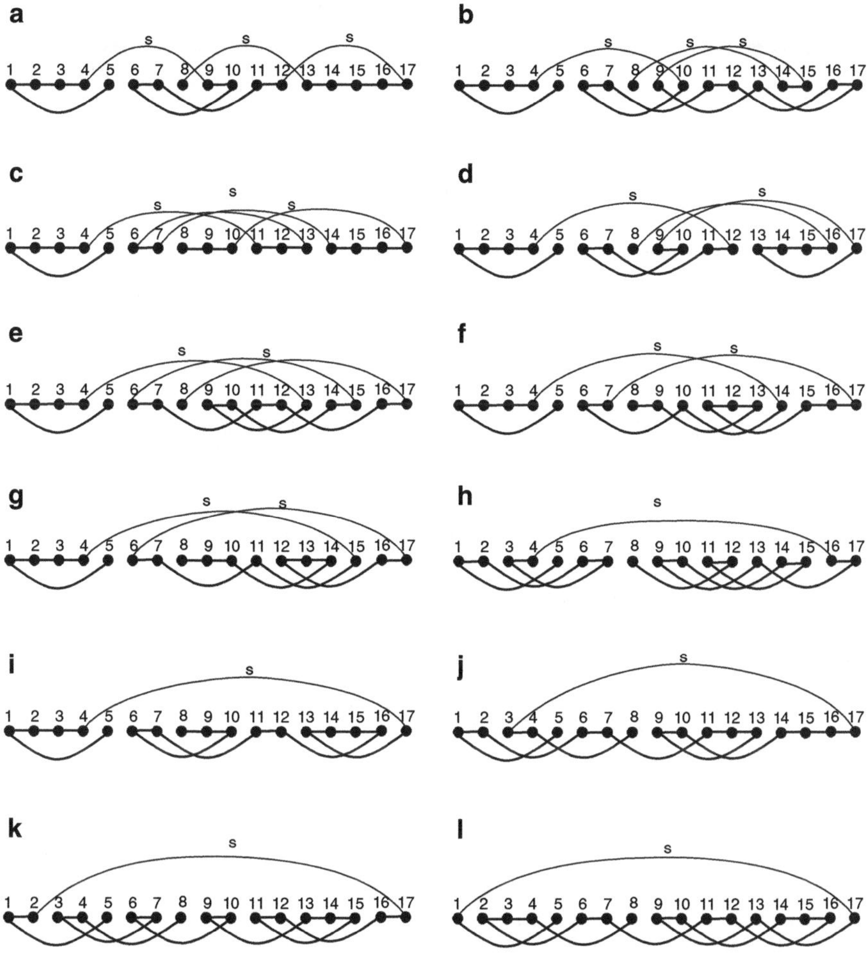

Fig. 39 Hamiltonian paths between 5 and 8 for different values of s, when $n = 17$. (**a**) $s = 5$. (**b**) $s = 6$. (**c**) $s = 7$. (**d**) $s = 8$. (**e**) $s = 9$. (**f**) $s = 10$. (**g**) $s = 11$. (**h**) $s = 12$. (**i**) $s = 13$. (**j**) $s = 14$. (**k**) $s = 15$. (**l**) $s = 16$

Subcase (*ii*). This subcase is symmetrical to $x \in \{1,2,3,5\}$ and $y \geq 6$. It was treated inside of (i) except for the cases $y = n-4, n-2, n-1, n$.

To obtain a Hamiltonian path from $x \in \{1,2,3,5\}$ to $y \in \{n-4, n-2, n-1, n\}$, we first collect the four Hamiltonian paths in $T_8\langle 1,4\rangle$ from $x \in \{1,2,3,5\}$ to 8; see Fig. 40. Symmetrically, we have paths in $T_n\langle 1,4\rangle$ from $y \in \{n-4, n-2, n-1, n\}$ to $n-7$, of vertex set $\{n-7, n-6, \ldots, n\}$. Joining 8 to $n-7$ by the direct path (8,9,...,$n-7$) gives the desired Hamiltonian path in $T_n\langle 1,4\rangle$ from x to y.

Subcase (*iii*). This subcase is symmetrical with $y \leq 5$, treated inside of (i).

□

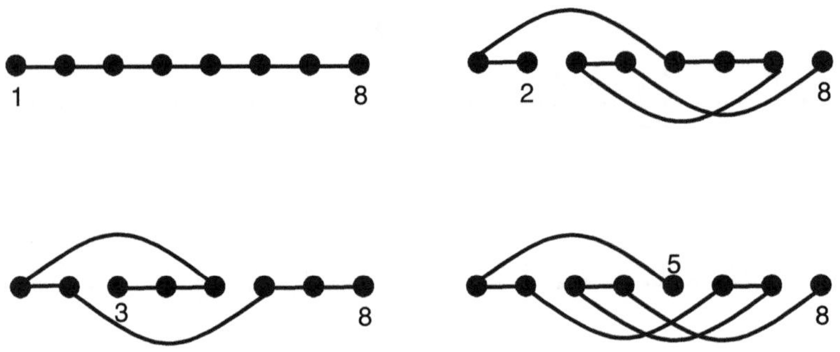

Fig. 40

To see whether $T_n\langle 1, 4, s\rangle$ is Hamiltonian connected or not, for $6 \leq n \leq 14$, see the following table:

	Hamiltonian connected when s is
$T_6\langle 1, 4, s\rangle$	
$T_7\langle 1, 4, s\rangle$	
$T_8\langle 1, 4, s\rangle$	5, 7
$T_9\langle 1, 4, s\rangle$	5, 8
$T_{10}\langle 1, 4, s\rangle$	5, 6, 7, 9
$T_{11}\langle 1, 4, s\rangle$	5, 7, 8, 10
$T_{12}\langle 1, 4, s\rangle$	5, 6, 7, 8, 9, 11
$T_{13}\langle 1, 4, s\rangle$	for all s
$T_{14}\langle 1, 4, s\rangle$	5, 6, 7, 8, 9, 10, 11, 13

Missing values for s mean that the corresponding Toeplitz graph is not Hamiltonian connected. This was verified by using a computer.

References

1. van Dal, R., Tijssen, G., Tuza, Z., van der Veen, J.A.A., Zamfirescu, Ch., Zamfirescu, T.: Hamiltonian properties of Toeplitz graphs. Discret. Math. **159**, 69–81 (1996)
2. Euler, R.: Characterizing bipartite Toeplitz graphs. Theor. Comput. Sci. **263**, 47–58 (2001)
3. Euler, R., LeVerge, H., Zamfirescu, T.: A characterization of infinite, bipartite Toeplitz graphs. In: Tung-Hsin, K. (ed.) Combinatorics and Graph Theory 95, Vol. 1. Academia Sinica, pp. 119–130. World Scientific, Singapore (1995)
4. Euler, R., Zamfirescu, T.: On planar Toeplitz graphs. Graphs Comb. **29**, 1311–1327 (2013)
5. Heuberger, C.: On Hamiltonian Toeplitz graphs. Discret. Math. **245**, 107–125 (2002)
6. Malik, S.: Hamiltonian cycles in directed Toeplitz graphs II. Ars Comb. (to appear)

7. Malik, S.: Hamiltonicity in directed Toeplitz graphs of maximum (out or in) degree 4. Util. Math. **89**, 33–68 (2012)
8. Malik, S., Qureshi, A.M.: Hamiltonian cycles in directed Toeplitz graphs. Ars Comb. **109**, 511–526 (2013)
9. Malik, S., Zamfirescu, T.: Hamiltonian connectedness in directed Toeplitz graphs. Bull. Math. Soc. Sci. Math. Roum. **53**(101) No. 2, 145–156 (2010)

Discriminants, Polytopes, and Toric Geometry

Ragni Piene

1 Elimination Theory

In the meeting of the Bourbaki group in 1969, it was decided to abandon Demazure's appendix to the volume Algèbre concerning resultants and discriminants:

Il faut éliminer la théorie de l'élimination.

J. Dieudonné (1969)

As a reaction to this decision, Abhyankar wrote a poem [1] at the International Congress of Mathematicians in Nice in 1970:

Eliminate, eliminate, eliminate
Eliminate the eliminators of elimination theory.

S.S. Abhyankar (1970)

Urged by Jean-Pierre Serre and supported by Pierre Cartier, Demazure finally published his Bourbaki manuscript in 2011, as a tribute to Jean-Pierre Serre "pour son 85-ième anniversaire" [6]. By then, it was of course obvious that the theory of elimination had not been eliminated! On the contrary, with the development of

This paper is a written version of the talk I gave at the 6th World Conference on 21st Century Mathematics at ASSMS in Lahore in March 2013. I would like to thank the organizers, especially Professor A. D. R. Chouadry, as well as the staff and students of ASSMS. Thanks are also due to my "toric collaborators" Alicia Dickenstein and Sandra Di Rocco.

R. Piene (✉)
CMA/Department of mathematics, University of Oslo, P.O. Box 1053 Blindern,
NO-0316 Oslo, Norway
e-mail: ragnip@math.uio.no

P. Cartier et al. (eds.), *Mathematics in the 21st Century*, Springer Proceedings in Mathematics & Statistics 98, DOI 10.1007/978-3-0348-0859-0_9

computers and computer algebra programs that made it easier to actually perform computations, elimination theory came back in fashion.

2 Resultants

Consider two polynomials in one variable

$$f(x) = a_m x^m + \cdots + a_0 \text{ and } g(x) = b_n x^n + \cdots + b_0$$

of degrees m and n, with $n \geq m$.

One wants to know which condition(s) the coefficients $a_0, \ldots, a_m, b_0, \ldots, b_n$ must satisfy in order for $f(x)$ and $g(x)$ to have a common root. One way of answering this problem is to consider the condition $\prod(\alpha_i - \beta_j) = 0$, where the α_i and β_j are the roots of f and g, respectively. This product is in fact a polynomial in $a_0, \ldots, a_m, b_0, \ldots, b_n$ and is an "eliminant" of f and g: the variable x has been eliminated.

Using linear algebra, it is not hard to see that one also gets an eliminant by considering the *Sylvester matrix*, which is the $(m+n) \times (m+n)$-matrix

$$\begin{pmatrix} a_m & a_{m-1} & a_{m-2} & \cdots & \cdots \\ 0 & a_m & a_{m-1} & a_{m-2} & \cdots \\ \vdots & & & \vdots & \\ b_n & b_{n-1} & b_{n-2} & \cdots & \cdots \\ 0 & b_n & b_{n-1} & b_{n-2} & \cdots \\ \vdots & & & \vdots & \end{pmatrix}$$

The determinant of this matrix is called the *resultant* of f and g and is denoted $\mathrm{Res}(f, g)$.

This matrix was studied by Sylvester and by Cayley. They also studied a smaller matrix called the Bézout matrix, obtained as follows. Consider the polynomial in two variables x, y:

$$\sum c_{ij} x^i y^j := \frac{f(x)g(y) - f(y)g(x)}{x - y}.$$

Recall that we assumed $n \geq m$. By writing $\sum_{i,j} c_{ij} x^i y^j = \sum_i p_i(y) x^i$, where $p_i(y) = \sum_j c_{ij} y^j$, we see that if f and g have a common root ξ, then $p_i(\xi) = 0$ for $i = 0, 1, \ldots, n-1$. So a necessary condition for f and g to have a common root is that the determinant of the $n \times n$ Bézout matrix

$$B(f, g) := (c_{ij})$$

is 0. (For a proof of the sufficiency, see, e.g., [13].)

More generally, if $f_0(x_1, \ldots, x_m), \ldots, f_k(x_1, \ldots, x_m)$ are $k+1$ polynomials in m variables, by successive pairwise elimination of the variables (taking a sequence of resultants), one obtains a set of polynomials in the coefficients of the f_i's. The ideal in the coefficient ring generated by this set will be called the resultant of the given polynomials and denoted by $\mathrm{Res}(f_0, \ldots, f_k)$.

The geometric interpretation of elimination is found in the concept of linear projections of projective varieties. Assume $X \subset \mathbb{P}^m$ is a projective variety defined by $X = Z(F_0, \ldots, F_k)$, where the $F_i(x_0, \ldots, x_r, x_{r+1}, \ldots, x_m)$ are homogeneous polynomials. Set $L := Z(x_{r+1}, \ldots, x_m)$. Then the projection $\overline{X} \subset \mathbb{P}^r$ of X from the linear space L is given by $\overline{X} = Z(\mathrm{Res}(F_0, \ldots, F_k))$, where now $\mathrm{Res}(F_0, \ldots, F_k)$ is the ideal obtained by eliminating the variables $x_{r+1}, \ldots, x_m$ from the F_i's.

3 Discriminants

The *discriminant* of a polynomial $f(x)$ in one variable is by definition the resultant $\mathrm{Res}(f, f_x)$ of f and its first derivative $f_x := \partial f/\partial x$. Thus, the discriminant is a polynomial in the coefficients of f. If the coefficients of f are such that the discriminant is zero, then the polynomial has at least one multiple root. The discriminant of a polynomial $f(x_1, \ldots, x_m)$ in several variables is the resultant of f and all its partial derivatives $f_{x_1}, \ldots, f_{x_m}$. Since we are eliminating m variables from $m+1$ equations, we expect that the discriminant $\Delta(f) := \mathrm{Res}(f, f_{x_1}, \ldots, f_{x_m})$ is one polynomial. This is in fact the general case, but need not happen always.

Example 1. Consider $f(x_1, x_2, x_3) = a_0 + a_1x_1 + a_2x_2 + a_3x_3 + a_4x_1x_2 + a_5x_1x_3$. Then

$$\Delta(f) = \mathrm{Res}(f, f_{x_1}, f_{x_2}, f_{x_3}) = (a_0a_4 - a_1a_2, a_0a_5 - a_1a_3, a_2a_5 - a_3a_4).$$

4 The Cayley Trick

Consider two polynomials $f(x)$ and $g(x)$ in one variable. Introduce a new variable y and set

$$h(x, y) := f(x) + yg(x).$$

We want to compare $\mathrm{Res}(f, g)$ and the discriminant $\Delta(h)$. By definition, $\Delta(h) = \mathrm{Res}(h, h_x, h_y)$. Now we have $h_x = f_x + yg_x$ and $h_y = g$. So the set of equations $h = h_x = h_y = 0$ is equivalent to $f = g = f_x + yg_x = 0$.

Assume ξ is a common root of f and g. Then $h(\xi, y) = 0$ and $h_y(\xi, y) = g(\xi) = 0$ for all y. By taking $\eta := -\frac{f_x(\xi)}{g_x(\xi)}$, we see that (ξ, η) is a common zero of h,

h_x, and h_y (at least provided $g_x(\xi) \neq 0$, i.e., ξ is a simple root of g). Conversely, if (ξ, η) is a common zero of h, h_x, and h_y, then clearly ξ is a common root of f and g. So we see that $\mathrm{Res}(f, g)$ and $\Delta(h)$ are "equivalent." This observation was called the *Cayley Trick* in [11, 2.5, p. 260].

More generally, consider $k + 1$ polynomials in m variables $f_i(x_1, \ldots, x_m)$, $i = 0, \ldots, k$, and set

$$h(x_1, \ldots, x_m, y_1, \ldots, y_k) := f_0(x_1, \ldots, x_m) + y_1 f_1(x_1, \ldots, x_m) + y_k f_k(x_1, \ldots, x_m).$$

The resultant ideal $\mathrm{Res}(f_0, \ldots, f_k)$ is obtained by eliminating (if possible) the variables $x_1, \ldots, x_m$ from the f_j's. The discriminant of h is obtained by eliminating the x_i's and y_j's from the $m + k + 1$ polynomials $h, h_{x_1}, \ldots, h_{x_m}, h_{y_1}, \ldots, h_{y_k}$. As in the case when $k = 1$ and $m = 1$, we see that $\Delta(h)$ and $\mathrm{Res}(f_0, \ldots, f_k)$ are equivalent. Hence, the Cayley Trick works also in this case (see also [16, Ex. 4.4, p. 204]).

5 Cayley Polytopes

Let $P_0, \ldots, P_k \subset \mathbb{R}^{n-k}$ be convex lattice polytopes, and let $e_0, \ldots, e_k$ denote the vertices of the unit k-simplex $\Delta_k \subset \mathbb{R}^k$.

Any polytope affinely equivalent to the polytope

$$P = \mathrm{Conv}\{e_0 \times P_0, \ldots, e_k \times P_k\} \subset \mathbb{R}^k \times \mathbb{R}^{n-k} = \mathbb{R}^n,$$

is called a *Cayley polytope*. We write

$$P = P_0 \star \cdots \star P_k.$$

If at least one of the P_j has "full" dimension $n-k$, then the dimension of P is equal to n.

A Cayley polytope is "hollow": it has no interior lattice points. The hollowness is measured by the *codegree*. The codegree of P is defined as

$$\mathrm{codeg}(P) = \min\{m \mid mP \text{ has interior lattice points}\}.$$

The *degree* of P is defined as

$$\deg(P) = n + 1 - \mathrm{codeg}(P)$$

Example 2. Let Δ_n be the standard n-simplex. We have

$$\mathrm{codeg}(\Delta_n) = n + 1 \text{ and } \mathrm{codeg}(2\Delta_n) = \left\lceil \frac{n+1}{2} \right\rceil.$$

Example 3. Let

$$P = P_0 \star \cdots \star P_k$$

be a Cayley polytope. Then

$$\operatorname{codeg}(P) \geq k + 1.$$

In [2], Batyrev and Nill asked the following question, often referred to as the Cayley polytope conjecture: does there exist an integer $N(d)$ such that any polytope P of degree d and $\dim P \geq N(d)$ is a Cayley polytope?

An answer was given by Haase, Nill, and Payne [14]: for general polytopes, there is such an integer $N(d)$, and $N(d) \leq (d^2 + 19d - 4)/2$. It was suspected that this was not the best bound and that $N(d)$ is linear in d. Indeed, we showed in [10] that $N(d) = 2d + 1$ for a smooth polytope (with an additional assumption). Observe that $n \geq 2d + 1$ is equivalent to $\operatorname{codeg}(P) \geq \frac{n+3}{2}$.

Theorem 1 ([7, 9]). *Let P be a smooth lattice polytope of dimension n. The following are equivalent*

(1) $\operatorname{codeg}(P) \geq \frac{n+3}{2}$
(2) $P = P_0 \star \cdots \star P_k$ *is a smooth Cayley polytope with* $k + 1 = \operatorname{codeg}(P)$ *and* $k > \frac{n}{2}$.
(3) P *is defective, with defect* $2k - n > 0$.

In [10] we proved this theorem under an additional assumption in (1), namely, that P is what we called $\mathbb{Q}$-normal (meaning that the tau value of P, $\tau(P)$, is equal to the $\mathbb{Q}$-codegree $\operatorname{codeg}_{\mathbb{Q}}(P)$). This assumption was later shown to be unnecessary [7]. The proof in [10] is essentially algebro-geometric, based on adjunction theory and using nef-value maps (cf. [3]), as well as the theory of toric fibrations (cf. [22]). The proof in [7] that $\operatorname{codeg}(P) \geq \frac{n+3}{2}$ implies $\tau(P) = \operatorname{codeg}_{\mathbb{Q}}(P)$ is purely combinatorial and uses Ehrhart polynomials and Ehrhart reciprocity.

6 Lattice Polytopes and Toric Embeddings

A convex lattice polytope $P \subset \mathbb{Z}^n$ gives a toric embedding $(\mathbb{C}^*)^n \to \mathbb{P}^N$, where $N = \#(P \cap \mathbb{Z}^n) - 1$. The map is defined by sending a point $\mathbf{x} = (x_1, \ldots, x_n)$ to $(\ldots, \mathbf{x}^{\mathbf{a}}, \ldots)$, where $\mathbf{a} \in P \cap \mathbb{Z}^n$.

Example 4. The polytope P_0:

corresponds to the toric embedding $\mathbb{C}^* \to \mathbb{P}^2$ given by $x \mapsto (1 : x : x^2)$; its closure X_{P_0} is a conic.

The polytope P_1:

corresponds the toric embedding $\mathbb{C}^* \to \mathbb{P}^3$ given by $x \mapsto (1 : x : x^2 : x^3)$; its closure X_{P_1} is a twisted cubic curve.

The Cayley sum $P = P_0 \star P_1$:

corresponds to the embedding

$$(\mathbb{C}^*)^2 \to \mathbb{P}^6$$

given by

$$(x, y) \mapsto (1 : x : x^2 : y : xy : x^2y : x^3y);$$

its closure X_P is a rational normal scroll of type $(2, 3)$.

The toric variety associated to a Cayley polytope $P = P_0 \star \cdots \star P_k$ is a *ruled variety*, in fact a toric fibration [5, Section 3] if the polytopes P_i are strictly combinatorially equivalent [10, Section 3]. Let Y be the (toric) base of the fibration and $\pi: X_P \to Y$ the fibration, with fibers equal to $\mathbb{P}^k$'s linearly embedded. In particular, X_P is covered by k-planes. More generally, a polytope P is an $(k + 1)$-Cayley polytope if and only if the associated toric variety X_P is covered by k-planes [15, Thm. 1.1].

7 Hyperplane Sections and Discriminants

Let P be a polytope and $X_P \subset \mathbb{P}^N$ the corresponding toric variety. The *dual variety* $X_P^\vee \subset (\mathbb{P}^N)^\vee$ is the closure of the set of hyperplanes $H \in (\mathbb{P}^N)^\vee$ that are tangent to X_P at a smooth point. A hyperplane $H \subset \mathbb{P}^N$ is tangent to X_P if the hyperplane section $X_P \cap H$ is singular. If the hyperplane section is given by $f(x_1, \ldots, x_n) = 0$, then it is singular if the coefficients satisfy the relations given by the discriminant $\Delta(f)$. If the discriminant is one equation, then this equation defines the dual variety, which is a hypersurface. If there are more than one equation, then the dual variety has smaller dimension. A variety is called *defective* if its dual variety is not a hypersurface. A polytope P is defective if X_P is defective. The *defect* of a defective variety X is the positive integer $\operatorname{codim} X^\vee - 1$.

Let $P_0, \ldots, P_k \subset \mathbb{R}^m$ be polytopes and $X_{P_i} \subset \mathbb{P}^{N_i}$ the corresponding toric varieties. The Cayley polytope $P = P_0 \star \cdots \star P_k \subset \mathbb{R}^n$, where $n = m + k$, gives a toric embedding $X_P \subseteq \mathbb{P}^N$, where $N = \sum(N_i + 1) - 1$.

A hyperplane section of X_P is given by

$$h(x_1, \ldots, x_m, y_1, \ldots, y_k) := f_0(x_1, \ldots, x_m) + y_1 f_1(x_1, \ldots, x_m) + \cdots + y_k f_k(x_1, \ldots, x_m) = 0,$$

where $f_i(x_1, \ldots, x_m) = 0$ is a hyperplane section of X_{P_i}.

The hyperplane section is singular if the discriminant $\Delta(h)$ of h is zero, i.e., if $h = \partial h/\partial x_i = \partial h/\partial y_j = 0$. By the Cayley Trick, this condition is the same as saying that $\operatorname{Res}(f_0(x_1, \ldots, x_m), \ldots, f_k(x_1, \ldots, x_m))$ is zero. The geometric interpretation of the Cayley Trick is simply that a hyperplane is tangent to X_P if and only if it contains a ruling.

If $k \leq n - k$, then $\Delta(h)$ is a polynomial in the coefficients of h and defines a hypersurface in the dual space, the dual variety $X_P^\vee \subseteq (\mathbb{P}^N)^\vee$ of X_P.

If $k > n - k$, the system $f_0 = \cdots = f_k = 0$ has more equations than variables. Hence, "the discriminant" of h consists of more than one polynomial, and the dual variety is not a hypersurface.

We conclude: If $k > n - k$, then the Cayley polytope $P = P_0 \star \cdots \star P_k$ is defective.

Example 5. Consider the Cayley polytope $P = P_0 \star P_1 \star P_2$, where $P_0 = P_1 = P_2 = \Delta_1$. Here $n = 3$, and $k = 2$, and P is defective, with defect 1. Indeed, the variety X_P is the Segre embedding of $\mathbb{P}^2 \times \mathbb{P}^1$ in $\mathbb{P}^5$, which is well known to be defective (see, e.g., [23]). Moreover, a hyperplane section of X_P (in the affine torus) is given as the zero set of the polynomial of Example 1.

In [15, Cor. 4.2], it is shown that if X_P has dual defect $r > 0$, then P is a Cayley polytope of the form $P = P_0 \star \cdots \star P_r$. Note that the converse is not true: it is well known that a smooth surface is never defective. In particular, a smooth Cayley polytope $P = P_0 \star P_1$ of dimension 2 is not defective.

8 The Degree of the Dual Variety

Gelfand, Kapranov, and Zelevinski showed [12, Thm. 2.8, Ch. 9] that if X_P is smooth and not defective, then there is a combinatorial formula for the degree of the dual variety:

$$\deg X_P^\vee = \sum_{F \subseteq P} (-1)^{\operatorname{codim} F} (\dim F + 1) \operatorname{Vol}_{\mathbb{Z}}(F),$$

where the sum is taken over all faces F of P and $\operatorname{Vol}_{\mathbb{Z}}$ denotes the lattice volume.

The formula follows from the fact that the degree of the dual variety of X_P is a polynomial in the Chern classes of X_P and the line bundle L_P giving the toric embedding; moreover, the Chern numbers of X_P and L_P can be expressed combinatorially in terms of the polytope P. For example,

$$\deg X_P = c_1(L_P)^n = \mathrm{Vol}_{\mathbb{Z}}(P),$$
$$c_n(T_{X_P}) = \#\text{ vertices of } P,$$

and

$$c_i(T_{X_P})c_1(L_P)^{n-i} = \sum_{\operatorname{codim} F_i = i} \mathrm{Vol}_{\mathbb{Z}}(F_i),$$

where the sum is taken over all faces of P of codimension i.

The degree of the dual variety is called the *class* of the variety, or the 0th *rank*. More generally, the ith *rank* r_i of a projective variety is a projective invariant that can be defined as the class of the subvariety obtained as the intersection of the variety with i hyperplanes (see, e.g., [20] and [17]). If $X \subset \mathbb{P}^N$ is smooth, then the ith rank is the degree of the $(n-i)$th Chern class of the bundle of principal parts (or jet bundle) of the tautological line bundle on X, namely, $\mathscr{P}^1_X(1)$. Hence, the ranks can be computed in terms of the Chern numbers of $(X, \mathscr{O}_X(1))$ from the exact sequence

$$0 \to \Omega^1_X \otimes \mathscr{O}_X(1) \to \mathscr{P}^1_X(1) \to \mathscr{O}_X(1) \to 0.$$

In the toric case, if X_P is smooth, then the ranks of $X_P^\vee$ are given by

$$r_i = c_{n-i}\left(\mathscr{P}^1_{X_P}(L_P)\right) c_1(L_P)^i$$

and hence have a combinatorial expression. If X_P is not defective, then $\deg X_P^\vee = r_0$. If X_P is defective with defect δ, then $r_0 = \cdots = r_{\delta-1} = 0$, and $\deg X_P^\vee = r_\delta$ — and vice versa.

Example 6. With P as in Example 5, we get $r_0 = 0$ and

$$r_1 = c_2\left(\mathscr{P}^1_{X_P}(L_P)\right) c_1(L_P) = 6\,\mathrm{Vol}_{\mathbb{Z}}(P) - 3\sum_F \mathrm{Vol}_{\mathbb{Z}}(F) + \sum_E \mathrm{Vol}_{\mathbb{Z}}(E),$$

where the first sum is over the facets of P and the second over the edges. Hence, we can compute

$$r_1 = 6 \cdot 3 - 3 \cdot 8 + 9 = 3.$$

9 Higher-Order Dual Varieties

The *kth dual variety* $X^{(k)}$ of a projective variety $X \subset \mathbb{P}^m$ is defined as

$$X^{(k)} = \overline{\{H \in \mathbb{P}^{m\vee} \mid H \text{ is tangent to } X \text{ to the order } k\}}$$
$$= \overline{\{H \in \mathbb{P}^{m\vee} \mid H \supseteq \mathbb{T}^k_{X,x} \text{ for some } x \in X_{\text{smooth}}\}},$$

where $\mathbb{T}^k_{X,x}$ denotes the kth osculating space to X at x (see [10, 21]).

Clearly $X^{(1)} = X^\vee$ and $X^{(k-1)} \supseteq X^{(k)}$, and one can also see that $X^{(k)} \subseteq \text{Sing}(X^\vee)$ for $k \geq 2$.

The dimension of $\mathbb{T}^k_{X,x}$ is $\leq \binom{n+k}{k} - 1$, where $n = \dim X$. If equality holds for (almost) all x, we say that X is (generically) k-regular. Therefore, the expected dimension of $X^{(k)}$ is $n + m - \binom{n+k}{k}$. We say that X is k*-defective* if $\dim X^{(k)} < n + m - \binom{n+k}{k}$.

For toric threefolds, we have the following result for the case $k = 2$.

Theorem 2 ([10]). *Let $(X, L) = (X_P, L_P)$ be a smooth, 2-regular toric threefold embedding corresponding to a lattice polytope P. Then X is 2-defective if and only if $(X, L) = (\mathbb{P}^3, \mathscr{O}(2))$. Moreover:*

(1) $\deg X^{(2)} = 120$ *if* $(X, L) = (\mathbb{P}^3, \mathscr{O}_{\mathbb{P}^3}(3))$
(2) $\deg X^{(2)} = 6(8(a+b+c)-7)$ *if* $(X, L) = (\mathbb{P}(\mathscr{O}_{\mathbb{P}^1}(a) \oplus \mathscr{O}_{\mathbb{P}^1}(b) \oplus \mathscr{O}_{\mathbb{P}^1}(c)), 2\xi)$, *where ξ denotes the tautological line bundle,*
(3) In all other cases,

$$\deg X^{(2)} = 62V - 57F + 28E - 8v + 58V_1 + 51F_1 + 20E_1,$$

where V, F, E (resp. V_1, F_1, E_1) denote the (lattice) volume, area of facets, length of edges of P (resp. the adjoint polytope Conv(interior P)*), and v = #{vertices of P}.*

Example 7. Let P be a cube with edge lengths 2. The corresponding toric variety is $(X_P, L_P) = (\mathbb{P}^1 \times \mathbb{P}^1 \times \mathbb{P}^1, \mathscr{O}(2, 2, 2))$. In this case we have $V = 3!8 = 48$, $F = 6 \cdot 2 \cdot 4 = 48$, $E = 12 \cdot 2 = 24$, $v = 8$, and $V_1 = F_1 = E_1 = 0$, since interior $P = \{(1, 1, 1)\}$ is a point. Hence,

$$\deg X^{(2)} = 62V - 57F + 28E - 8v = 848.$$

10 k-Self-Dual Toric Varieties

A lattice point configurations is a finite set $\mathscr{A} = \{a_0, \ldots, a_N\} \subset \mathbb{Z}^n$. Its convex hull $P = \text{Conv}(\mathscr{A})$ is a lattice polytope, and we can consider $\mathscr{A}$ as a subset of the lattice points $P \cap \mathbb{Z}^n$ in P. The configuration $\mathscr{A}$ defines a toric embedding $X_{\mathscr{A}} \subset \mathbb{P}^N$, and we can consider $X_{\mathscr{A}}$ as a linear (toric) projection of X_P.

Let $\mathscr{A} = \{a_0, \ldots, a_N\} \subset \mathbb{Z}^n$ be a lattice point configuration and $X_{\mathscr{A}} \subset \mathbb{P}^N$ the corresponding toric embedding.

Form the matrix A by adding a row of 1s to the matrix $(a_0|\cdots|a_N)$. Denote by $\mathbf{v}_0 = (1, \ldots, 1), \mathbf{v}_1, \ldots, \mathbf{v}_n \in \mathbb{Z}^{N+1}$ the row vectors of A.

For any $\alpha \in \mathbb{N}^{n+1}$, denote by $\mathbf{v}_\alpha \in \mathbb{Z}^{N+1}$ the vector obtained as the coordinatewise product of α_0 times the row vector $\mathbf{v}_0$ times ... times α_n times the row vector $\mathbf{v}_n$.

For any k, we define the matrix $A^{(k)}$ as follows. Order the vectors $\{\mathbf{v}_\alpha : |\alpha| \leq k\}$ (for instance, by degree and then by lexicographic order with $0 > 1 > \ldots > n$), and let $A^{(k)}$ be the $\binom{n+k}{k} \times (N+1)$ integer matrix with these rows.

The configuration $\mathscr{A}$ is *non-pyramidal* (nap) if the configuration of columns in A is not a pyramid (i.e., no basis vector e_i of $\mathbb{R}^{N+1}$ lies in the rowspan of the matrix).

The configuration $\mathscr{A}$ is $k$nap if the configuration of columns in $A^{(k)}$ is not a pyramid.

Example 8. The following matrix is a pyramid:

$$A = \begin{pmatrix} 1 & 1 & 1 & 1 & 1 \\ 0 & 1 & 2 & 3 & 4 \\ 0 & 0 & 1 & 0 & 0 \end{pmatrix}$$

Note that this matrix corresponds to a cone over a quartic rational normal curve.

We say that $X_{\mathscr{A}}$ is *k-self-dual* if there exists a linear isomorphism $\phi: \mathbb{P}^N \cong (\mathbb{P}^N)^\vee$ such that $\phi(X_{\mathscr{A}}) = X_{\mathscr{A}}^{(k)}$. The following theorem, and its proof, generalizes [4].

Theorem 3 ([8]). *Let $\mathscr{A}$ be a lattice configuration.*

(1) $X_{\mathscr{A}}$ is k-self-dual if and only if $\dim X_{\mathscr{A}} = \dim X_{\mathscr{A}}^{(k)}$ *and $\mathscr{A}$ is $k$nap.*
(2) If $\mathscr{A}$ is $k$nap and $\dim \operatorname{Ker} A^{(k)} = 1$, *then $X_{\mathscr{A}}$ is k-self-dual.*
(3) If $\mathscr{A}$ is $k$nap and k-self-dual, and $\dim \operatorname{Ker} A^{(k)} = r > 1$, *then $\mathscr{A}$ is r-Cayley.*

Example 9. The lattice configuration

$$\mathscr{A} = \{(0,0), (1,0), (1,1), (0,2)\}$$

gives a toric variety parameterized as

$$X_{\mathscr{A}}: (x, y) \mapsto \left(1 : x : xy : y^2\right).$$

Its dual variety is given by the parameterization

$$(X_{\mathscr{A}})^\vee: (x, y) \mapsto \left(-y^2 : 2x^{-1}y^2 : -2x^{-1}y : 1\right),$$

We see that $(X_{\mathscr{A}})^\vee \cong X_{\mathscr{A}^\vee}$, where

$$\mathscr{A}^\vee = \{(0,2), (-1,2), (-1,1), (0,0)\}.$$

The dual variety is projectively equivalent to the original variety; hence, the variety is self-dual.

The corresponding polytopes are equivalent, but not equal:

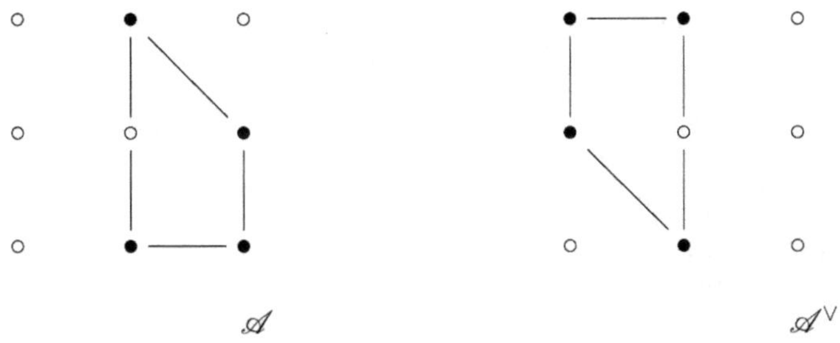

In his B.A. Thesis [18], Perkinson's student Mulliken gave a definition of (higher) self-duality which is stronger than ours: he required that the toric embedding giving the kth dual variety should be *equal* to the original toric embedding. Therefore, only *centrally symmetric* lattice configurations can possibly give self-dual varieties. In his terminology, the previous example is not self-dual.

Example 10. Consider the configuration $\mathscr{A} := P \cap \mathbb{Z}^3$, where P is the Cayley polytope of Example 5. Then

$$A = \begin{pmatrix} 1 & 1 & 1 & 1 & 1 & 1 \\ 0 & 1 & 0 & 1 & 0 & 1 \\ 0 & 0 & 1 & 1 & 0 & 0 \\ 0 & 0 & 0 & 0 & 1 & 1 \end{pmatrix}$$

We see that A is not a pyramid, and Ker A has rank 1. So $X_{\mathscr{A}}$ is self-dual (cf. [23]).

As observed by Perkinson [19], any nontrivial linear relation between the rows of $A^{(k)}$ corresponds to a polynomial of degree k that vanish on $\mathscr{A}$. One can use this fact to construct examples of k-self-dual varieties, and the method gives an interesting connection to diophantine geometry by considering polynomials with many integer solutions.

Example 11. Consider three quadrics $Q_1, Q_2, Q_3 \in \mathbb{Z}[x_1, x_2, x_3]$ with

$$Q_1 \cap Q_2 \cap Q_3 = \{a_0, \dots, a_7\} \subset \mathbb{Z}^3 \subset \mathbb{R}^3.$$

Then $\mathscr{A} = \{a_0 \dots, a_7\}$ gives a 2-self-dual threefold. This follows from Theorem 3. Indeed, corresponding to the three quadrics, there are three nontrivial linear relations between the rows of the (10×8)-matrix $A^{(2)}$. So the rank of $A^{(2)}$ is $10 - 3 = 7$, which is one less than the maximal rank. Therefore its kernel has rank 1.

References

1. Abhyankar, S.S.: Polynomials and Power Series. Mathematical Intelligencer, vol. 3 (1972). Reprinted in: Christensen, C., Sathaye, A., Sundaram, G., Bajaj, C. (eds.): Algebra, Arithmetic and Geometry with Applications, pp. 783–784. Springer, Berlin/Heidelberg (2004)
2. Batyrev, V., Nill, B.: Multiples of lattice polytopes without interior lattice points. Moscow Math. J. **7**(2), 195–207 (2007)
3. Beltrametti, M.C., Sommese, A.J.: The Adjunction Theory of Complex Projective Varieties. de Gruyter Expositions in Mathematics, vol. 16. Walter de Gruyter, Berlin (1995)
4. Bourel, M., Dickenstein, A., Rittatore, A.: Self-dual projective toric varieties. J. London Math. Soc. **84**(2), 514–540 (2011)
5. Casagrande, C., Di Rocco, S.: Projective Q-factorial toric varieties covered by lines. Commun. Contemp. Math. **10**(3), 363–389 (2008)
6. Demazure, M: Résultant, discriminant. Enseign. Math. (2) **58**(3–4), 333–373 (2012)
7. Dickenstein, A., Nill, B.: A simple combinatorial criterion for projective toric manifolds with dual defect. Math. Res. Lett. **17**(3), 435–438 (2010)
8. Dickenstein, A., Di Rocco, S., Piene, R.: Higher order duality and toric embeddings. Ann. Inst. Fourier **64**, 1:375–400 (2014)
9. Dickenstein, A., Di Rocco, S., Piene, R.: Classifying regular lattice polytopes via toric fibrations. Adv. Math. **222**(1), 240–254 (2009)
10. Dickenstein, A., Di Rocco, S., Piene, R.: Higher order duality and toric embeddings. Ann. Inst. Fourier **64** (2014) (to appear). arXiv:1111.4641
11. Gel'fand, I.M., Kapranov, M.M., Zelevinsky, A.V.: Generalized Euler integrals and A-hypergeometric functions. Adv. Math. **84**, 255–271 (1990)
12. Gel'fand, I.M., Kapranov, M.M., Zelevinsky, A.V.: Discriminants, resultants, and multidimensional determinants. Birkhäuser, Boston (1994)
13. Griffiths, H.B.: Cayley's version of the resultant of two polynomials. Am. Math. Mon. **88**(5), 328–338 (1981)
14. Haase, C., Nill, B., Payne, S.: Cayley decompositions of lattice polytopes and upper bounds for h^*-polynomials. J. Reine Angew. Math. **637**, 207–216 (2009)
15. Ito, A.: Algebro-geometric characterization of Cayley polytopes. arXiv:1202.6663v1.
16. Kapranov, M.M., Sturmfels, B., Zelevinsky, A.V.: Chow polytopes and general resultants. Duke Math. J. **67**(1), 189–218 (1992)
17. Kleiman, S.L.: Tangency and Duality. CMS Conference Proceedings, vol. 6, pp. 163–225. American Mathematical Society, Providence (1986).
18. Mulliken, J.: Lotsa dots: self-dual affine monomial varieties. B.A. Thesis, Reed College (1995)
19. Perkinson, D.: Inflections of toric varieties. Mich. Math. J. **48**, 483–515 (2000)
20. Piene, R.: Polar classes of singular varieties. Ann. Sci. Écolle Norm. Sup. (4) **11**(2), 247–276 (1978)
21. Piene, R.: A note on higher order dual varieties, with an application to scrolls. In Singularities, Part 2, (Arcata, Calif., 1981), pp. 335–342. Proceedings of the Symposium on Pure Mathematics, vol. 40. American Mathematical Society, Providence (1983)
22. Reid, M.: Decomposition of toric morphisms. In: Artin, M., Tate, J. (eds.) Arithmetic and Geometry, vol. II. Progress in Mathematics, 36, pp. 395–418. Birkhäuser, Boston (1983)
23. Vallès, J.: Variétés de type Togliatti. C. R. Acad. Sci. Paris, Ser. I **343**, 411–414 (2006)

Some Classical Problems in Number Theory via the Theory of $K3$ Surfaces

Hironori Shiga

1 From the Elliptic Modular Function to the $K3$ Modular Function

The theory of the elliptic modular function plays an important role in many situations in number theory. The elliptic modular function is obtained as a one-to-one correspondence between the parameter space of the family of elliptic curves (given by the Weierstrass normal form) and its period domain (i.e., the complex upper half plane). The $K3$ surface is considered to be a two-dimensional counterpart of the elliptic curve. So, if we consider a family of algebraic $K3$ surfaces with some normal form, we can obtain its modular function. We call it a $K3$ modular function (see [18, 19], some mathematical physicists call it a mirror map for $K3$ surfaces).

In this article the author wishes to explain the notion of the $K3$ modular function and one explicit example with several applications to the classical problems in number theory.

Many of the results have already been obtained by the author. In the present paper we give an alternate viewpoint, and several new aspects are developed. The readers are expected to consult the references for the detailed arguments.

H. Shiga (✉)
Graduate School of Science, Chiba University, Inage-ku Yayoi-cho 1-33, Chiba 263-8522, Japan

Abdus Salam School of Mathematical Sciences, Lahore, Pakistan
e-mail: shiga@math.s.chiba-u.ac.jp

P. Cartier et al. (eds.), *Mathematics in the 21st Century*, Springer Proceedings in Mathematics & Statistics 98, DOI 10.1007/978-3-0348-0859-0_10

2 Review of the Elliptic Modular Functions

2.1 The Elliptic Modular Function $\lambda(\tau)$

Start from the elliptic curve $w^2 = z(z-1)(z-\lambda)$ with $\lambda \in \boldsymbol{C} - \{0,1\}$. Consider the elliptic integrals

$$\begin{cases} \eta_1(\lambda) = \int_{-\infty}^{0} \frac{dz}{\sqrt{z(z-1)(z-\lambda)}} \\ \eta_2(\lambda) = \int_{1}^{+\infty} \frac{dz}{\sqrt{z(z-1)(z-\lambda)}} \end{cases}. \tag{1}$$

For the case $\lambda \in \boldsymbol{H} = \{\mathrm{Im}\,\lambda > 0\}$, we define

$$\Phi(\lambda) = \frac{\eta_2(\lambda)}{\eta_1(\lambda)}.$$

We can choose the branch of $\Phi(\lambda)$ so that the image $F_1 = \Phi(\boldsymbol{H})$ becomes Fig. 1.

If we make an analytic continuation of $\Phi(\lambda)$ in $\boldsymbol{C} - \{0,1\}$, we obtain infinitely many Schwarz reflection images of F_1. They tesselate the upper half plane $\boldsymbol{H}$. The Schwarz reflections form the modular group

$$\Gamma(2) =< \begin{pmatrix} 1 & 2 \\ 0 & 1 \end{pmatrix}, \begin{pmatrix} 1 & 0 \\ 2 & 1 \end{pmatrix} >= \left\{ \begin{pmatrix} a & b \\ c & d \end{pmatrix} \in SL(2,\boldsymbol{Z}) : \begin{pmatrix} a & b \\ c & d \end{pmatrix} \equiv \begin{pmatrix} 1 & 0 \\ 0 & 1 \end{pmatrix} \mod 2 \right\}.$$

Hence, the inverse map of the multivalued map $\Phi(\lambda)$ becomes a single-valued function on $\boldsymbol{H}$ that is invariant under the action of $\Gamma(2)$. It is denoted by $\lambda(\tau)$.

Putting $\tilde{q} = \exp[\pi i \tau]$ we define the following Jacobi theta constants:

$$\vartheta \begin{bmatrix} 0 \\ 0 \end{bmatrix}(\tau) = \vartheta_{00}(\tau) = \sum_{n\in \boldsymbol{Z}} \tilde{q}^{n^2} = 1 + 2\tilde{q} + 2\tilde{q}^4 + 2\tilde{q}^9 + \cdots,$$

$$\vartheta \begin{bmatrix} 0 \\ 1 \end{bmatrix}(\tau) = \vartheta_{01}(\tau) = \sum_{n\in \boldsymbol{Z}} (-1)^n \tilde{q}^{n^2} = 1 - 2\tilde{q} + 2\tilde{q}^4 - 2\tilde{q}^9 + \cdots.$$

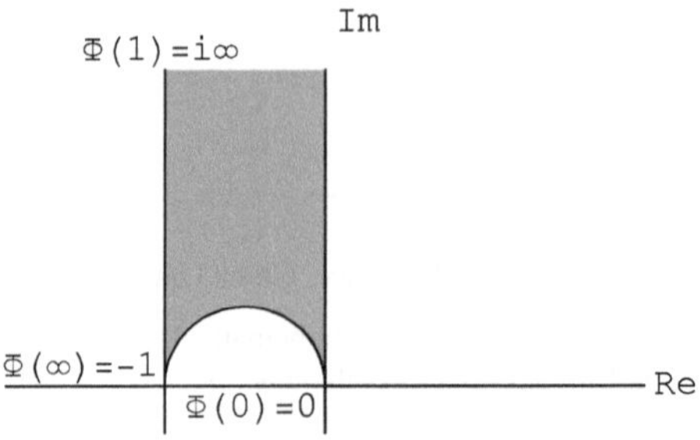

Fig. 1 The image F_1

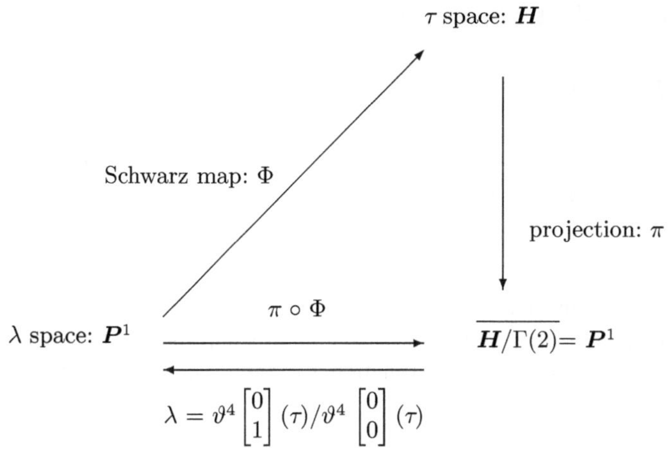

Diagram 1 Schwarz map Φ and its inverse

They are modular forms of weight 2 with respect to $\Gamma(2)$. Moreover we have the following expression.

Theorem 1 (Theta representation of $\lambda(\tau)$).

$$\lambda(\tau) = \frac{\vartheta \begin{bmatrix} 0 \\ 1 \end{bmatrix}^4 (\tau)}{\vartheta \begin{bmatrix} 0 \\ 0 \end{bmatrix}^4 (\tau)}. \tag{2}$$

The above story is summarized by Diagram 1.

2.2 *The Elliptic Modular Function $j(\tau)$*

The elliptic modular function $j(\tau)$ is defined by

Definition 1.

$$j(\tau) = 2^8 \frac{\left(1 - \lambda + \lambda^2\right)^3}{\lambda^2 (1 - \lambda)^2}. \tag{3}$$

The function $j(\tau)$ is invariant under the usual action of $SL(2, \boldsymbol{Z})$. It gives a bijective correspondence between the isomorphism classes of elliptic curves $y^2 = 4x^3 - g_2 x - g_3$ and $\boldsymbol{H}/SL(2, \boldsymbol{Z})$ via the invariant $j = 12^3 g_2^3 / \left(g_2^3 - 27 g_3^2\right)$.

3 Two Classical Theorems

3.1 Schneider's Theorem

Theorem 2 (Schneider, 1937). *Let us consider the elliptic modular function* $j(\tau)$ *under the condition* $\tau \in \overline{\boldsymbol{Q}}$ *(the field of algebraic numbers). Let* $y^2 = 4x^3 - g_2x - g_3$ *be an elliptic curve with* $j = j(\tau)$*. Suppose that* $j(\tau)$ *is an algebraic number. Then* τ *is an imaginary quadratic number.*

Remark 1. The converse of the above theorem is true also. That is a consequence of the fundamental theorem of the classical complex multiplication theory. So, we can imagine that the elliptic modular function is closely related to the number theory of imaginary quadratic fields.

Let us consider the elliptic curve $E = E(\tau) = \boldsymbol{C}/(\boldsymbol{Z} + \tau\boldsymbol{Z})$. The modulus τ is imaginary quadratic if and only if the extended endomorphism algebra $\mathrm{End}_0(E)$ is isomorphic to some imaginary quadratic field ($= \boldsymbol{Q}(\tau)$). It is equivalent to say $\mathrm{End}_0(E)$ has an element other than the trivial one in $\boldsymbol{Q}$. An element $\phi \in \mathrm{End}_0(E)$ is induced from a homomorphism $\tilde{\phi} : nE = \boldsymbol{C}/(n\boldsymbol{Z} + n\tau\boldsymbol{Z}) \to E$ for some positive integer n. So it is determined by a subalgebraic group (i.e., an algebraic subvariety with an abelian group structure) of $E \times E$. Set the Weierstrass $\wp$ function:

$$\wp(z) = \wp(z; 1, \tau) = \frac{1}{z^2} + \sum_{(m,n)\neq(0,0)} \left(\frac{1}{(z - (m\tau + n))^2} - \frac{1}{(m\tau + n)^2} \right).$$

The elliptic curve $E(\tau)$ has a realization $y^2 = 4x^3 - g_2x - g_3 = 4(x - e_1)(x - e_2)(x - e_3)$ by the embedding map

$$\rho : z \to (x, y) = \left(\wp(z), \wp'(z)\right).$$

Note that $E(\tau') = E(-1/\tau)$ is isomorphic to $E(\tau)$. So, if we have an algebraic relation between $\wp(z) = \wp(z; 1, \tau)$ and $\wp^*(z) = \wp(z; 1, -1/\tau)$ over $\overline{\boldsymbol{Q}}$, we can find an (extra) endomorphism of $\mathrm{End}_0(E)$. Observing that $\wp^*(z) = \wp(z; 1, -1/\tau) = \tau^2\wp(\tau z)$, the problem is reduced to the following theorem.

Theorem 3 (see Schneider, 1937, [17]). *Let* τ *be an algebraic number in* $\boldsymbol{H}$*. Suppose* $g_2(\tau), g_3(\tau), \wp\left(\frac{1}{2}\right)$*, and* $\wp^*\left(\frac{1}{2}\right) = \tau^2\wp\left(\frac{\tau}{2}\right)$ *are all algebraic numbers, then we have an algebraic relation between* $\wp(z) = \wp(z; 1, \tau)$ *and* $\wp^*(z) = \wp(z; 1, -1/\tau)$ *over* $\overline{\boldsymbol{Q}}$*.*

This theorem is great because we can deduce the functional equation of two functions $\wp$ and $\wp^*$ only from the algebraicity at one point. For the proof of this theorem, we need typical arguments of transcendental number theory which involve analytic methods. So we omit it (see [17] for detailed arguments).

Remark 2. The condition of the algebraicity of $g_2(\tau)$ and $g_3(\tau)$ is equivalent to saying $j(\tau)$ is algebraic (note that we are allowed to make a scale change of x, y coordinates). For the branch points of E, we have $e_1 = \wp\left(\frac{1}{2}\right), e_3 = \wp\left(\frac{\tau}{2}\right), and e_2 = \wp\left(-\frac{1+\tau}{2}\right)$. So the algebraicity of $\wp\left(\frac{1}{2}\right)$ and $\wp^*\left(\frac{1}{2}\right) = \tau^2\wp\left(\frac{\tau}{2}\right)$ is automatically satisfied. Then we have the following restatement of the above theorem.

[Restatement of Theorem 3]

Let τ be an algebraic number in $\boldsymbol{H}$. Suppose $j(\tau)$ is an algebraic number. Then we have an algebraic relation between $\wp(z) = \wp(z; 1, \tau)$ and $\wp^*(z) = \wp(z; 1, -1/\tau)$ over $\overline{\boldsymbol{Q}}$.

3.2 *The Gauss Arithmetic Geometric Mean Theorem*

Let a, b be a pair of positive numbers with $a \geq b > 0$.

Set $a_0 = a,\ b_0 = b$. And we define a_n, b_n recursively by

$$a_{n+1} = \frac{a_n + b_n}{2},\ b_{n+1} = \sqrt{a_n b_n}.$$

Then we have a common limit $\lim\limits_{n\to\infty} a_n = \lim\limits_{n\to\infty} b_n$. We denote this limit by $M(a,b)$.

The Gauss hypergeometric differential equation $E(a,b,c)$ is given by

$$x(1-x)y'' + (c - (a+b+1)x)y' - aby = 0.$$

(Here, the a and b in $E(a,b,c)$ should not be confused with those of $M(a,b)$.) It is a (complex linear) Fuchsian differential equation with regular singularities at $x = 0, 1, \infty$. The integrals (1) (as functions of λ) are fundamental solutions of $E\left(\frac{1}{2}, \frac{1}{2}, 1\right)$. The Gauss hypergeometric series

$$F\left(\frac{1}{2}, \frac{1}{2}, 1; x\right) = 1 + \sum_{n=1}^{\infty} \frac{\left(\left(\frac{1}{2}\right)\cdots\left(\frac{1}{2}+(n-1)\right)\right)^2}{(n!)^2} x^n \tag{4}$$

is the unique holomorphic solution at $x = 0$ up to a constant factor.

Theorem 4 (Gauss' AGM theorem, 1799).

$$\frac{1}{M(1,x)} = F\left(\frac{1}{2}, \frac{1}{2}, 1; 1-x^2\right) = \frac{1}{\pi}\int_1^{\infty} \frac{dz}{\sqrt{z(z-1)(z-(1-x^2))}}\quad (0<x<1). \tag{5}$$

We can see it in the "Tagesbuch" of Gauss ([7][98], ..., [102], 30 Mai, ..., 23 Dez, 1799) with exceptional comments of surprise.

We find the following strange formula in the work of Jacobi ([9] p. 235 = Fundamenta Nova section 65)

Theorem 5 (Jacobi's theta formula). *Under the relation (2) we have*

$$\vartheta_{00}^2(\tau) = F\left(\frac{1}{2}, \frac{1}{2}, 1; 1-\lambda\right) = \frac{1}{\pi}\int_1^\infty \frac{dz}{\sqrt{z(z-1)(z-(1-\lambda))}} \quad (0<\lambda<1). \tag{6}$$

Theorem 6 (Isogeny formula (see [6][8] p. 141)).

$$\begin{cases} \vartheta_{00}^2(2\tau) = \frac{1}{2}\left(\vartheta_{00}^2(\tau) + \vartheta_{01}^2(\tau)\right), \\ \vartheta_{01}^2(2\tau) = \vartheta_{00}(\tau)\vartheta_{01}(\tau). \end{cases} \tag{7}$$

We may summarize the story:

Theta representation + Jacobi's formula + isogeny formula = Gauss AGM theorem.

In fact, start from a given number x $(0<x<1)$. We may put $x = \vartheta_{01}^2(\tau)/\vartheta_{00}^2(\tau)$ for some $\tau \in \boldsymbol{H}$. Regarding that the isogeny formula realizes the AGM procedure, we have

$$\begin{aligned} \vartheta_{00}^2(\tau)M(1,x) &= M\left(\vartheta_{00}^2(\tau), \vartheta_{01}^2(\tau)\right) = \lim_{n\to\infty}\frac{1}{2}\left(\vartheta_{00}^2(2^n\tau) + \vartheta_{01}^2(2^n\tau)\right) \\ &= \vartheta_{00}^2(i\infty) = 1. \end{aligned}$$

Hence $\vartheta_{00}^2(\tau)M(1,x) = 1$. Due to Jacobi's formula, it is the AGM theorem.

4 The $K3$ Modular Function

4.1 $K3$ Surface

An algebraic $K3$ surface S is characterized as a simply connected algebraic surface with unique holomorphic 2-form φ (up to a constant factor). The second homology group $H_2(S,\boldsymbol{Z})$ is always a free $\boldsymbol{Z}$ module of rank 22 and isomorphic to $M_0 = E_8(-1) \oplus E_8(-1) \oplus U \oplus U \oplus U$ as a lattice via the naturally defined intersection form, where E_8 is the uniquely determined positive definite integral even unimodular lattice of rank 8, U is the hyperbolic lattice given by the matrix $\begin{pmatrix} 0 & 1 \\ 1 & 0 \end{pmatrix}$, and $E_8(-1)$ is the sign reversed lattice of E_8. The sublattice generated by the divisors is called the Neron-Severi lattice denoted by $\mathrm{NS}(S)$, and its orthogonal complement is said to be the transcendental lattice denoted by $\mathrm{Tr}(S)$. Let $\{\gamma_1, \ldots, \gamma_{22}\}$ be a basis of $H_2(S, \boldsymbol{Z})$ with intersection matrix M_0. The ratio of integrals

$$\eta = \left(\int_{\gamma_1} \varphi : \cdots : \int_{\gamma_{22}} \varphi \right) \in \boldsymbol{P}^{21}$$

is said to be the period of S. It satisfies the Riemann-Hodge relation

$$\eta {M_0}^t \eta = 0, \eta {M_0}^t \overline{\eta} > 0 \quad \text{(positive definite)}.$$

Note that $\int_\gamma \omega = 0 \Longleftrightarrow \gamma \in \mathrm{NS}(S)$. Roughly speaking, the period determines the complex structure of S like in the case of the elliptic curve. So $\mathrm{Tr}(S)$ is important. To investigate its exact meaning, we need a detailed study of "The Torelli type theorem in an explicit style."

4.2 Our Family of $K3$ Surfaces

We consider a special family $\mathcal{F}$ of elliptic $K3$ surfaces given by the following affine singular equations with two complex parameters λ_1, λ_2:

$$\begin{cases} S(\lambda) = S(\lambda_1, \lambda_2) : z^2 = y^3 + x^2(x-1)^2(x-\lambda_1)^2(x-\lambda_2)^2 \\ \cong \left\{ z^2 = y^3 + x^2(x-\xi_0)^2(x-\xi_1)^2(x-\xi_2)^2 \right\}, \\ \text{where} \\ \xi \in \Xi = \left\{ [\xi_0 : \xi_1 : \xi_2] \in \mathbb{P}^2(\mathbb{C}) : \xi_0\xi_1\xi_2(\xi_0 - \xi_1)(\xi_1 - \xi_2)(\xi_2 - \xi_0) \neq 0 \right\}, \\ \lambda = (\lambda_1, \lambda_2) = (\xi_1/\xi_0, \xi_2/\xi_0). \end{cases} \tag{8}$$

Always we consider their compact minimal nonsingular models. The elliptic fibration is given by the projection $\pi : (x, y, z) \mapsto x$. Excepting $x = a_i$ with $a_0 = 0, a_1 = \lambda_1, a_2 = \lambda_2, a_3 = 1$, and $a_4 = \infty$, always $\pi^{-1}(x)$ is an elliptic curve. At $x = a_i$ we have some degenerations. We call them singular fibers. The elliptic surface $S(\lambda)$ has singular fibers of type IV for $\pi^{-1}(a_i)$ $(i = 0, 1, 2, 3)$, which are composed of three rational curves meeting at one point, and a singular fiber $\pi^{-1}(\infty) = \pi^{-1}(a_4)$ that is a desingularization of the isolated surface singularity E_6. So we have 14 divisors $\theta_1, \ldots, \theta_{14}$ which are independent of the general fiber $F = \pi^{-1}(x)$. Adding the section σ at infinity and the general fiber F, the divisors $\{\theta_1, \ldots, \theta_{14}, \sigma, F\}$ generate the sublattice $\mathrm{NS}(S)$ of rank 16 for a generic member of $\mathcal{F}$. As we shall see later, we have the transcendental lattice of rank 6 with the intersection matrix.

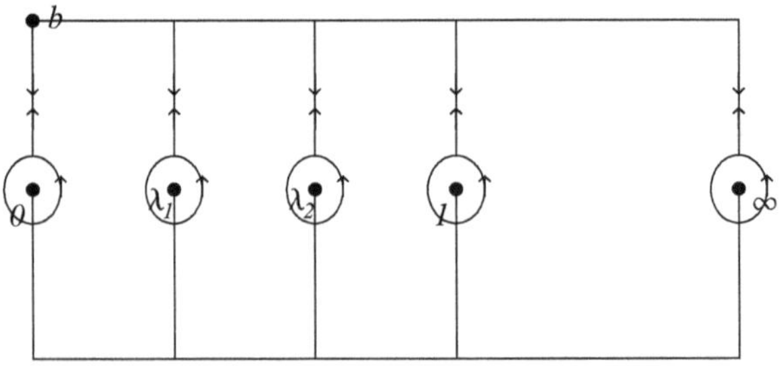

Fig. 2 Singular fiber of $S(\lambda)$

$$A = \begin{pmatrix} 0 & -1 & -1 & 0 & -1 & -1 \\ -1 & 0 & -1 & 2 & 0 & -1 \\ -1 & -1 & 0 & 2 & 2 & 0 \\ 0 & 2 & 2 & 0 & -1 & -1 \\ -1 & 0 & 2 & -1 & 0 & -1 \\ -1 & -1 & 0 & -1 & -1 & 0 \end{pmatrix}. \tag{9}$$

4.3 *Period Map and Period Domain*

For the moment, we fix a reference surface $S_0 = S(1/3, 2/3)$ and consider the parameter λ restricted in a small neighborhood δ of $(1/3, 2/3)$.

On the base x-space of S_0, we set a base point $(x =)b = i$. We may choose a basis $\{\gamma_1, \gamma_2\}$ of $H_1(\pi^{-1}(b), \boldsymbol{Z})$ so that we have the intersection number $\gamma_1 \cdot \gamma_2 = 1$ with $r^2(\gamma_1) = \gamma_2$ for the fiber preserving automorphism $r : (x, y, z) \mapsto (x, \omega y, z)$ of S_0. Here we put $\omega = e^{2\pi i/3}$. We use the same notation also in the sequel.

We take a closed arc δ_i on the x-plane that starts from b and returns back to b after going around a_i in the positive sense. If we deform the fiber $\pi^{-1}(x)$ along δ_i, we obtain a linear transformation M_i, considered to be a left action of $H_1(\pi^{-1}(b), \boldsymbol{Z})$ with respect to the basis $\{\gamma_1, \gamma_2\}$.

Proposition 1. *It holds*

$$M_0 = M_1 = M_2 = M_3 = \begin{pmatrix} 0 & 1 \\ -1 & -1 \end{pmatrix}, M_4 = \begin{pmatrix} -1 & -1 \\ 1 & 0 \end{pmatrix} = M_0^{-1}. \tag{10}$$

Proof. We consider the case δ_0. Near the origin $x = 0$ the fiber $\pi^{-1}(x)$ takes the form

$$\pi^{-1}(x) : z^2 - \left(y^3 + x^2 u_0(x)\right) = z^2 - \left(y - x^{2/3}u(x)\right)\left(y - \omega x^{2/3}u(x)\right)$$
$$\times \left(y - \omega^2 x^{2/3}u(x)\right) = 0,$$

where $u_0(x)$ and $u(x)$ are some unit functions. If $x = x_0$ is fixed, this is a double cover over the y-plane branching at $y = v_0, \omega v_0, \omega^2 v_0$ and ∞, where $v_0 = x_0^{2/3} u(x_0)$. Let us make the variation $x = x_0 e^{i\theta}$ $(\pi/2 \leq \theta \leq \pi/2 + 2\pi)$. As a consequence, the branch points are transformed by the rotation

$$v_0 \to \omega^2 v_0 \to \omega v_0.$$

By the construction of the system $\{\gamma_1, \gamma_2\}$, this monodromy procedure induces the transformation

$$\gamma_1 \to \gamma_2\,(= r^2(\gamma_1)) \to -\gamma_1 - \gamma_2 (= r(\gamma_1)).$$

Hence we obtain M_0. We can get $M_1, M_2, and M_3$ by the same method. M_4 is calculated by the relation $M_0 M_1 M_2 M_3 M_4 = \text{id}$. □

For an oriented closed arc α on the domain x-space-$\{a_0, \ldots, a_4\}$, by making a deformation of a 1-cycle $\gamma(x)$ in the fiber $\pi^{-1}(x)$, we obtain a 2-chain $\bigcup_{x\in\alpha} \gamma(x)$. If $\gamma(x)$ returns back to the starting cycle at the terminal, it becomes to be a (torus-shaped) 2-cycle on S. By observing the list (10), we can determine a basis $\{\Gamma_1, \ldots, \Gamma_6\}$ of $\mathrm{Tr}(S)$ for the reference surface S_0 as the following Fig. 3. So, we can calculate their intersection numbers $\Gamma_i \cdot \Gamma_j$.

Proposition 2. *The intersection matrix* $(\Gamma_i \cdot \Gamma_j)_{1\leq i,j\leq 6}$ *is given by A of (9).*

Proof. We make a set of cut lines L which connects five branch points $0, \lambda_1, \lambda_2, 1, and \infty$ in the lower half plane of x-sphere $\boldsymbol{P}^1(x)$ given in Fig. 3. $\boldsymbol{P}^1(x) - L$ is simply connected. So we can determine the meaning of $\gamma_i(x)$ $(i = 1, 2)$ on $\boldsymbol{P}^1(x) - L$. We define the orientation of the 2-cycle given by $\Gamma = \bigcup_{x\in\alpha} \gamma(x)$ as the ordered pair of the orientation of the 1-cycle γ and that of the arc α. We denote $\Gamma_i = \bigcup_{x\in\alpha_i} \gamma_1(x)$ $(i = 1, 2)$. Let us calculate the intersection number $\Gamma_1 \cdot \Gamma_2$.

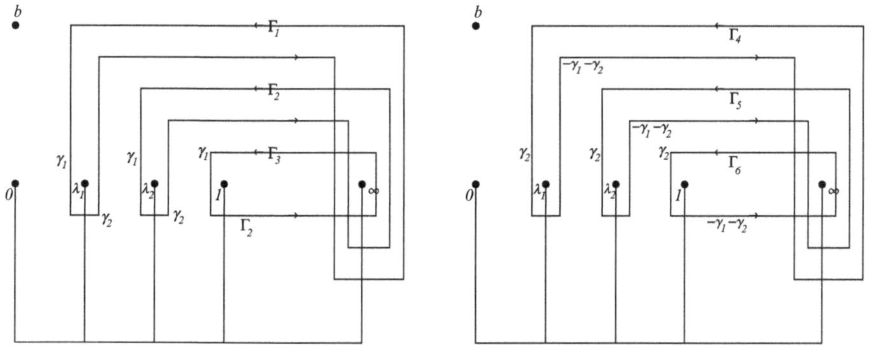

Fig. 3 Transcendental cycles on S_0

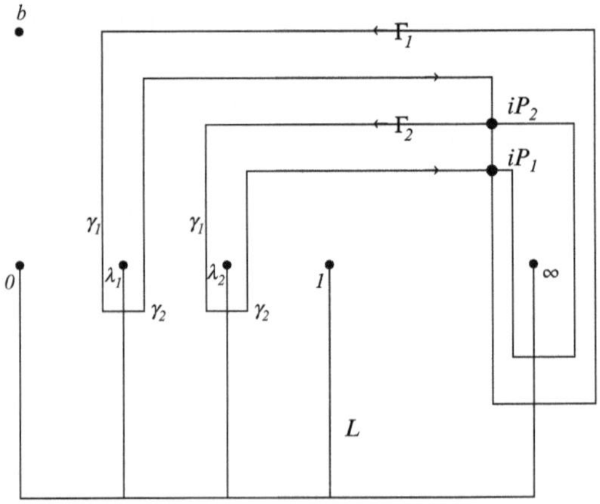

Fig. 4 Intersection of Γ_1 and Γ_2

We have two intersections iP_1, iP_2 of their projections onto $\boldsymbol{P}^1(x)$ (see Fig. 4).

At iP_1, Γ_1 and Γ_2 have the same 1-cycle γ_2. So at this point we have the intersection multiplicity

$$I_{iP_1}(\Gamma_1, \Gamma_2) = 0.$$

At iP_2, Γ_1 has the cycle γ_2, and Γ_2 has the cycle γ_1. At this point the arc α_1 intersects α_2 with clockwise orientation (i.e., the converse of the orientation of the x-sphere). So we have their intersection multiplicity $I_{iP_2}(\alpha_1, \alpha_2) = -1$. Consequently, we have the intersection multiplicity

$$I_{iP_2}(\Gamma_1, \Gamma_2) = -I_{iP_2}(\alpha_1, \alpha_2) \times (\gamma_2 \cdot \gamma_1) = -(-1) \times (-1) = -1.$$

Hence, we have $\Gamma_1 \cdot \Gamma_2 = -1$. By the same procedure we obtain other intersection numbers. □

We denote the period of S_0 by

$$\check{\eta}_{i-1} = \int_{\Gamma_i} \varphi \ (i = 1, 2, \ldots, 6). \tag{11}$$

We may define the period for $S(\lambda), \lambda \in \delta$ by the small deformation of the surface S_0.

By considering the natures of γ_1, γ_2 and the construction of Γ_i, we have $\Gamma_{i+3} = r^2(\Gamma_i)$. Hence we have

$$\check{\eta}_{i+3} = \omega^2 \check{\eta}_i \ (i = 0, 1, 2). \tag{12}$$

As a consequence, the equality $\check{\eta} A^t \check{\eta} = 0$ of the Riemann-Hodge relation becomes trivial. The inequality $\check{\eta} A^t \overline{\check{\eta}} > 0$ is reduced to

$$\begin{cases} (\check{\eta}_0, \check{\eta}_1, \check{\eta}_2)\, B_1{}^t \left(\overline{\check{\eta}_0}, \overline{\check{\eta}_1}, \overline{\check{\eta}_2}\right) > 0 \\ B_1 = \begin{pmatrix} 1 & -1 & -\omega \\ -1 & -1 & -1 \\ -\omega^2 & -1 & -1 \end{pmatrix} \end{cases} . \tag{13}$$

By the transformation of the homogeneous coordinates,

$$\begin{pmatrix} \eta_0 \\ \eta_1 \\ \eta_2 \end{pmatrix} = \begin{pmatrix} 0 & 0 & \omega^2 \\ \omega & 1 & \omega^2 \\ \omega & 0 & 0 \end{pmatrix} \begin{pmatrix} \check{\eta}_0 \\ \check{\eta}_1 \\ \check{\eta}_2 \end{pmatrix} \tag{14}$$

the period relation is described by

$$(\eta_0, \eta_1, \eta_2) B_0{}^t\, (\overline{\eta}_0, \overline{\eta}_1, \overline{\eta}_2) > 0,\ B_0 = \begin{pmatrix} 0 & 0 & 1 \\ 0 & 1 & 0 \\ 1 & 0 & 0 \end{pmatrix}.$$

By taking affine coordinates $u = \eta_1/\eta_0, v = \eta_2/\eta_0$, we obtain a normalized period domain (i.e., a hyperball $\boldsymbol{B}^2$)

$$\mathscr{D} = \left\{(u, v) \in \boldsymbol{C}^2 : 2\,\mathrm{Re}\,(v) + |u|^2 < 0\right\} \subset \boldsymbol{C}^2. \tag{15}$$

Now let the parameter λ move on Ξ. So we get a multivalued analytic map

$$\Phi_P = \Phi : \Xi \to \mathscr{D}, \Phi(\lambda_1, \lambda_2) = (\eta_1 : \eta_2 : \eta_3)$$

by making the analytic continuation along any arc (starting from the reference point $\lambda = (1/3, 2/3)$) in Ξ. We call Φ_P the period map for $\mathcal{F}$. The image of Φ_P is contained in the domain $\mathscr{D}$.

4.4 Modular Group and Theta Representation

The Appell hypergeometric differential equation (see [1, 24], also) $E\left(\frac{1}{3}, \frac{1}{3}, \frac{1}{3}, 1\right)$ is given by

$$\begin{cases} 9\lambda(1-\lambda)(\lambda-\mu)r = 3\left(5\lambda^2 - 4\lambda\mu - 3\lambda + 2\mu\right)p + 3\mu(1-\mu)q + (\lambda-\mu)u \\ 3(\lambda-\mu)s = p - q \\ 9\lambda(1-\mu)(\mu-\lambda)t = 3\lambda(1-\lambda)p + 3\left(5\mu^2 - 4\lambda\lambda - 3\mu + 2\lambda\right)q + (\mu-\lambda)u, \end{cases}$$

where $r = \frac{\partial^2 u}{\partial \lambda^2}, s = \frac{\partial^2 u}{\partial \lambda \partial \mu}, t = \frac{\partial^2 u}{\partial \mu^2}, p = \frac{\partial u}{\partial \lambda}$, and $q = \frac{\partial u}{\partial \mu}$. At any point $(\lambda, \mu) \in \Xi$, the space of the solutions is a three-dimensional complex vector space. The Appell hypergeometric function

$$F_1\left(\frac{1}{3}, \frac{1}{3}, \frac{1}{3}, 1; \lambda, \mu\right) = \frac{1}{\Gamma(1/3)\Gamma(2/3)} \int_1^{\infty} \frac{dt}{\sqrt[3]{t(t-1)(t-\lambda)(t-\mu)}} \tag{16}$$

gives the solution (unique up to a constant factor) in the neighborhood of the origin.

According to [19] (Prop. 4–1), the period $\eta(\lambda_1, \lambda_2)$ of $S(\lambda)$ satisfies the Appell hypergeometric differential equation $E\left(\frac{1}{3}, \frac{1}{3}, \frac{1}{3}, 1\right)$. So, the period map Φ is considered to be the ratio of its three independent solutions.

Set a family of algebraic curves of genus three given by

$$C(\lambda) = C(\lambda_1, \lambda_2) : y^2 = t(t-1)(t-\lambda_1)(t-\lambda_2),\ \lambda = (\lambda_1, \lambda_2) \in \Xi.$$

As we see in [16, 20], we can define the corresponding period map $\Phi_0 : \Xi \to \boldsymbol{B}$ by using the ratio of periods $\left(\int_{A_1} \frac{dt}{y} : -\omega^2 \int_{B_1} \frac{dt}{y} : \int_{A_2} \frac{dt}{y}\right)$ for some 1-cycles A_1, A_2, B_1 on $C(\lambda)$. Observing (16), we know that Φ_0 is also given by the ratio of three independent solutions of $E\left(\frac{1}{3}, \frac{1}{3}, \frac{1}{3}, 1\right)$. As a consequence our period map Φ coincides with Φ_0. So, we can use the properties of Φ_0 as those of Φ.

Set

$$\Gamma = U\left(B_0, \boldsymbol{Z}[\omega]\right) = \left\{g \in \mathrm{GL}_3\left(\boldsymbol{Z}[\omega]\right) : {}^t\overline{g} B_0 g = B_0\right\},$$

and consider its congruence subgroup

$$\Gamma\left(\sqrt{-3}\right) = \left\{g \in \Gamma : g \equiv I_3 \mod \left(\sqrt{-3}\right)\right\}.$$

For $g = \begin{pmatrix} p_0 & q_0 & r_0 \\ p_1 & q_1 & r_1 \\ p_2 & q_2 & r_2 \end{pmatrix} \in \Gamma$, we define its action on $\boldsymbol{B} = \mathscr{D}$

$$g(u, v) = \left(\frac{p_1 + q_1 u + r_1 v}{p_0 + q_0 u + r_0 v}, \frac{p_2 + q_2 u + r_2 v}{p_0 + q_0 u + r_0 v}\right). \tag{17}$$

Our hyperball $\boldsymbol{B}$ has a modular embedding into the Siegel upper half space $\mathfrak{S}_3$, $\Omega : \boldsymbol{B} \to \mathfrak{S}_3$

$$\Omega(u, v) = \begin{pmatrix} \frac{u^2 + 2\omega^2 v}{1-\omega} & \omega^2 u & \frac{\omega u^2 - \omega^2 v}{1-\omega} \\ \omega^2 u & -\omega^2 & u \\ \frac{\omega u^2 - \omega^2 v}{1-\omega} & u & \frac{\omega^2 u^2 + 2\omega^2 v}{1-\omega} \end{pmatrix}. \tag{18}$$

In fact, the embedding is obvious. The map Ω is compatible with the homomorphism $\rho : \Gamma \to Sp_6(\boldsymbol{Z})$ given by

$$\rho(g) = \left(\begin{array}{ccc|ccc} a_{22}-b_{22} & a_{23}-b_{23} & -b_{22} & b_{21} & b_{23} & a_{21}-b_{21} \\ a_{32}-b_{32} & a_{33}-b_{33} & -b_{32} & b_{31} & b_{33} & a_{31}-b_{31} \\ b_{22} & b_{23} & a_{22} & -a_{21} & -a_{23} & b_{21} \\ \hline -b_{12} & -b_{13} & -a_{12} & a_{11} & a_{13} & -b_{11} \\ -b_{32} & -b_{33} & -a_{32} & a_{31} & a_{33} & -b_{31} \\ a_{12}-b_{12} & a_{13}-b_{13} & -b_{12} & b_{11} & b_{13} & a_{11}-b_{11} \end{array}\right)$$

where $g = (g_{ij}),\ g_{ij} = a_{ij} + \omega b_{ij}$. Namely, we have $\rho(g)\cdot\Omega(u,v) = \Omega(g\cdot(u,v))$.

For rational characteristics $a = (a_1,\ldots,a_g), b = (b_1,\ldots,b_g) \in \mathbb{Q}^g$, we define the Riemann theta constant

$$\vartheta\begin{bmatrix} a \\ b \end{bmatrix}(\Omega) = \sum_{n\in\mathbb{Z}^g} \exp\left[\pi i(n+a)\Omega^t(n+a) + 2\pi i(n+a)^t b\right]$$

where Ω is a variable on $\mathfrak{S}_g$ (see [15]). We use only the following:

$$\vartheta_k(u,v) = \vartheta\begin{bmatrix} 0 & 1/6 & 0 \\ k/3 & 1/6 & k/3 \end{bmatrix}(\Omega(u,v)) = \sum_{\mu\in\mathbb{Z}[\omega]} \omega^{2kTr(\mu)} H(\mu u)\overline{q}^{N(\mu)}, \quad (19)$$

where $k \in \{0,1,2\}$, $Tr(\mu) = \mu + \bar{\mu}$, $\mathrm{N}(\mu) = \mu\bar{\mu}$ and

$$H(u) = \exp\left[\frac{\pi}{\sqrt{3}}u^2\right]\vartheta\begin{bmatrix} 1/6 \\ 1/6 \end{bmatrix}\left(u, -\omega^2\right), \quad \overline{q} = \exp\left[\frac{2\pi}{\sqrt{3}}v\right].$$

According to [20], we have the following properties of $\Phi = \Phi_0$:

Fact 1. (0) *Periods $\eta_i(\lambda_1,\lambda_2)(i = 0,1,2)$ give a system of fundamental solutions of Appell's system of hypergeometric differential equations $E\left(\frac{1}{3},\frac{1}{3},\frac{1}{3},1\right)$. Its projective monodromy group is $\Gamma\left(\sqrt{-3}\right)/\langle \mathrm{id}, \omega\mathrm{id}, \omega^2\mathrm{id}\rangle$.*

(1) *The period map Φ can be extended to a biholomorphic correspondence from the compactification $\mathbb{P}^2$ of the ξ-space to the Satake compactification $\overline{\boldsymbol{B}/\Gamma\left(\sqrt{-3}\right)}$ of $\boldsymbol{B}/\Gamma\left(\sqrt{-3}\right)$ (see Fig. 5).*

(2):

Theorem 7 (Theta representation theorem for our "λ" function). *The inverse of Φ is given by*

$$[\xi_0,\xi_1,\xi_2] = \Lambda\left([\eta_0,\eta_1,\eta_2]\right) = \left[\vartheta_0(u,v)^3, \vartheta_1(u,v)^3, \vartheta_2(u,v)^3\right]. \quad (20)$$

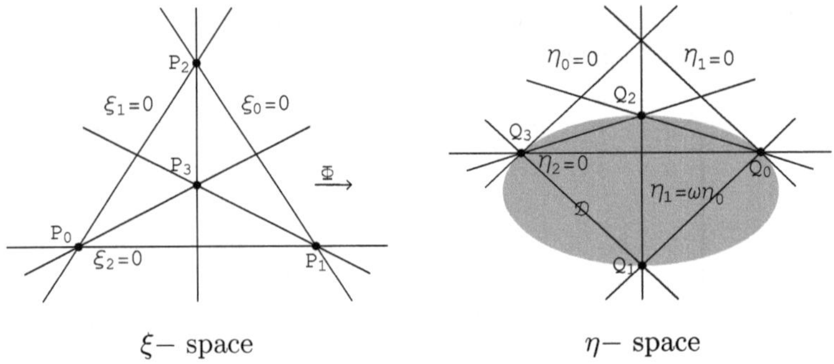

Fig. 5 Correspondence between ξ space and $\boldsymbol{B}$

5 Applications to Number Theory

5.1 Schneider Type Theorem

Theorem 8. *Let* $(u, v) \in \boldsymbol{B}^2$ *be a variable in* $\overline{\boldsymbol{Q}}^2$. *The parameter* $\lambda \in \Xi$ *of the* $K3$ *surface* $S(\lambda)$ *is algebraic if and only if* $\Omega(u, v) \in \mathfrak{S}_3$ *is a CM point, namely,* (u, v) *is an isolated fixed point of* $PU(B_0, \boldsymbol{Q}(\omega))$.

Remark 3. As a general theory established by Shimura, we can construct the Hilbert class field over the cubic Galois extension of $\boldsymbol{Q}\left(\sqrt{-3}\right)$ by the adjunction of the special value of the above $K3$ modular function. But, until now we don't have any explicit example of this.

Remark 4. The above theorem is a simple application of more general theorem given in [21, 22] (see also [5]). The proof of the generalized Schneider theorem relies on the key theorem of Wüstholz [23].

5.2 AGM Theorem in 2 Variables

After Gauss there are some pioneering trials for the generalization of the AGM theorem (e.g., [2–4, 14]). As far as we know, we cannot find any explicit AGM system of several variables with the description of the consequent AGM function. Here we can state just one example. After this theorem, we have several variants of the AGM theorem in several variables. The most remarkable results are given by Matsumoto-Terasoma [12] and [13].

Start from three positive real numbers a, b, c, and make

$$\begin{cases} a' = \frac{1}{3}(a+b+c), \\ b'^3 + c'^3 = \frac{1}{3}\left(a^2b + b^2c + c^2a + ab^2 + bc^2 + ca^2\right), \\ b'^3 - c'^3 = \frac{1}{3\sqrt{-3}}(a-b)(b-c)(c-a). \end{cases}$$

Set $\psi : (a,b,c) \mapsto (a',b',c')$. We know that we find $\psi^2(a,b,c)$ to be a triple of positive numbers again. By the iteration of this procedure, we define

$$M_3(a,b,c) := \lim_{n\to\infty} \psi^n(a,b,c).$$

This is our new "AGM" function. The Appell hypergeometric function $F_1(a,b,b',c;\lambda_1,\lambda_2)$ is defined by the series

$$F_1(a,b,b',c;\lambda_1,\lambda_2) = \sum_{m,n\geq 0} \frac{(a,m+n)(b,m)(b',n)}{(c,m+n)m!n!}\lambda_1^m\lambda_2^n \tag{21}$$

$$(a,n) = \begin{cases} a(a+1)\cdots(a+n-1) & \text{for } n>0, \\ 1 \quad \text{for } n=0. \end{cases}$$

Theorem 9 (Three terms AGM theorem, [10]).

$$\frac{1}{M_3(1,x,y)} = F_1\left(\frac{1}{3},\frac{1}{3},\frac{1}{3},1;1-x^3,1-y^3\right)$$
$$= \frac{1}{\Gamma(1/3)\Gamma(2/3)}\int_1^\infty \frac{dt}{\sqrt[3]{t(t-1)(t-(1-1x^3))(1-(1-y^3))}}, \quad (|x|<1,|y|<1).$$

Theorem 10 (Isogeny formula, [10]).

$$\begin{cases} \vartheta_0\left(\sqrt{-3}u, 3v\right) = \frac{1}{3}\left(\vartheta_0 + \vartheta_1 + \vartheta_2\right), \\ \vartheta_1^3\left(\sqrt{-3}u, 3v\right) + \vartheta_2^3\left(\sqrt{-3}u, 3v\right) = \frac{1}{3}\left(\vartheta_0^2\vartheta_1 + \vartheta_1^2\vartheta_2 + \vartheta_2^2\vartheta_0 + \vartheta_0\vartheta_1^2 \right. \\ \qquad \left. + \vartheta_1\vartheta_2^2 + \vartheta_2\vartheta_0^2\right), \\ \vartheta_1^3\left(\sqrt{-3}u, 3v\right) - \vartheta_2^3\left(\sqrt{-3}u, 3v\right) = \frac{1}{3\sqrt{-3}}\left(\vartheta_0 - \vartheta_1\right)\left(\vartheta_1 - \vartheta_2\right)\left(\vartheta_2 - \vartheta_0\right). \end{cases}$$

By using our theta representation of the $K3$ modular function,

$$(\lambda_1,\lambda_2) = \left(\frac{\vartheta_1(u,v)^3}{\vartheta_0(u,v)^3}, \frac{\vartheta_2(u,v)^3}{\vartheta_0(u,v)^3}\right) \tag{22}$$

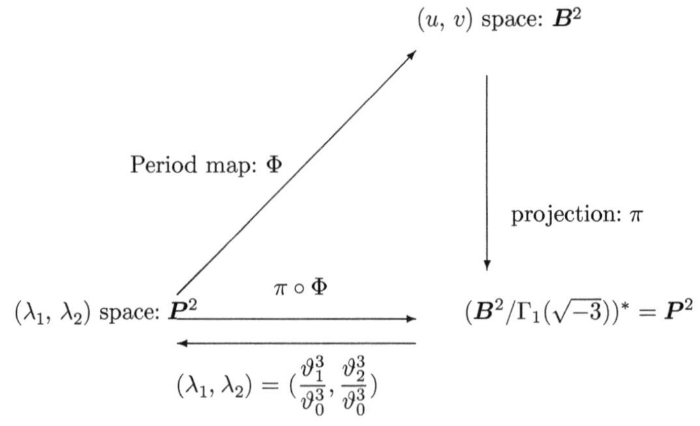

Diagram 2 Period map Φ and its inverse

We can obtain

Theorem 11 (A Jacobi type formula in two variables, [11]).

$$\vartheta_0(u,v) = C_0 F_1\left(\frac{1}{3}, \frac{1}{3}, \frac{1}{3}, 1; 1-\lambda_1, 1-\lambda_2\right), \qquad C_0 = \vartheta\begin{bmatrix}\frac{1}{6}\\ \frac{1}{6}\end{bmatrix}(-\omega^2). \tag{23}$$

The proof for our AGM theorem is given by the quite similar argument

"Theta representation + Isogeny formula + Jacobi's theta formula = AGM theorem" as in the classical case. We can summarize the above situation by Diagram 2.

References

1. Appell, P.: Sur les Fonctions hypergéométriques de plusieurs variables les polynomes d'Hermite et autres fonctions sphériques dans l'hyperspace. Gauthier-Villars, Paris (1925)
2. Borchardt, C.W.: Über das arithmetisch-geometrisch Mittel aus vier Elementen. Monatsber. Akad. Wiss. Berlin **53**, 611–621 (1876)
3. Borwein, J.M., Borwein, P.B.: A cubic counterpart of Jacobi's identity and the AGM. Trans. A.M.S. **323**, 691–701 (1991)
4. Borwein, J.M., Borwein, P.B.: Pi and the AGM-A Study in Analytic Number Theory and Computational Complexity. Wiley, New York (1987)
5. Cohen, P.B.: Humbert surfaces and transcendence properties of automorphic functions. Rocky Mountain J. Math. **26**, 987–1001 (1996). Symposium of Diophantine Problems, Boulder, (1994)
6. Gauss, C.F.: Hundert Theoreme über die neuen Transzendenten, 1818, Werke IIIer band. Georg Olms Verlag, Hildesheim/New York (1973)
7. Gauss, C.F.: Mathematisches Tagebuch 1796–1814. Ostwalds Kalassiker der exakte Wissenschaften 256. Geest und Portig, Leibzig (1976)

8. Igusa, J.: Theta Functions. Springer, Heidelberg/New York (1972)
9. Jacobi, C.G.: Gesammelte Werke I. Chelsea, New York (1969)
10. Koike, K., Shiga, H.: Isogeny formulas for the Picard modular form and a three terms arithmetic geometric mean. J. Number Theory **124**, 123–141 (2007)
11. Matsumoto, K., Shiga, H.: A variant of Jacobi type formula for Picard curves. J. Math. Soc. Jpn. **62**, 1–15 (2010)
12. Matsumoto, K., Terasoma, T.: Arithmetic-geometric means for hyperelliptic curves and Calabi-Yau varieties. Int. J. Math. **21**(7), 939–949 (2010)
13. Matsumoto, K., Terasoma, T.: Thomae type formula for $K3$ surfaces given by double covers of the projective plane branching along six lines. J. Reine U. Angew. Math. **669**, 121–150 (2012)
14. Mestre, J.: Moyenne de Borchardt et integrales elliptiques. C. R. Acad. Sci. Paris Ser. I Math. **313**, 273–276 (1991)
15. D. Mumford, Tata Lectures on Theta I. Birkhäuser, Boston/Basel/Stuttgard (1983)
16. Picard, E.: Sur les fonctions de deux variables indépendantes analogues aux fonctions modulaires. Acta Math. **2**, 114–135 (1883)
17. Schneider, T.: Arithmetische Untersuchungen elliptischer Integrale. Math. Ann. **113**, 1–13 (1937)
18. Shiga, H.: One attempt to the $K3$ modular function I. Ann. Sc. Norm. Sup. Pisa Ser. IV **VI**, 609–635 (1979)
19. Shiga, H., One attempt to the $K3$ modular function II. Ann. Sc. Norm. Sup. Pisa Ser. IV **VIII**, 157–182 (1981)
20. Shiga, H.: On the representation of the Picard modular function by θ constants I-II, Pub. R.I.M.S. Kyoto Univ. **24** 311–360 (1988)
21. Shiga, H.: On the transcendency of the values of the modular function at algebraic points. Soc. Math. Fr. Asterisque **209**, 293–305 (1992)
22. Shiga, H., Wolfart, J.: Criteria for complex multiplication and transcendence properties of automorphic functions. J. Reine Angew. Math. Bd **463**, 1–25 (1995)
23. Wüstholz, G.: Algebraische Punkte auf analytischen Untergruppen algebraischer Gruppen. Ann. Math. **129**, 501–507 (1989)
24. Yoshida, M.: Fuchsian differential equations. Aspects of Mathematics, vol. E11. Friedr. Vieweg & Sohn, Braunschweig (1987)

Poisson Smooth Structures on Stratified Symplectic Spaces

Petr Somberg, Hông Vân Lê, and Jiři Vanžura

1 Introduction

Many classical problems on various classes of topological spaces reduce to the quest for its appropriate functional structure. Examples of topological spaces we are interested in comprise stratified spaces equipped with an additional structure of geometric origin. Due to the lack of canonical notion of the sheaf of smooth (or analytic) functions on such spaces, one has to define a smooth structure with all derived smooth (or analytical) notions such that the obtained smooth structure satisfies good formal properties.

In this paper we study smooth structures on stratified spaces, developing the ideas in [16]. We observe that many properties of smooth structures on pseudomanifolds with isolated conical singularities also hold for a larger class of locally trivial spaces with singularities with cone as typical fiber, if one poses a mild, natural, local

Funding This work was supported by RVO: 67985840 and by Grant of ASCR Nr100190701 (H.V. L. and J. V.) and by MSM 0021620839 and GACR 20108/0397 (P.S.). A part of this paper was written while H.V.L. was visiting the ASSMS, GCU, Lahore-Pakistan. She thanks ASSMS for their hospitality and financial support.

H.V.L. thanks Dmitry Panyushev and Nguyen Tien Zung for discussions, which have helped to shape the final form of this paper.

P. Somberg
Mathematical Institute, Charles University, Sokolovska 83, 18000 Praha 8, Czech Republic

H. Vân Lê (✉)
Institute of Mathematics of ASCR, Zitna 25, 11567 Praha 1, Czech Republic
e-mail: hvle@math.cas.cz

J. Vanžura
Institute of Mathematics of ASCR, Zizkova 22, 61662 Brno, Czech Republic

P. Cartier et al. (eds.), *Mathematics in the 21st Century*, Springer Proceedings in Mathematics & Statistics 98, DOI 10.1007/978-3-0348-0859-0_11

condition on these smooth structures (Definition 3). In this extension we have to take care of (possibly disconnected) regular strata of different dimensions. A large part of our note concerns with compatible smooth structures on stratified symplectic spaces. Stratified symplectic spaces appear abundantly in geometry and mathematical physics [13–15]. They are subjects of intensive study in symplectic geometry since the 1990s [2, 5, 13, 25, 27]. In [1, 7] and many other papers in this direction, the authors considered complex manifolds $M_{\mathbb{C}}^{2n}$ with a holomorphic symplectic form ω^2, which can be turned into a symplectic form $\tilde{\omega}^2$ on differentiable manifolds $M_{\mathbb{C}}^{2n}$ of real dimension $4n$ by setting $\tilde{\omega}^2 := \mathrm{Re}\,(\omega^2) + \mathrm{Im}\,(\omega^2)$. The theory of smooth structures on general stratified symplectic spaces has been introduced by Sjamaar and Lerman [25], then developed by many others [2, 13, 27] (in Remarks 2.3 and 4 and Example 5, we compare our notion of a smooth structure on a stratified symplectic space with that given by Sjamaar-Lerman). Pflaum introduced smooth structures on stratified spaces by means of a maximal atlas (Remark 2.3), and with reference to a smooth structure, he developed extensions to stratified spaces of standard differential geometric concepts [27]. Huebschmann refined the notion of symplectic stratified spaces by introducing the notion of stratified polarization [13]. Our axiomatic approach to the notion of a smooth structure uses heavily the notion of C^∞-algebras defined in [21]. We refer the reader to [11] for the notion of C^∞-differentiable space developed by Gonsalez and de Salas, which also grew from the notion of C^∞-algebra, but they did not treat stratified spaces. In our study we discover the existence of a class of symplectic stratified spaces which can be supplied with a Poisson and weakly symplectic smooth structure. This class is large enough to encompass basic examples of symplectic singular spaces considered in [25] (Example 5 and Proposition 4) as well as many important singularities on the closure of adjoint orbits of nilpotent elements of complex Lie algebras (Example 6 and Proposition 5).

The structure of this paper is as follows. In Sect. 2 we introduce the notion of a smooth structure on a stratified space (Definition 3) and compare our concept of a smooth structure with some other concepts (Remarks 2.2–3). We prove that a smooth stratified space possesses several important properties, e.g., the existence of smooth partitions of unity (Lemma 1, Proposition 1, and Corollary 1), which will be needed in later sections (Remarks 6.2 and 7). In Sect. 3 we study natural smooth structures on stratified symplectic spaces (Definition 6). We show that, under mild conditions, stratified symplectic spaces (X, ω) equipped with a Poisson smooth structure possess a variety of basic properties of smooth symplectic manifolds, e.g., the existence and uniqueness of a Hamiltonian flow, the isomorphism between the Brylinski-Poisson homology and the de Rham homology, and the existence of a Lefschetz decomposition (Theorems 2, 1, 3, and Proposition 7). We also show many examples of weakly symplectic smooth structures and Poisson smooth structures (Examples 5 and 6).

2 Stratified Spaces and Their Smooth Structures

In this section we recall the notion of a stratified space following Goresky's and MacPherson's concept [10, p.36], [25, §1] (Definition 2). Then we introduce the notion of a smooth structure on a stratified space (Definition 3). Our concept of a smooth structure on a stratified space is a natural extension of our concept of a smooth structure on a pseudomanifold with isolated conical singularities given in [16, §2], using the key notion of a product smooth structure (Remark 2.2). (The notion of a product smooth structure has been introduced by Mostow in [23, §3] and has been used by many authors). We prove several important properties of a smooth structure on a stratified space, e.g., the existence of smooth partitions of unity and its consequences (Lemma 1, Proposition 1, and Corollary 1) and the existence of a locally smoothly contractible, resolvable smooth structure on pseudomanifolds with edges (Lemma 4). We show that a resolvable smooth structure satisfying a mild condition is not finitely generated (Proposition 2). We introduce the notion of a smooth differential form and a smooth Zariski vector field (Definition 4). We compare our concept of a smooth structure with some other concepts (Remark 2.3).

2.1 *Stratified Spaces*

We begin with the notion of a decomposed space.

Definition 1 ([10, p. 36], [25, Definition 1.1]). Let X be a Hausdorff and paracompact topological space, and let $\mathcal{S}$ be a partially ordered set with ordering denoted by $\leq$. *An $\mathcal{S}$-decomposition of X* is a locally finite collection of disjoint locally closed manifolds $S_i \subset X$ (one for each $i \in \mathcal{S}$) called *strata* such that:

(1) $X = \cup_{i \in \mathcal{S}} S_i$
(2) $S_i \cap \bar{S}_j \neq \emptyset \Longleftrightarrow S_i \subset \bar{S}_j \Longleftrightarrow i \leq j$

We define *the depth of a stratum S* as follows:

$$\mathrm{depth}_X S := \sup\{n| \text{ there exist strata } S = S_0 < S_1 < \cdots < S_n\}.$$

We define *the depth of X* to be the number $\mathrm{depth}\, X := \sup_{i \in \mathcal{S}} \mathrm{depth}_X S_i$. *The dimension of X* is defined to be the maximal dimension of its strata.

As in [25] we consider only finite-dimensional decomposed spaces X. We call a stratum $S \subset X$ *singular*, if there is another stratum $S' \subset X$ such that $S \subset \bar{S}'$. Otherwise S is called *regular*. We call

$$X^{\mathrm{reg}} := \{\cup S | S \text{ is a regular stratum of } X\}$$

the regular part (component) of X. Clearly $\overline{X^{\mathrm{reg}}} = X$.

Now we specify a subclass of decomposed spaces consisting of locally trivial spaces with cone as typical fiber. Recall that a cone cL over a topological space L is the topological space $L \times [0, \infty)/L \times \{0\}$. Let $[z, t]$ denote the image of (z, t) in a cone cL under the projection $\pi : L \times [0, \infty) \to cL$. We denote by $cL(\varepsilon)$ the open subset $\{[z, t] \in cL \| \ t < \varepsilon\}$.

Definition 2 ([9], [25, Definition 1.7]). A decomposed space X is called *a stratified space* if the strata of X satisfy the following condition defined recursively. Given a point x in a stratum S, there exist an open neighborhood $U(x)$ of x in X; an open ball B_x around x in S; a compact stratified space L, called the link of x; and a homeomorphism $\phi_x : U(x) \to B_x \times cL(1)$ that preserves the decomposition.

Remark 1. 1. It is well known that a Whitney stratified subspace in $\mathbb{R}^n$ is a stratified space in sense of Definition 2 [19].
2. In the literature there are different concepts of a stratified space; see, e.g., [27, Chapter 1] for a discussion.

Example 1. 1. Among important examples of stratified spaces of depth 1 are pseudomanifolds with edges; see, e.g., [24]. Let us recall the definition of a pseudomanifold with edges. Suppose that M is a compact connected smooth manifold with boundary ∂M, and suppose that ∂M is the total space of a smooth locally trivial bundle $\pi : \partial M \to N$ over a closed smooth base N whose fiber is a closed smooth manifold. The topological space X obtained by gluing M to N with help of π (i.e., the points in each fiber $\pi^{-1}(s)$ are identified with $s \in N$) is called *a pseudomanifold with edges* corresponding to the pair (M, π). The natural surjective map $M \to X$ which is the identity on $M \setminus \partial M$ is denoted by $\bar{\pi}$. In general, N need not be connected, and the connected components of N are called *edges* of X. Clearly, $X = (X \setminus N) \cup N$ is a decomposed space; moreover $X \setminus N$ is an open connected stratum of X. Now we will prove that N satisfies the condition in Definition 2. For $s \in N$ let L be the fiber $\pi^{-1}(s) \subset \partial M$ and B be an open neighborhood of s in N such that $\pi^{-1}(B) = B \times \pi^{-1}(s)$. Let $\pi^{-1}(N)_\varepsilon$ be a collar neighborhood of the boundary component $\pi^{-1}(N) \subset \partial M$ in M provided with a trivialization $(p, t) : \pi^{-1}(N)_\varepsilon \to \pi^{-1}(N) \times [0, \varepsilon)$, where p is a smooth retraction $\pi^{-1}(N)_\varepsilon \to \pi^{-1}(N)$. Set $U(s) := \bar{\pi}(p^{-1} \circ \pi^{-1}(B))$. We define a trivialization $\phi_s : U(s) \to B \times cL(1)$ by $\phi_s(x) = (\pi \circ p(x), [t, p(x)])$. This shows that X is a stratified space of depth 1. An important class of pseudomanifolds with edges consists of *pseudomanifolds X with isolated conical singularities*, if its edges X_i of X are just points s_i of X.
2. If L is a compact stratified space, then the cone cL is also a stratified space.
3. If L_1 and L_2 are stratified spaces, then $L_1 \times L_2$ is a stratified space. In particular, a product of cones $cL_1 \times cL_2 = c\,(L_1 \times L_2 \times [0, 1])$ has the following decomposition: $c(L_1 \times L_2 \times [0, 1]) = \{pt\} \cup L_1 \times (0, 1) \cup L_2 \times (0, 1) \cup L_1 \times L_2 \times (0, 1)$.

2.2 Smooth Structure on Stratified Spaces and Its Simplest Properties

We now introduce the notion of a smooth structure on a stratified space X, which is a natural extension of our notion of a smooth structure on a pseudomanifold with isolated conical singularities (pseudomanifold w.i.c.s.) in [16] (Remark 2.2).

As the notion of a stratified space, X is defined inductively on the depth of X, the notion of a smooth structure on X is also defined inductively on the depth on X.

Since X is locally modelled as a product $B \times cL(1)$, we define the notion of a smooth structure on the product $B \times cL(1)$ inductively on the depth of L. More generally, we introduce the notion of a smooth structure on a stratified space X of depth k, recursively on k. The key notion is a product smooth structure (Definition 3.2).

For a stratum $S \subset X$ we consider the following associated spaces.

- $C^\infty_u(S)$ – the space of all usual smooth functions on S.
- $C^\infty_{u,0}(S) := C^\infty_u(S) \cap C^0_0(S)$, where $C^0_0(S)$ is the space of all continuous functions with compact support.
- For $f \in C^0_0(S)$, let $j_*(f) \in C^0_0(X)$ denote the unique extension of f such that $j_*(f) = 0$ if $x \notin S$.

Definition 3 (cf. [16, Definition 2.2]). *A smooth structure* on a stratified space X of depth k is a choice of a $\mathbb{R}$-subalgebra $C^\infty(X)$ of the algebra $C^0(X)$ of continuous real-valued functions on X that satisfy the following properties.

1. $C^\infty(X)$ is a germ-defined C^∞-ring.
2. For any $x \in X$ there exists a local trivialization $\phi_x : U(x) \to B_x \times cL(1)$ which is a local diffeomorphism of stratified spaces, i.e., $C^\infty(U) = \phi^*_x(C^\infty(B \times cL(1)))$, where $C^\infty(B \times cL(1))$ is *a product smooth structure*. In other words, $C^\infty(B \times cL(1))$ is the germ-defined C^∞-ring whose sheaf $SC^\infty(B \times cL(1))$ is generated by $\pi_1^*(SC^\infty(B))$ and $\pi_2^*(SC^\infty(cL(1)))$, where π_1 and π_2 are the projections from $B \times cL(1)$ to B and $cL(1)$, respectively (cf. [23, §3]).
3. A smooth structure $C^\infty(cL(1))$ on the cone over a compact stratified space L must satisfy the following two additional properties: (3a) $C^\infty(cL(1))|_{L\times(0,1)} \subset C^\infty(L \times (0,1))$ and (3b) $j_*[\pi_1^*(C^\infty_{u,0}(0,1))] \subset C^\infty(c(L(1)))$, where $\pi_1 : L \times (0,1) \to (0,1)$ is the projection.

Lemma 1. *Any smooth structure $C^\infty(X)$ on a stratified space X satisfies the following properties.*

1. *$C^\infty(X)_{|S} \subset C^\infty_u(S)$ for each stratum S of X.*
2. *(cf. [16, Lemma 2.1]) $C^\infty(X)$ is partially invertible in the following sense. If $f \in C^\infty(X)$ is nowhere vanishing, then $1/f \in C^\infty(X)$.*

Proof. We prove the first assertion of Lemma 1 using induction on the dimension of X. Clearly this assertion is valid if $\dim X = 1$. Since $C^\infty(X)$ is germ defined, it suffices to prove Lemma 1.1 locally. Hence, w.l.o.g., by Definition 3.2 we can assume that $X = B \times cL(1)$ and $C^\infty(X)$ is a product smooth structure.

First, we consider the case $S = B \times [L, 0]$. By Definition 3.2, any smooth function on X is locally written as $G(f_1, \dots f_n, h_1, \dots h_k)$ where $G \in C^\infty(\mathbb{R}^{n+k})$, $f_i \in C^\infty(B)$, $h_i \in C^\infty(cL(1))$. Since the restriction of h_i to B is constant, $G(f_1, \dots, f_n, h_1, \dots, h_k)_{|S}$ is a smooth function on B. Now assume that $S = B \times S_L \times (0,1)$, where S_L is a stratum of L. Using the condition (3a) in Definition 3, we can assume that $X = B \times L \times (0,1)$ and $C^\infty(X)$ is generated by $\pi_1^*(SC^\infty(B))$, $\pi_2^*(SC^\infty(L))$, $\pi_3^*(SC_u^\infty(0,1))$, where $\pi_1, \pi_2, and \pi_3$ are the projections from X to B, L, $(0,1)$, respectively. Thus, any function in $C^\infty(X)$ is locally of the form $G(f_i, h_j, t)$ where $f_i \in C^\infty(B)$, $h_j \in C^\infty(L)$ and $t \in (0,1)$. Using the induction assumption, noting that $\dim L \le \dim X - 1$, we have $h_i|_{S_L} \in C_u^\infty(S_L)$. Hence $G|_{B \times S_L \times (0,1)} \in C_u^\infty(B \times S_L \times (0,1))$. This completes the proof of the first assertion of Lemma 1.

To prove the last assertion of Lemma 1, it suffices to show that locally $1/f$ is a smooth function. Since $f \neq 0$, shrinking a neighborhood U of x if necessary, we can assume that there is an open interval $(-\varepsilon, \varepsilon)$ which has no intersection with $f(U)$. Now there exists a function $\psi : \mathbb{R} \to \mathbb{R}$ such that:

(a) $\psi|_{f(U)} = \mathrm{Id}$.
(b) $(-\varepsilon/2, \varepsilon/2)$ does not intersect with $\psi(\mathbb{R})$.

Clearly $G : \mathbb{R} \to \mathbb{R}$ defined by $G(x) := \psi(x)^{-1}$ is a smooth function. Note that $1/f(y) = G(f(y))$ for any $y \in U$. This completes the proof of the last assertion of Lemma 1. □

Remark 2. 1. Denote by i the canonical inclusion $X^{\mathrm{reg}} \to X$. Since $X = \overline{X^{\mathrm{reg}}}$, the kernel of $i^* : C^\infty(X) \to C^0(X^{\mathrm{reg}})$ is zero. Lemma 1.1 implies that $i^*(C^\infty(X))$ is a subalgebra of $C_u^\infty(X^{\mathrm{reg}})$. Roughly speaking, we can regard $C^\infty(X)$ as a subalgebra of $C_u^\infty(X^{\mathrm{reg}})$.

2. The condition (3b) in Definition 3 is a relaxing of the condition 3 of Definition 2.2 in [16] for pseudomanifolds w.i.c.s. that requires $j_*(C_0^\infty M^{\mathrm{reg}}) \subset C^\infty(M^{\mathrm{reg}})$. In fact, in [16] (and in the present note) we need only the (weaker) condition (3b) of Definition 3 for the existence of partition of unity and nothing more.
3. Our definition of a smooth structure on a stratified space is a refinement of the definition due to Sjammar and Lerman [25], which requires a smooth structure to satisfy only Lemma 1.1. Pflaum introduces smooth structures by means of a maximal atlas; thus a smooth structure appears as an equivalence class of a system of local embeddings into suitable $\mathbb{R}^n$ [27].

We are going to prove the existence of smooth partitions of unity, which is important for later applications (Remarks 6 and 7).

Proposition 1 (cf. [16, Proposition 2.1]). *Let $\{U_i\}_{i\in I}$ be a locally finite open covering of X such that each U_i has a compact closure $\bar{U}_i$. Then there exists a smooth partition of unity $\{f_i\}_{i\in I}$ subordinate to $\{U_i\}_{i\in I}$.*

Proof. This is a local statement; hence it suffices to prove for the case $X = B \times cL(1)$. Since the smooth structure on $B \times cL(1)$ is a product smooth structure, it is not hard to reduce Proposition 1 to the case B is a point, i.e., $X = cL(1)$ is a cone over a stratified space L. For the case L is a smooth manifold, by Remark 2.2, we have proved the corresponding assertion in [16, Proposition 2.1]. The proof of [16, Proposition 2.1] can be repeated word by word for the case L is a stratified space, using the last two conditions of Definitions 3; so we omit its proof. □

The following Corollary is an immediate consequence of the existence of partition of unity; see, e.g., [16, Lemma 2.11].

Corollary 1. *Smooth functions on X separate points on X.*

Next, we introduce the notion of the cotangent bundle and the notion of the Zariski tangent bundle of a stratified space X, in the same way as we did in [16], which are similar to the notions introduced in [25], [27, B.1]. Note that the germs of smooth functions $C^\infty_x(X)$ is a local $\mathbb{R}$-algebra with the unique maximal ideal $\mathfrak{m}_x$ consisting of functions that vanish at x. Set $T^*_x(X) := \mathfrak{m}_x/\mathfrak{m}_x^2$. Since the following exact sequence

$$0 \to \mathfrak{m}_x \to C^\infty_x \xrightarrow{j} \mathbb{R} \to 0 \tag{1}$$

splits, where j is the evaluation map, $j(f_x) = f_x(x)$ for any $f_x \in C^\infty_x$, the space $T^*_x X$ can be identified with the space of Kähler differentials of $C^\infty_x(X)$. The Kähler derivation $d : C^\infty_x(X) \to T^*_x X$ is defined as follows:

$$d(f_x) := (f_x - j^{-1}(f_x(x))) + \mathfrak{m}_x^2, \tag{2}$$

where $j^{-1} : \mathbb{R} \to C^\infty_x$ is the left inverse of j; see, e.g., [18, Chapter 10], or [27, Proposition B.1.2]. We call $T^*_x X$ *the cotangent space* of X at x. Its dual space $T^Z_x X := \mathrm{Hom}(T^*_x X, \mathbb{R})$ is called *the Zariski tangent space* of X at x. The union $T^*X := \cup_{x\in X} T^*_x X$ is called *the cotangent bundle* of X. The union $T^Z X := \cup_{x\in X} T^Z_x X$ is called *the Zariski tangent bundle* of X.

Let us denote by $\Omega^1_x(X)$ the $C^\infty_x(X)$-module $C^\infty_x(X) \otimes_{\mathbb{R}} \mathfrak{m}_x/\mathfrak{m}_x^2$. We call $\Omega^1_x(X)$ *the germs of 1-forms at* x. Set $\Omega^k_x(X) := C^\infty_x(X) \otimes_{\mathbb{R}} \Lambda^k(\mathfrak{m}_x/\mathfrak{m}_x^2)$. Then $\oplus_k \Omega^k_x(X)$ is an exterior algebra with the following wedge product:

$$(f \otimes_{\mathbb{R}} dg_1 \wedge \cdots \wedge dg_k) \wedge (f' \otimes_{\mathbb{R}} dg_{k+1} \wedge \cdots \wedge dg_l) := (f \cdot f') \otimes_{\mathbb{R}} dg_1 \wedge \cdots \wedge dg_l, \tag{3}$$

where $f, f' \in C^\infty_x$ and $dg_i \in T^*_x M$.

Note that the Kähler derivation $d : C^\infty_x(X) := \Omega^0_x(X) \to \Omega^1_x(X)$ extends to the unique derivation $d : \Omega^k_x(X) \to \Omega^{k+1}_x(X)$ satisfying the Leibniz property. Namely, we set

$$d(f \otimes 1) := 1 \otimes df,$$

$$d(f \otimes \alpha \wedge g \otimes \beta) := d(f \otimes \alpha) \wedge g \otimes \beta + (-1)^{\deg \alpha} f \otimes \alpha \wedge d(g \otimes \beta).$$

Definition 4. 1. (cf. [23, §2]) A section $\alpha : X \to \Lambda^k T^*(X)$ is called *a smooth differential k-form*, if for each $x \in X$ there exists a neighborhood $U(x) \subset X$ of x such that $\alpha(x)$ can be represented as $\sum_{i_0 i_1 \dots i_k} f_{i_0} df_{i_1} \wedge \dots \wedge df_{i_k}$ for some $f_{i_0}, \dots, f_{i_k} \in C^\infty(X)$.
2. A section $V : X \to \Lambda^k T^Z X$ will be called *a smooth Zariski k-vector field*, if for any $\alpha \in \Omega^k(X)$ the value $V(\alpha)$ is a smooth function on X.

Denote by $\Omega(X) = \oplus_k \Omega^k(X)$ the space of all smooth differential forms on X. We identify the germ at x of a k-form $\sum_{i_0 i_1 \dots i_k} f_{i_0} df_{i_1} \wedge \dots \wedge df_{i_k}$ with the element $\sum_{i_0 i_1 \dots i_k} f_{i_0} \otimes df_{i_1} \wedge \dots \wedge df_{i_k} \in \Omega^k_x(X)$. Clearly the Kähler derivation d extends to a map, also denoted by d, that sends $\Omega(X)$ to $\Omega(X)$.

Now we set

$$\Omega_u(X^{\text{reg}}) := \Omega(X^{\text{reg}}, C^\infty_u(X^{\text{reg}})).$$

Remark 2.1 implies immediately.

Lemma 2. *The kernel* $i^* : \Omega(X) \to \Omega_u(X^{\text{reg}})$ *is zero. Roughly speaking, we can regard* $\Omega(X)$ *as a subspace in* $\Omega_u(X^{\text{reg}})$.

2.3 Examples of Smooth Structures on Stratified Spaces

Example 2. Assume that X is a realization of a compact polytope in $\mathbb{R}^n$, i.e., X is a stratified space such that each stratum S of dimension k of X is an open disk in some affine subspace of dimension k in $\mathbb{R}^n$. Then X has a natural smooth structure induced from the standard smooth structure on $\mathbb{R}^n$, i.e., $SC^\infty(X) := SC^\infty(\mathbb{R}^n)|_X$. Indeed, by construction $C^\infty(X)$ is a germ-defined C^∞-ring; hence the condition 1 in Definition 3 is trivially satisfied. Now let us prove the validity of the second condition on the product smooth structure inductively on the dimension of X. Note that the validity of the second condition is trivially satisfied, if $\dim X = 0$. Since X is a realization of a polytope, the tubular neighborhood of any point $x \in S^k$ in $\mathbb{R}^n$ has the form $B_x \times cL(1)$, where $L(1)$ is the intersection of X with the sphere S^{n-k-1} of a small enough radius centered at x on a hyperplane through x in $\mathbb{R}^n$ that is orthogonally complement to S, and B_x is an open ball around x in S. By the dimension induction assumption, $L(1) \subset S^{n-k-1}$ has a natural smooth structure induced from the embedding $L(1) \to S^{n-k-1} \subset \mathbb{R}^n$, which satisfies the conditions of Definition 3, since the projection from a punctured sphere $S^{n-k-1} \setminus \{pt\}$ to $\mathbb{R}^{n-k-1}$, where $\{pt\} \notin L(1)$, sends $L(1)$ to a realization of a polytope of lower dimension in $\mathbb{R}^{n-k-1}$, and this projection is a diffeomorphism between $S^{n-k-1} \setminus \{pt\}$ and $\mathbb{R}^{n-k-1}$. To study the smooth structure on $cL(1)$, we need the following.

Lemma 3. *Assume that $A \subset \mathbb{R}^n$ and $B \subset \mathbb{R}^m$. Then the sheaf $SC^\infty(\mathbb{R}^n \times \mathbb{R}^m)|_{A\times B}$ is generated by $\pi_1^*(SC^\infty(\mathbb{R}^n)|_A)$ and $\pi_2^*(SC^\infty(\mathbb{R}^m)|_B)$, where π_1 and π_2 are the projection of $\mathbb{R}^{n+m}$ onto $\mathbb{R}^n$ and $\mathbb{R}^m$, respectively.*

Proof. Lemma 3 is a consequence of the simple fact that $SC^\infty(\mathbb{R}^n \times \mathbb{R}^m)$ is generated by $SC^\infty(\mathbb{R}^n)$ and $SC^\infty(\mathbb{R}^m)$. □

Lemma 3 implies that the induced smooth structure on the product $B_x \times cL(1)$ satisfies the condition 2 of Definition 3. The condition (3a) of Definition 3 trivially holds, and the condition (3b) of Definition 3 also holds for the smooth structure on the cone $cL(1) \subset \mathbb{R}^{n-k} \subset \mathbb{R}^n$, using partition of unity on $\mathbb{R}^{n-k}$. This proves that the induced smooth structure on X satisfies all the conditions of Definition 3.

Example 3. Assume that $(X_i, C^\infty(X_i))$ are stratified spaces provided with smooth structures. Then it is easy to verify that $(\Pi X_i, \Pi(C^\infty(X_i))$ is a stratified manifold provided with a smooth structure.

Example 4. Assume that we have a continuous surjective map $M \xrightarrow{\pi} X$ from a smooth manifold M with corner to a stratified space X of depth 1 such that for each stratum $S_i \subset X$ the triple $(\pi^{-1}(S_i), \pi_i, S_i)$ is a differentiable fibration; moreover for each $x \in X^{\rm reg}$ the preimage $\pi^{-1}(x)$ consists of a single point. Clearly π induces a stratified space structure on M. The $\mathbb{R}$-subalgebra $C^\infty(X) := \{f \in C^0(X) | \pi^* f \in C^\infty(M)\}$ will be called *a resolvable smooth structure*. We are going to show that a resolvable smooth structure satisfies the conditions in Definition 3. First, $C^\infty(X)$ is a germ-defined C^∞-ring, since $C^\infty(M)$ possesses this property. Next, the existence of a local smooth trivialization ϕ_x for each $x \in X$, which satisfies the last two conditions of Definition 3, is a consequence of the existence of a differentiable fibration $(\pi^{-1}(S_i), \pi_i, S_i)$ and the fact that π induces a stratified space structure on M. The space M will be called *a resolution* of X.

In what follows we study some properties of a resolvable smooth structure on a stratified space of depth 1.

We say that $C^\infty(M)$ is *locally smoothly contractible*, if for any $x \in M$ there exists an open neighborhood $U(x) \ni x$ together with a smooth homotopy $\sigma : U(x) \times [0, 1] \to U(x)$ joining the identity map with the constant map $U(x) \mapsto x$ [23, §5].

A C^∞-ring $C^\infty(X)$ is called *finitely generated*, if there are finite elements $f_1, \dots, f_k \in C^\infty(X)$ such that any $h \in C^\infty(X)$ can be written as $h = G(f_1, \dots, f_k)$, where $G \in C^\infty(\mathbb{R}^k)$.

Lemma 4. *Every pseudomanifold X with edges has a resolvable smooth structure, which is locally smoothly contractible.*

Proof. By definition (see Example 1.1), there exist a compact smooth manifold M with boundary ∂M and a surjective map $\bar{\pi} : M \to X$. In Example 2.3 we have shown that such an X has a resolvable smooth structure $C^\infty(X) := \{f \in C^0(X) | \bar{\pi}^* f \in C^\infty(M)\}$. We will show that $C^\infty(X)$ is locally smoothly contractible. Let S_i be a singular stratum of X, and $\bar{\pi}^{-1}(S_i) = \partial M_i \subset \partial M$. Let $V(\partial M_i)$ be a collar open neighborhood of ∂M_i in M. Then $U(S_i) := \bar{\pi}(V(\partial M_i))$ is an open neighborhood of S_i in X. Let us consider the following commutative diagram

$$\begin{array}{ccc} I \times V(\partial M_i) & \xrightarrow{\tilde{F}} & V(\partial M_i) \\ \Big\downarrow{\scriptstyle (Id \times \bar{\pi})} & & \Big\downarrow{\scriptstyle \bar{\pi}} \\ I \times U(S_i) & \xrightarrow{F} & U(S_i) \end{array}$$

where $\tilde{F}$ is a smooth retraction from $V(\partial M_i)$ to ∂M_i, constructed using the fibration $[0, 1) \to V(\partial M_i) \to \partial M_i$. We set

$$F(t, x) := \bar{\pi}\left(\tilde{F}\left(t, \bar{\pi}^{-1}(x)\right)\right).$$

Since $\tilde{F}_{|\partial M_i} = \mathrm{Id}$, the map F is well defined. Clearly F is a smooth homotopy, since $\tilde{F}$ is a smooth homotopy. This proves Proposition 4. □

Proposition 2. *A resolvable smooth structure on a stratified space X of depth 1 obtained from a smooth manifold M with corner is not finitely generated as a C^∞-ring, if there exists $x \in X$ such that* $\dim \pi^{-1}(x) \geq 1$, *where $\pi : M \to X$ is the associated projection.*

Proof. Assume the opposite, i.e., $C^\infty(X)$ is generated by $g_1, \ldots, g_n \in C^\infty(X)$. Then $G := (g_1, \ldots, g_n)$ defines a smooth embedding $X \to \mathbb{R}^n$. Hence $C^\infty(X) = C^\infty(\mathbb{R}^n)/I$, where I is an ideal of $C^\infty(\mathbb{R}^n)$ of smooth functions on $\mathbb{R}^n$ vanishing on $G(X)$ [21, p. 21, Proposition 1.5]. In particular, the cotangent space $T_x^* X$ is a finite dimensional linear space for all $x \in X$. We will show that this assertion leads to a contradiction.

Let S be a stratum of X such that $\dim(\pi^{-1}(S)) \geq \dim S + 1$. Let $x \in S$ and $U(x)$ a small open neighborhood of x in X. Let $f \in C^\infty(U(x))$, equivalently $\pi^*(f) \in C^\infty(\pi^{-1}(U(x)))$. Let $\chi : \pi^{-1}(U(x)) \to \mathbb{R}_+^p \times \mathbb{R}^{n-p} \subset \mathbb{R}^n$ be a coordinate map on $\pi^{-1}(U(x)) \subset M$. By definition of manifolds with corner, we have $\left(\chi^{-1}\right)^* \pi^*(f) \in C^\infty\left(\tilde{U}\right)$ for some open set $\tilde{U} \subset \mathbb{R}^n$ containing $\chi\left(\pi^{-1}(U(x))\right)$. Denote by $\tilde{S}$ the preimage $\chi \circ \pi^{-1}(S \cap U(x))$, which is a submanifold of $\tilde{U}(x)$. Let us denote the restriction of $\pi \circ \chi^{-1}$ to $\tilde{S}$ by $\tilde{\pi}$. Then the triple $\left(\tilde{S}, \tilde{\pi}, S \cap U\right)$ is a smooth fibration, whose fiber $\tilde{\pi}^{-1}(y)$ is a smooth manifold of dimension at least 1. We note that $\left(\chi^{-1}\right)^* \pi^*(f)$ belongs to the subalgebra $C^\infty\left(\tilde{U}, \tilde{S}, \tilde{\pi}\right)$ consisting of smooth functions on $\tilde{U}$ that are constant along fiber $\tilde{\pi}^{-1}(y)$ for all $y \in S \cap U(x)$. Since the depth of X is 1, $C^\infty\left(\tilde{U}, \tilde{S}, \tilde{\pi}\right)$ is identified with the set of smooth functions on U.

Shrinking $U(x)$ we can assume that $\tilde{S} = \tilde{U} \cap \mathbb{R}^k$ and $\tilde{\pi} : \tilde{S} \to S \cap U$ is the restriction of a linear projection $\bar{\pi} : \mathbb{R}^k \to \mathbb{R}^l$, where $l = \dim S$, $k = \dim \tilde{S}$, and $S \cap U = \mathbb{R}^l \cap U$. Here we assume that U is an open set in $\mathbb{R}^n$. Let $\mathbb{R}^{n-k}$ with coordinate $\tilde{x} = (\tilde{x}^1, \dots, \tilde{x}^{n-k})$ be a complement to $\mathbb{R}^k$ in $\mathbb{R}^n \supset \tilde{U}$, and let $\mathbb{R}^{k-l} \subset \mathbb{R}^k$ with coordinate $\tilde{y} = (\tilde{y}^1, \dots, \tilde{y}^{k-l})$ be the set $\bar{\pi}^{-1}(0)$. We also equip the subspace $\mathbb{R}^l$ with coordinate $\tilde{z} = (\tilde{z}^1, \dots, \tilde{z}^l)$. The condition $\dim \pi^{-1}(x) \geq 1$ in Proposition 4 is equivalent to the equality $k - l \geq 1$; in other words, y is an essential variable. Furthermore, a point $\tilde{s} \in \tilde{S} \subset \mathbb{R}^n$ has (local) coordinates with $\tilde{x} = 0$.

Lemma 5. *A function $g \in C^\infty(\tilde{U})$ belongs to $C^\infty(\tilde{U}, \tilde{S}, \tilde{\pi})$ if and only if g has the form*

$$(\tilde{x}^1, \dots, \tilde{x}^{n-k}, \tilde{y}, \tilde{z}) = \tilde{x}^1 g_1(\tilde{x}, \tilde{y}, \tilde{z}) + \dots + \tilde{x}^{n-k} g_{n-k}(\tilde{x}, \tilde{y}, \tilde{z}) + c(\tilde{z}),$$

where $g_i \in C^\infty(\tilde{U})$ and $c(\tilde{z})$ is a smooth function on U depending only on variable $\tilde{z}$.

Proof. We write for $g \in C^\infty(\tilde{U}, \tilde{S}, \tilde{\pi})$

$$g(\tilde{x}, \tilde{y}, \tilde{z}) - g(0, \tilde{y}, \tilde{z}) = \int_0^1 \frac{dg(t\tilde{x}, \tilde{y}, \tilde{z})}{dt} dt = \int_0^1 \sum_{i=1}^{n-k} \frac{\partial g(t\tilde{x}^1, \dots, t\tilde{x}^{n-k}, \tilde{y}, \tilde{z})}{\partial(t\tilde{x}^i)} \tilde{x}_i \, dt.$$

Setting

$$g_i := \int_0^1 \frac{\partial g(t\tilde{x}^1, \dots, t\tilde{x}^{n-k}, \tilde{y}, \tilde{z})}{\partial(t\tilde{x}^i)} dt,$$

we obtain $g(\tilde{x}, \tilde{y}, \tilde{z}) = \sum_{i=1}^{n-k} \tilde{x}^i g_i(\tilde{x}, \tilde{y}, \tilde{z}) + g(0, \tilde{y}, \tilde{z})$. Since $g(0, \tilde{y}, \tilde{z})$ depends only on $\tilde{z}$, we obtain the "only if" part of Lemma 5 immediately. The "if" part is trivial. This proves Lemma 5. □

Now let us complete the proof of Proposition 2. Take a point $s \in S$ and a point $\tilde{s} \in \tilde{\pi}^{-1}(s)$ such that $\tilde{x}(\tilde{s}) = \tilde{y}(\tilde{s}) = \tilde{z}(\tilde{s}) = 0$. Since X has depth 1, Lemma 5 implies that the maximal ideal $\mathfrak{m}_s$ is a linear space generated by functions of the form $\tilde{x}^i g_{i,\alpha}(\tilde{x}, \tilde{y}, \tilde{z}), i = \overline{1, n-k}$. Let us consider the sequence $S := \{\tilde{x}^1 \tilde{y}^1, \dots, \tilde{x}^1 (\tilde{y}^1)^m \in \mathfrak{m}_s\}$, $m \to \infty$. If $\dim T_s^* X = \dim \mathfrak{m}_s/\mathfrak{m}_s^2 = n$, there exists a subsequence $\tilde{x}^1(\tilde{y}^1)^{k_1}, \dots, \tilde{x}^1(\tilde{y}^1)^{k_n}$ such that $\tilde{x}^1(\tilde{y}^1)^m$ is a linear combination of $\tilde{x}^1(\tilde{y}^1)^{k_j}$ modulo $\mathfrak{m}_s^2$ for any m, which is impossible. This completes the proof of Proposition 2. □

Remark 3. Proposition 2 partially answers the question 2 we posed in [16, §5]. We observe that there are many quotient smooth structures which are finitely generated, i.e., a quotient by a smooth group action. In this case the dimension of the fiber over singular strata (e.g. the dimension of a singular orbit) is smaller than or equal to the dimension of the generic fiber (the dimension of a generic orbit, respectively).

3 Symplectic Stratified Spaces and Compatible Smooth Structures

In this chapter we introduce the notion of a stratified symplectic space (X, ω) (Definition 5), which is close to that introduced by Sjamaar and Lerman [25] (Remark 4). We also introduce the notion of a weakly symplectic smooth structure and the notion of a Poisson smooth structure on (X, ω) (Definition 6). We give examples of weakly symplectic smooth structures and Poisson smooth structures (Propositions 4, 5 and Examples 6.1–3). We prove the existence and uniqueness of a Hamiltonian flow associated with a smooth function H on a symplectic stratified space X, which is equipped with a Poisson smooth structure (Theorem 1). We compare our result with a result by Sjamaar and Lerman in [25, §3] (Remark 6). We prove that the Brylinski-Poisson homology of a symplectic stratified space X provided with a Poisson weakly symplectic smooth structure is isomorphic to the de Rham cohomology of X, if the regular strata of X have the same dimension (Theorem 2.) Then we show that, under a mild condition, a stratified symplectic space (X, ω) provided with a Poisson weakly symplectic smooth structure $C^\infty(X)$ enjoys many nice properties related to the existence of a Lefschetz decomposition (Lemma 10, Proposition 7, and Theorem 3).

3.1 Symplectic Stratified Spaces

Definition 5. A stratified space X is called *symplectic*, if every stratum S_i is provided with a symplectic form ω_i. The collection $\omega := \{\omega_i\}$ is called *a stratified symplectic form*, or simply *a symplectic form*, if no misunderstanding can occur.

Remark 4. Definition 5 coincides with the first topological condition in [25, Definition 1.12.(i)] of a symplectic stratified space X introduced by Sjamaar and Lerman. (The other conditions [25, Definitions 1.12.(ii), 1.12.(iii)] require the existence of a compatible smooth structure on X, which is also called by other authors [13] a stratified symplectic Poisson algebra). Thus any stratified symplectic space in Sjamaar's and Lerman's definition is a stratified symplectic space in our definition.

3.2 Weakly Symplectic Smooth Structures and Poisson Smooth Structures

On each symplectic stratum (S_i, ω_i) we define the bivector G_{ω_i} to be the section of the bundle $\Lambda^2 TS_i$ such that $G_{\omega_i}(x) = \partial y_1 \wedge \partial x_1 + \cdots + \partial y_n \wedge \partial x_n$ if $\omega_i(x) = \sum_{j=1}^n dx^j \wedge dy^j$ [3, §1.1]. If we regard ω_i as an element in $\mathrm{End}(TS_i, T^*S_i)$ and

G_{ω_i} as an element in $\mathrm{End}(T^*S_i, TS_i)$, then G_{ω_i} is the inverse of ω_i. The bivector G_{ω_i} defines a Poisson bracket on $C^\infty(S_i)$ by setting $\{f, g\}_{\omega_i} := G_{\omega_i}(df \wedge dg)$.

Definition 6. Let (X, ω) be a symplectic stratified space and $C^\infty(X)$ be a smooth structure on X.

1. A smooth structure $C^\infty(X)$ is said to be *weakly symplectic*, if there is a smooth 2-form $\tilde{\omega} \in \Omega^2(X)$ such that the restriction of $\tilde{\omega}$ to each stratum S_i coincides with ω_i. In this case we also say that $\tilde{\omega}$ is *compatible with* $C^\infty(X)$.
2. A smooth structure $C^\infty(X)$ is called *Poisson*, if there is a Poisson structure $\{,\}_\omega$ on $C^\infty(X)$ such that $(\{f, g\}_\omega)_{|S_i} = \{f_{|S_i}, g_{|S_i}\}_{\omega_i}$ for any stratum $S_i \subset X$.

Remark 5. 1. Lemma 2 implies that there exists at most one 2-form $\tilde{\omega} \in \Omega^2(X)$ which is compatible with a given smooth structure $C^\infty(X)$.
2. We claim that the condition 2 in Definition 6 is equivalent to the existence of a smooth Zariski bivector field $\tilde{G}_\omega \in \Gamma(\Lambda^2 T^Z(X))$ such that

$$\tilde{G}_\omega(\alpha)_{|S_i} = G_{\omega_i}(\alpha_{|S_i}) \tag{4}$$

for any stratum $S_i \subset X$. Indeed, the existence a section $\tilde{G}_\omega$ satisfying (4) defines a Poisson structure on $C^\infty(X)$ by setting $\{f, g\}(x) := \tilde{G}_\omega(df \wedge dg)(x)$. Conversely, assume that there is a Poisson structure $\{,\}_\omega$ on $C^\infty(X)$ whose restriction to each stratum S_i coincides with the given Poisson structure on S_i. We claim the bivector $\tilde{G}_{\omega_i}$ is a smooth Zariski bivector field. Since the space of smooth differential forms is germ defined, it suffices to show the above claim locally. Note that on some neighborhood U we can write $\Omega^2(X) \ni \alpha = \sum_i f_i dg_i \wedge dh_i$, where $f_i, g_i, h_i \in C^\infty(U)$. Since the smooth structure is Poisson, we get

$$\tilde{G}_\omega(\alpha) = \sum_i \tilde{G}_\omega(f_i dg_i \wedge dh_i) = \sum_i f_i\{g_i, h_i\} \in C^\infty(U). \tag{5}$$

This proves our claim.
3. The condition 2 of Definition 6 agrees with the condition (iii) in Definition 1.12 of [25] by Sjamaar and Lerman of a stratified symplectic Poisson algebra. It also agrees with our Definition of a Poisson smooth structure on a conical symplectic pseudomanifold in [16, §4].

Now we are going to consider important examples of weakly symplectic smooth structures and Poisson smooth structures.

Example 5. We assume that a compact Lie group G acts on a connected symplectic manifold (M, ω) with proper moment map $J : M \to \mathfrak{g}^*$. Let $Z = J^{-1}(0)$. The quotient space $M_0 = Z/G$ is called a symplectic reduction of M. If 0 is a singular value of J, then Z is not a manifold, and M_0 is called *a singular symplectic reduction*. It is known that M_0 is a stratified symplectic space in Sjamaar's and Lerman's definition [25], and hence in our definition, see Remark 4. Let us recall the description of M_0 by Sjamaar and Lerman. For a subgroup H of G, denote by $M_{(H)}$ the set of all points whose stabilizer is conjugate to H, the stratum of M of orbit type (H).

Lemma 6 ([25, Theorem 2.1]). *Let (M, ω) be a Hamiltonian G-space with moment map $J : M \to \mathfrak{g}^*$. The intersection of the stratum $M_{(H)}$ of orbit type (H) with the zero level set Z of the moment map is a manifold, and the orbit space*

$$(M_0)_{(H)} = (M_{(H)} \cap Z)/G$$

has a natural symplectic structure $(\omega_0)_{(H)}$ whose pullback to $Z_{(H)} := M_{(H)} \cap Z$ coincides with the restriction to $Z_{(H)}$ of the symplectic form ω on M. Consequently the stratification of M by orbit types induces a decomposition of the reduced space $M_0 = Z/G$ into a disjoint union of symplectic manifolds $M_0 = \cup_{H \subset G}(M_0)_{(H)}$.

Since J is proper, by Theorem 5.9 in [25], the regular part M_0^{reg} is connected. Sjamaar and Lerman also defined a "canonical" smooth structure on M_0 as follows. Set $C^\infty(M_0)_{\mathrm{can}} := C^\infty(M)^G/I^G$, where I^G is the ideal of G-invariant functions vanishing on Z [25, Example 1.11]. We will show that $C^\infty(M_0)_{\mathrm{can}}$ is also a smooth structure in the sense of Definition 3. Denote by π the natural projection $Z \to Z/G$. Set $C^\infty(Z) := C^\infty(M)|_Z$. Since Z is closed, $C^\infty(Z) = C^\infty(M)/I_Z$, where I_Z is the ideal of smooth functions on M vanishing on Z.

We claim that the space $C^\infty(Z)^G$ of G-invariant smooth functions on Z can be identified with the space $C^\infty(M_0)_{\mathrm{can}} = C^\infty(M)^G/I^G$. Clearly $C^\infty(M)^G/I^G$ is a subspace of G-invariant smooth functions on Z. On the other hand, any smooth function f on M can be modified to a G-invariant smooth function $f_G \in C^\infty(M)$ by setting

$$f_G(x) := \int_G f(g \cdot x)\mu_g$$

for a G-invariant measure μ_g on G normalized by the condition $\mathrm{vol}(G) = 1$. So if $g \in C^\infty(Z)^G$, then g is the restriction of a G-invariant function on M. In other words, we have an injective map $C^\infty(Z)^G \to C^\infty(M)^G/I^G$. Hence follows the identity $C^\infty(Z)^G = C^\infty(M_0)_{\mathrm{can}}$. It follows that $C^\infty(M_0)_{\mathrm{can}}$ is the quotient of the smooth structure obtained from $C^\infty(Z)$ via the projection $\pi : Z \to M_0$. In particular, $C^\infty(M_0)_{\mathrm{can}}$ is a germ-defined C^∞-ring, since $C^\infty(Z)$ is a germ-defined C^∞-ring.

Proposition 3. *$C^\infty(M_0)_{\mathrm{can}}$ is a smooth structure in the sense of Definition 3.*

Proof. Note that $C^\infty(M_0)_{\mathrm{can}}$ satisfies the first condition in Definition 3. To show that $C^\infty(M_0)_{\mathrm{can}}$ satisfies the other conditions in Definition 3, we use Theorem 6.7 in [25] which asserts that there is a proper smooth embedding $(M_0, C^\infty(M_0)_{\mathrm{can}}) \xrightarrow{i} (\mathbb{R}^n, C^\infty(\mathbb{R}^n))$ such that the image $i(M_0)$ is a stratified Whitney subspace X of $\mathbb{R}^n$. In particular $C^\infty(M_0)_{\mathrm{can}} = i^*\,(C^\infty(\mathbb{R}^n))$.

We describe a neighborhood of a point $p \in X$ in $\mathbb{R}^n$ following [25, p. 410]. Let S denote the stratum of X that contains p. Let N' be a submanifold in $\mathbb{R}^n$ that is transversal to each stratum of X, intersects S in the single point p, and satisfies $\dim N' + \dim S = n$. Let $B_\delta(p) \subset \mathbb{R}^n$ denote the ball of radius δ. By Whitney's condition B, if δ is sufficiently small, then the sphere $\partial B_\delta(p)$ will be transversal to to each stratum in $X \cap N'$. Fix such a $\delta > 0$. Next we consider the *normal slice* $N(p) := N' \cap X \cap B_\delta(p)$ and *the link* $L(p) := N' \cap X \cap \partial B_\delta(p)$ of the stratum S at the point p. These spaces are canonical Whitney stratified spaces, since they are transversal intersections of Whitney stratified spaces. Furthermore, S has an open neighborhood T_S in X with local trivial fibration $\pi : T_S \to S$ such that $\pi^{-1}(p)$ is homeomorphic to $cL(p)$. Using Lemma 3, this implies that the induced smooth structure on X satisfies the condition 2 of Definition 3. The last two conditions (3a) and (3b) of Definition 3 also hold for X, since locally we have the same description of X as that in Example 2. This completes the proof of Proposition 3. (We also refer the reader to a more explicit, algebraic description of the local structure of $C^\infty(M_0)_{\text{can}}$ given in Theorem 5.1 [25] and its proof). □

Proposition 4. *The smooth structure $C^\infty(M_0)_{\text{can}}$ is both weakly symplectic and Poisson.*

Proof. We observe that $C^\infty(M_0)_{\text{can}}$ is weakly symplectic, since by Proposition 6 the pullback $\pi^*(\omega_0)$ is equal to the restriction of the symplectic form ω to Z. Furthermore, the Poisson property of $C^\infty(M_0)_{\text{can}}$ follows from [25, Proposition 3.1], where they showed that $C^\infty(M_0)_{\text{can}}$ is closed under the Poisson bracket. This completes the proof of Proposition 4. □

Let us consider another important class of stratified symplectic spaces, which are the closure of nilpotent orbits in a complex semisimple Lie algebra $\mathfrak{g}$. This class has been examined by Panyshev [26], Huebschmann [13], Fu [8], and many other under different perspectives.

Example 6. For $x \in \mathfrak{g}$ let $x = x_s + x_n$ be the Jordan decomposition of x, where $x_n \neq 0$ is a nilpotent element, x_s is a semisimple, and $[x_s, x_n] = 0$. Denote by G the adjoint group of $\mathfrak{g}$ and by $\mathcal{Z}_G(x_s)$ the centralizer of x_s in G. The adjoint orbit $G(x)$ is a fibration over $G(x_s)$ whose fiber over x_s is the $\mathcal{Z}_G(x_s)$-orbit of x_n. Since $G(x_s)$ is a closed orbit, a neighborhood U of a point $x \in \overline{G(x)}$ is isomorphic to the product $B \times \overline{\mathcal{Z}_G(x_s) \cdot x_n}$, where B is an open neighborhood of x_s in $G(x_s)$. It is known that the closure $\overline{\mathcal{Z}_G(x_s) \cdot x_n}$ is a finite union of $\mathcal{Z}_G(x_s)$-orbits of nilpotent elements in the Lie subalgebra $\mathcal{Z}_\mathfrak{g}(x_s)$ [6, chapter 6], so the closure $\overline{G(x)}$ is a finite union of adjoint orbits in $\mathfrak{g}$ provided with the Kostant-Kirillov symplectic structure. Thus $\overline{G(x)}$ is a decomposed space, whose strata are symplectic manifolds. Moreover $\overline{G(x)}^{\text{reg}} = G(x)$ is connected.

1. Now assume that x_n is a minimal nilpotent element in $\mathcal{Z}_\mathfrak{g}(x_s)$. Then $\overline{G(x)}$ is a stratified symplectic space of depth 1, since $\overline{\mathcal{Z}_G(x_s) \cdot x_n} = \mathcal{Z}_G(x_s) \cdot x_n \cup \{0\}$, [6, §4.3]. The embedding $\overline{G(x)} \to \mathfrak{g}$ provides $\overline{G(x)}$ with a natural finitely generated C^∞-ring

$$C_1^\infty(\overline{G(x)}) := \left\{ f \in C^0(\overline{G(x)}) | \, f = \tilde{f}_{|\overline{G(x)}} \text{ for some } \tilde{f} \in C^\infty(\mathfrak{g}) \right\}.$$

Clearly, $C_1^\infty(\overline{G(x)})$ satisfies the first condition of Definition 3. The last two conditions in Definition 3 also hold, since $\overline{G(x)}$ is a fibration over $G(x_s)$ whose fiber is the cone $\overline{\mathcal{Z}_{G(x_s)}(x_n)}$ containing the origin $\{0\} \in \mathfrak{g}$. Thus $C_1^\infty(\overline{G(x)})$ is a smooth structure according to Definition 3.

The smooth structure $C_1^\infty(\overline{G(x)})$ is Poisson that is inherited from the Poisson structure on $\mathfrak{g}$. It is also weakly symplectic, since the symplectic form on $\overline{G(x)}$ is the restriction of the smooth 2-form $\omega_x(v, w) = \langle x, [v, w] \rangle$ on $\mathfrak{g}$.

2. We still assume that x_n is minimal in $\mathcal{Z}_{\mathfrak{g}}(x_s)$. In [26, Lemma 2] Panyushev showed that $\overline{G(x)}$ possesses an algebraic (Springer's) resolution of the singularity at $\{0\} \in \overline{G(x)} \subset \mathfrak{g}$. We will show that this resolution brings a resolvable smooth structure $C_2^\infty(\overline{G(x)})$. First we recall the construction in [26]. Let

   - h be a characteristic of x_n (i.e., $h \in \mathcal{Z}_{\mathfrak{g}}(x_s)$ is a semisimple element and $(h, x_n, y_n) \subset \mathcal{Z}_{\mathfrak{g}}(x_s)$ is an $\mathfrak{sl}_2$-triple).
   - $\mathcal{Z}_{\mathfrak{g}}(x_s)(i) := \left\{ s \in \mathcal{Z}_{\mathfrak{g}}(x_s) | [h, s] = is \right\}$.
   - $\mathfrak{n}_2(x_s) := \oplus_{i \geq 2} \mathcal{Z}_{\mathfrak{g}}(x_s)(i)$.
   - $P(x_s)$ denote a parabolic subgroup with the Lie algebra $lP(x_s) := \oplus_{i \geq 0} \mathcal{Z}_{\mathfrak{g}}(x_s)(i)$.
   - N_- – the connected Lie subgroup of $\mathcal{Z}_G(x_s)$ with Lie algebra $lN_-(x_s) := \oplus_{i<0} \mathcal{Z}_{\mathfrak{g}}(x_s)(i)$.

   It is known that $\overline{P(x_s) \cdot x_n} = \mathfrak{n}_2(x_s)$ and $\mathcal{Z}_{\mathcal{Z}_{G(x_s)}}(x_n) \subset P(x_s)$. Hence there exists a natural map

$$\tau : \mathcal{Z}_G(x_s) *_{P(x_s)} \mathfrak{n}_2(x_s) \to \overline{\mathcal{Z}_G(x_s) \cdot x_n}, \qquad g * n \mapsto gn,$$

   which is a resolution of the singularity of the cone $\overline{\mathcal{Z}_G(x_s) \cdot x_n}$. The resolution of the $\overline{G(x)}$ is obtained by considering the fibration F over the orbit $G(x_s)$ whose fiber over x_s is $\mathcal{Z}_G(x_s) *_{P(x_s)} \mathfrak{n}_2(x_s)$. The map τ extends to a map $\tilde{\tau} : F \to \overline{G(x)}$ as follows. Denote by (x_s, y) the point in the fiber over $x_s \in G(x_s)$ in F that is defined by $y \in \mathcal{Z}_G(x_s) *_{P(x_s)} \mathfrak{n}_2(x_s)$. Then we set $\tilde{\tau}(x_s, y) := \tau_{x_s}(y)$.

   The resolvable smooth structure $C_2^\infty(\overline{G(x)})$ is defined by $\tilde{\tau}$ as in Example 2.3. By Lemma 4 $C_2^\infty(\overline{G(x)})$ is locally smoothly contractible.

3. In addition, now we assume that $x_s = 0$, so $\mathcal{Z}_G(x_s) = G$, $P(x_s) = P$ and $\mathfrak{n}_2(x_s) = \mathfrak{n}_2$. In this case it has been shown in [26] that $\tau^*(\omega)$ is a smooth 2-form on $G *_P \mathfrak{n}_2$. It follows that $C_2^\infty(\overline{G(x)})$ is weakly symplectic. Panyushev also showed that $\tau^*(\omega)$ is symplectic if and only if x is even. (We refer the reader to [6] and [7] for a detailed description of nilpotent orbits.)

Lemma 7. *Assume that X is a stratified symplectic space with isolated conical singularities and $(\tilde{X}, \tilde{\omega}, \pi : \tilde{X} \to X)$ a smooth resolution of X such that $\tilde{\omega}$ is* *a symplectic form on* $\tilde{X}$ *and* $\pi^*(\omega_{|X^{\mathrm{reg}}}) = \tilde{\omega}_{|\pi^{-1}(X^{\mathrm{reg}})}$. *If for each singular point* $x \in X$ *the preimage* $\pi^{-1}(x)$ *is a coisotropic submanifold in* $\tilde{X}$, *then the obtained resolvable smooth structure* $C^\infty(X)$ *is Poisson.*

Proof. We define a Poisson bracket on $C^\infty(X)$ by setting $\{g, f\}_\omega(x) := \{\pi^* g, \pi^* f\}_{\tilde{\omega}}(\tilde{x})$, for $\tilde{x} \in \pi^{-1}(x)$. We will show that this definition does not depend on the choice of a particular $\tilde{x}$. By definition $\{\pi^* g, \pi^* f\}_{\tilde{\omega}}(\tilde{x}) := G_{\tilde{\omega}}(d\pi^* g, d\pi^* f)(\tilde{x})$. Since $\pi^* f$ and $\pi^* g$ are constant along the coisotropic submanifold $\pi^{-1}(x)$, we get $G_{\tilde{\omega}}(d\pi^* g, d\pi^* f)(\tilde{x}) = 0$. This proves Lemma 7. □

It has been showed in [1, §2], [26] that the preimage $\tau^{-1}(\overline{\mathcal{Z}_G(x_s) \cdot x_n} \setminus \mathcal{Z}_G(x_s) \cdot x_n)$ is a Lagrangian submanifold in $\mathcal{Z}_G(x_s) *_{P(x_s)} \mathfrak{n}_2(x_s)$ which is the cotangent bundle $T^*(\mathcal{Z}_G(x_s)/P(x_s))$ supplied with the natural symplectic structure. Using Lemma 7 we summarize our examination of Example 6 in the following.

Proposition 5. *Let* $x = x_n + x_s$ *where* x_n *is a minimal nilpotent element in* $\mathcal{Z}_{\mathfrak{g}}(x_s)$. *Then* $C_1^\infty(\overline{G(x)})$ *is a weakly symplectic and Poisson smooth structure. If* $x_s = 0$, *then* $C_2^\infty(\overline{G(x)})$ *is a weakly symplectic and Poisson smooth structure.*

3.3 The Existence of Hamiltonian Flows

Let (X, ω) be a stratified symplectic space and $C^\infty(X)$ a Poisson smooth structure on X. For any $H \in C^\infty(X)$ we define a linear operator $X_H : C^\infty(X) \to X^\infty(X)$ by

$$X_H(f) := \{f, H\}_\omega \text{ for } f \in C^\infty(X).$$

By definition, for given H, the value $X_H(f)(x)$ depends only on the value $df(x)$. Hence X_H is a section of the Zariski tangent bundle of X. We call X_H *the Hamiltonian vector field associated with* H.

Lemma 8. *The Hamiltonian vector field* X_H *is a smooth Zariski vector field on* X. *If* x *is a point in a stratum* S, *then* $X_H(x) \in T_x S$.

Proof. By definition of a Poisson structure, the function $X_H(f)$ is smooth for all $f \in C^\infty(X)$. Hence X_H is a smooth Zariski vector field. This proves the first assertion of Lemma 8. To prove the second assertion, it suffices to show that if the restriction of a function $f \in C^\infty(X)$ to a neighborhood $U_S(x) \subset S$ of a point $x \in S$ is zero, then $X_H(f)(x) = 0$. The last identity holds, since $X_H(f)(x)$ is equal to the Poisson bracket of the restriction of H and f to S. This completes the proof. □

Theorem 1 (cf. [25, §3]). *Given a Hamiltonian function $H \in C^\infty(X)$ and a point $x \in X$, there exists a unique smooth curve $\gamma : (-\varepsilon, \varepsilon) \to X$ such that for any $f \in C^\infty(X)$ we have*

$$\frac{d}{dt} f(\gamma(t)) = \{f, H\}. \tag{6}$$

Proof. Let $\Phi_t(x)$ be the flow that is generated by $X_H|_S$ on each stratum $S \subset X$. Since $C^\infty(X)$ is Poisson, the validity of (6) for $\gamma(t) := \Phi_t$ follows from Lemma 8. This proves the existence of a flow satisfying (6).

Now let us prove the uniqueness of the flow satisfying (6), using Sjamaar's and Lerman's argument in [25, §3]. Let $x \in X$ and $\gamma(t)$, $t \in (-\varepsilon_1, \varepsilon_1)$ be an integral curve of Eq. (6) with $\gamma_0(0) = x$. We will show that $\Phi_{-t}(\gamma(t)) = x$ for all $0 \leq t \leq \min(\varepsilon, \varepsilon_1)$. By Corollary 1 smooth functions on X separate points, therefore it suffices to show that for all $t \leq \min(\varepsilon, \varepsilon_1)$ and for all $f \in C^\infty(X)$, we have

$$f(\Phi_{-t}(\gamma_t(t))) = f(x). \tag{7}$$

As in [25, §3], using (6), we have

$$\frac{d}{dt} f(\Phi_{-t}(\gamma(t))) = \{H, f\}_\omega(\gamma(t)) + \{f, H\}_\omega(\gamma(t)) = 0.$$

This implies (7) and completes the proof of Theorem 1. □

Remark 6. 1. In Example 5 we proved that the smooth structure on a singular symplectic reduced space $(M^{2n}, \omega)//G$ defined by Sjamaar and Lerman in [25] is a Poisson smooth structure in sense of our definition. Thus, their result on the existence of a Hamiltonian flow on $(M^{2n}, \omega)//G$ in [25, §3] is a consequence of our Theorem 1.

2. In [25] Sjamaar and Lerman used a slightly different method for their proof of the existence of a Hamiltonian flow on the singular symplectic reduced space $(M^{2n}, \omega)//G$. They looked at the corresponding Hamiltonian flow on M and showed that this flow descends to a Hamiltonian flow on the reduced space.

3.4 Brylinski-Poisson Homology

In this subsection we extend the study of the Brylinski-Poisson homology of symplectic pseudomanifolds with isolated conical singularities in [16, §4] to the case of stratified symplectic spaces X equipped with a Poisson smooth structure.

Assume that $C^\infty(X)$ is a Poisson smooth structure. We consider *the canonical complex*

$$\to \Omega^{n+1}(X) \stackrel{\delta}{\to} \Omega^n(X) \to \cdots,$$

where δ is a linear operator defined as follows. Let $\alpha \in \Omega(X)$ and $\alpha = \sum_j f_0^j df_1^j \wedge df_p^j$ be a local representation of α as in Definition 4. Then we set (see [3, 12]):

$$\delta(f_0 df_1 \wedge \cdots \wedge df_n) := \sum_{i=1}^{n} (-1)^{i+1} \{f_0, f_i\}_\omega df_1 \wedge \cdots \wedge \widehat{df_i} \wedge \cdots \wedge df_n$$
$$+ \sum_{1 \le i < j \le n} (-1)^{i+j} f_0 d\{f_i, f_j\}_\omega \wedge df_1 \wedge \cdots \wedge \widehat{df_i} \wedge \cdots \wedge \widehat{df_j} \wedge \cdots \wedge df_n.$$

Lemma 9 (cf. [16, Lemma 4.3]).

1. *$\delta = i(G_\omega) \circ d - d \circ i(G_\omega)$. In particular, δ is well defined.*
2. *$\delta^2 = 0$.*

Proof. Recall that i denotes the canonical inclusion $X^{\mathrm{reg}} \to X$.

1. Let $\alpha \in \Omega(X)$, then $i^*\alpha \in \Omega_u(X^{\mathrm{reg}})$. Using

$$\delta \circ i^* = i^* \circ \delta, \; i^* \circ d = d \circ i^*,$$

   and the validity of the first assertion of Lemma 9 for any smooth Poisson manifold M [3, Lemma 1.2.1], we have

$$i^*(\delta\alpha) = \delta(i^*\alpha) = i^*(i(G_\omega) \circ d\alpha - d \circ i(G_\omega)\alpha). \tag{8}$$

   By Lemma 2, i^* is injective; hence the above equality implies the first assertion of Lemma 9.
2. The second statement of Lemma 9 is proved in the same way, using the injectivity of i^*. This completes the proof of Lemma 9.

□

The following Theorem 2 generalizes [16, Corollary 4.2].

Theorem 2. *Suppose (X, ω) is a stratified symplectic space equipped with a Poisson smooth structure $C^\infty(X)$ which is also weakly symplectic. If all regular strata of X have the same dimension $2n$, the Brylinski-Poisson homology of the complex $(\Omega(X), \delta)$ is isomorphic to the de Rham cohomology of X with reverse grading : $H_k(\Omega(X), \delta) = H^{2n-k}(\Omega(X), d)$. If, moreover, the smooth structure $C^\infty(X)$ is locally smoothly contractible, $H_k(\Omega(X), \delta)$ is equal to the singular cohomology $H^{2n-k}(X, \mathbb{R})$.*

Proof. Using the injectivity of i^*, we derive from (8) the following formulas for all k:

$$H_k(\Omega(X), \delta) = H_k(i^*(\Omega(X), \delta) \text{ and } H^k(\Omega(X), d) = H^k(i^*(\Omega(X), d). \tag{9}$$

Since $C^\infty(X)$ is weakly symplectic, there exists a symplectic form $\tilde{\omega} \in \Omega^2(X)$ such that $i^*(\tilde{\omega})|_{S_i} = \omega_i$ for any stratum S_i. Set vol $:= \tilde{\omega}^n/n!$. Let $\tilde{G}^k_\omega$ be the pairing: $\Lambda^k(T^*X) \times \Lambda^k(T^*X) \to C^\infty(X)$ associated with $\tilde{G}_\omega$ whose existence is shown in Remark 5.2. We define a symplectic star operator $*_\omega : \Omega^k(X) \to \Omega^{2n-k}(X)$ as follows (cf. [3, §2.1]).

$$*_\omega : \Omega^k(X) \to \Omega^{2n-k}(X),\ \beta \wedge *_\omega\alpha := \tilde{G}^k_\omega(\beta,\alpha) \wedge \frac{\tilde{\omega}^n}{n!},$$

for all $\alpha, \beta \in \Omega^k(X)$. In particular, on singular strata, the image of $*_\omega$ is zero.

For the sake of simplicity, we also denote by ω the restriction of ω to X^{reg}. The following proposition is proved by repeating the proof of Proposition 4.2 in [16] word by word, so we omit its proof.

Proposition 6. *We have* $*_\omega\left(i^*\left(\Omega^k(X)\right)\right) = i^*\left(\Omega^{2n-k}(X)\right)$.

The first assertion of Theorem 2 follows immediately from (9) and Proposition 6. The second assertion of Theorem 2 follows from [23]. This completes the proof of Theorem 2. □

3.5 A Lefschetz Decomposition

The notion of a Lefschetz decomposition on a symplectic manifold $\left(M^{2n},\omega\right)$ has been introduced by Yan in [29], where he gives an alternative proof the Mathieu Theorem on harmonic cohomology classes of $\left(M^{2n},\omega\right)$ in [20] using the Lefschetz decomposition. Roughly speaking, a Lefschetz decomposition on a symplectic manifold $\left(M^{2n},\omega\right)$ is an $\mathfrak{sl}_2$-module structure of $\Omega\left(M^{2n}\right)$. The Lie algebra $\mathfrak{sl}_2$ acting on $\Omega\left(M^{2n}\right)$ is generated by linear operators L, L^*, A defined as follows. L is the wedge multiplication by ω, $L^* := i(G_\omega)$, and $A = [L^*, L]$.

Now assume that (X,ω) is a stratified symplectic space provided with a Poisson smooth structure $C^\infty(X)$, which is also weakly symplectic. Then $\Omega(X)$ is stable under L, L^*. Hence we obtain immediately.

Lemma 10. *The space* $\Omega(X)$ *is an* $\mathfrak{sl}_2$*-module, where* $\mathfrak{sl}_2$ *is the Lie algebra generated by* $(L, L^*, A = [L, L^*])$.

The next Lemma concerns the Lefschetz decomposition of $\Omega_u(X^{\mathrm{reg}})$. Recall that $\alpha \in \Omega_u\left(X^{\mathrm{reg}}\right)$ is called *primitive*, if $L^*\alpha = 0$. We define the dimension function $\mathrm{d} : X \to \mathbb{Z}$ by setting $\mathrm{d}(x) := \dim S_x$, where S_x is the connected component of the stratum containing x.

Lemma 11. *1. For any* $\gamma \in \Omega^k_u(X^{\mathrm{reg}})$ *and any* $r \in \mathbb{Z}$, *we have*

$$[L^r, L^*]\gamma = \left(r\left(k - \frac{\mathrm{d}}{2}\right) + r(r-1)\right)L^{r-1}.$$

2. *There exists a function* $c : \mathbb{Z} \times \mathbb{Z} \to \mathbb{R}$ *such that for any primitive form* $\gamma \in \Omega(X^{\mathrm{reg}})$, *we have* $\gamma = c(\mathrm{d}, k) \cdot (L^*)^k \circ (L^k \gamma)$.

Proof. The first assertion of Lemma 11 for $r = 1$ is well known; see [29, Corollary 1.6]. For $r \geq 2$ we use the following formula:

$$[L^r, L^*] = L[L^{r-1}, L^*] + [L, L^*]L^{r-1},$$

which leads to the first assertion of Lemma 11 by induction.

2. The second assertion of Lemma 11 is proved by applying the first assertion recursively. □

For any $k \geq 0$ set

$$\mathcal{P}_k(X) := \left\{\alpha \in \Omega^k(X) | \, \alpha \text{ is primitive} \right\}.$$

Proposition 7. *Assume that* $C^\infty(X)$ *is both Poisson and weakly symplectic. If all regular strata of* X *have the same dimension* $2n$, *then we have the following Lefschetz decomposition for* $k \geq 0$

$$\Omega^k(X) = \mathcal{P}_k(X) \oplus L(\mathcal{P}_{k-2}(X)) \oplus \cdots .$$

Proof. Using the Lefschetz decomposition on symplectic manifold, for $\alpha \in \Omega^k(X)$ we decompose $i^*(\alpha) \in i^* \left(\Omega^k(X)\right)$ as

$$i^*(\alpha) = \alpha_p^k + L\left(\alpha_p^{k-2}\right) + \cdots + L^{[(n-k)/2]} \alpha_p^{n-k-2[(n-k)/2]}, \tag{10}$$

where α_p^j are primitive forms in $\Omega_u^j(X^{\mathrm{reg}})$. To prove Proposition 7, using the injectivity of i^*, it suffices to show that α_p^j are elements in $i^*(\Omega(X))$. Now let us consider the decomposition of $i^*(\alpha) \in \Omega_u^k(X^{\mathrm{reg}})$. We will show that all terms α_p^j can be obtained from a linear combination of $L^p(i^*(\alpha)), (L^*)^p \left(i^*(\alpha)\right), \, p \geq 0$, inductively on the degree j.

First we assume that k is even, i.e., $k = 2q$, hence $\alpha_p^0 \in C_u^\infty(X^{\mathrm{reg}})$. Applying to the both sides of (10) the operator L^{n-q}, we get

$$L^{n-q}\left(i^*(\alpha)\right) = \omega^n \cdot \alpha_p^0 \in i^*\left(\Omega^{2n}(X)\right). \tag{11}$$

By Proposition 6, (11) implies that $\alpha_p^0 \in i^*\left(C^\infty(X)\right)$, what is required to prove in the first induction step.

Now let us assume that $k = 2q + 1$. As in (11), we have

$$L^{n-q-1}\left(i^*(\alpha)\right) = \omega^{n-1} \cdot \alpha_p^1 \in i^*\left(\Omega^{2n-1}(X)\right). \tag{12}$$

Taking into account $L^*\alpha_p^1 = 0$, using Lemma 11 and (12), the term α_p^1 can be obtained from $L^{n-q-1}(i^*(\alpha))$ by applying the operator $c_{n,1} \cdot (L^*)^{n-1}$. Hence $\alpha_p^1 \in i^*(\Omega(X^{2n}))$, which completes the next induction step.

Repeating this procedure, we get all terms α_p^j, which belong to $i^*(\Omega(X))$. This completes the proof of Proposition 7. □

Since $[L, d] = 0$ holds on $\Omega_u(X^{\mathrm{reg}})$ and $i^*(\Omega(X))$ is stable under the action of d and L, the equality $[L, d] = 0$ also holds on $\Omega(X)$. In particular, the wedge product with $[\omega^k]$ maps $H^{l-k}(\Omega(X), d)$ to $H^{l+k}(\Omega(X), d)$ for any $l \in \mathbb{Z}$. A stratified symplectic space $\left(X^{2n}, \omega\right)$ of dimension $2n$ equipped with a Poisson weakly symplectic smooth structure $C^\infty\left(X^{2n}\right)$ is said *to satisfy the hard Lefschetz condition*, if the cup product

$$[\omega^k] : H^{n-k}\left(\Omega\left(X^{2n}\right), d\right) \to H^{n+k}\left(\Omega\left(X^{2n}\right), d\right)$$

is surjective for any $k \le n = \frac{1}{2}\dim X^{2n}$. A differential form $\alpha \in \Omega\left(X^{2n}\right)$ is called *harmonic*, if $d\alpha = 0 = \delta\alpha$. Let us abbreviate $H^*\left(\Omega\left(X^{2n}\right), d\right)$ by $H^*_{dR}\left(X^{2n}\right)$.

Theorem 3. *Let $\left(X^{2n}, \omega\right)$ be a stratified symplectic space and $C^\infty\left(X^{2n}\right)$ Poisson smooth structure which is also weakly symplectic. Assume that all regular strata of X have the same dimension $2n$. Then the following two assertions are equivalent:*

(1) *Any cohomology class in $H^*_{dR}\left(X^{2n}\right)$ contains a harmonic cocycle.*
(2) *$\left(X^{2n}, \omega\right)$ satisfies the hard Lefschetz condition.*

Proof. The proof of Theorem 3 for smooth symplectic manifolds by Yan in [29, Theorem 0.1] can be repeated word by word. For the convenience of the reader, we outline a proof here. Denote by $H^k_{hr}\left(X^{2n}\right)$ the space of all harmonic k-forms on $\left(X^{2n}, \omega\right)$, and let $H^*_{hr} = \oplus_{i=0}^{2n} H^i_{hr}\left(X^{2n}\right)$.

Now let us prove that the assertion (1) of Theorem 2 implies the assertion (2) of Theorem 2. We consider the following diagram:

$$\begin{array}{ccc} H^{n-k}_{hr}(X^{2n}) & \xrightarrow{L^k} & H^{n+k}_{hr}(X^{2n}) \\ \downarrow & & \downarrow \\ H^{n-k}_{dR}(X^{2n}) & \xrightarrow{L^k} & H^{n+k}_{dR}(X^{2n}). \end{array}$$

Let us recall that $i : X^{\mathrm{reg}} \to X^{2n}$ is the canonical inclusion. Since $[L, \delta] = -d$ [29, Lemma 1.2], which can be easily proved for (X^{2n}, ω) satisfying the condition of Theorem 3, Proposition 7 implies that $L^k : H^{n-k}_{hr}\left(X^{2n}\right) \to H^{n+k}_{hr}\left(X^{2n}\right)$ is an isomorphism. Since the vertical arrows in the diagram are surjective, we conclude that the second horizontal arrow in the diagram is also surjective. This proves (1) $\Longrightarrow$ (2).

Now let us prove that (2) $\Longrightarrow$ (1). Note that the condition (2) implies that [29, §3]

$$H_{dR}^{n-k}\left(X^{2n}\right) = \operatorname{Im} L + P_{n-k},$$

where $P_{n-k} := \left\{\alpha \in H_{dR}^{n-k}\left(X^{2n}\right) \mid L^{k+1}\alpha = 0 \in H_{dR}^{n+k+2}\left(X^{2n}\right)\right\}$.

Using induction argument, it suffices to prove that in each primitive cohomology class, $v \in P_{n-k}$ there is a harmonic cocycle. Let $v = [z]$, $z \in \Omega^{n-k}\left(X^{2n}\right)$. Since v is primitive we have $[z \wedge \omega^{k+1}] = 0 \in H_{dR}^{m+k+2}\left(X^{2n}\right)$. Hence, $z \wedge \omega^{k+1} = d\gamma$ for some $\gamma \in \Omega^{n+k+1}\left(X^{2n}\right)$. By Proposition 7 the operator $L^{k+1} : \Omega^{n-k-1}\left(X^{2n}\right) \to \Omega^{n+k+1}\left(X^{2n}\right)$ is onto, consequently there exists $\theta \in \Omega^{m-k-1}\left(X^{2n}\right)$ such that $\gamma = \theta \wedge \omega^{k+1}$. It follows that $(z - d\theta) \wedge \omega^{k+1} = 0$. Therefore $w = z - d\theta$ is primitive and closed. Since $[L^*, d] = \delta$ [29, Corollary 1.3], we obtain $\delta w = 0$. This completes the proof of Theorem 2. □

Remark 7. 1. If $C^\infty(X^{2n})$ is locally smoothly contractible, by [23, Theorem 5.2] the de Rham cohomology, $H^*(\Omega(X^{2n}), d)$ coincides with the singular cohomology $H^*(M, \mathbb{R})$), since X^{2n} admits smooth partitions of unity; see Proposition 1.
2. In [4, Proposition 5.4] Cavalcanti proved that the hard Lefschetz property on a compact symplectic manifold implies $\operatorname{Im}\delta \cap \ker d = \operatorname{Im} d \cap \operatorname{Im}\delta$; see also [22]. His Theorem can be proved word by word for stratified symplectic spaces satisfying the conditions of Proposition 7, since the main ingredient of the proof is Proposition 7.
3. It is interesting to know whether we can extend the symplectic cohomology theory developed in [17,28] to stratified symplectic spaces satisfying the conditions of Proposition 7, since the main ingredient of this theory is the existence of a Lefschetz decomposition.

4 Conclusions

In this paper we have worked out a natural refinement of the concept of a smooth structure on a stratified space, which is well suited for the study of stratified symplectic spaces. In this refined concept there are two natural classes of smooth structures on stratified symplectic spaces: weakly symplectic smooth structures and Poisson smooth structures. We show that Poisson smooth structures on stratified symplectic spaces X, especially those are also weakly symplectic, enjoy many properties of symplectic manifolds, if the regular strata of X are of the same dimension. We suggest to study stratified symplectic spaces, whose regular strata are of varying dimension, for future works.

References

1. Beauville, A.: Symplectic singularities. Invent. Math. **139**, 541–549 (2000)
2. Bates, L., Lerman, E.: Proper group actions and stratified symplectic spaces. Pac. J. Math. **181**, 201–229 (1997)
3. Brylinksi, J.C.: A differential complex for Poisson manifolds. JDG **28**, 93–114 (1988)
4. Cavalcanti, G.R.: New aspects of the dd^c-lemma. PhD thesis, University of Oxford (2005)
5. Chen, W., Ruan, Y.: Orbifold Gromov-Witten theory. In: Orbifolds in Mathematics and Physics (Madison, 2001). Contemporary Mathematics, vol. 310, pp. 25–85 American Mathematical Society, Providence (2002)
6. Collingwood, D.H., McGovern, W.M.: Nilpotent Orbits in Semisimple Lie Algebras. Van Nostrand Reinhold, New York (1993)
7. Fu, B.: A survey on symplectic singularities and symplectic resolutions. Ann. Math. Blaise Pascal **13**(2), 209–236 (2006)
8. Fu, B.: Symplectic resolution for nilpotent orbits. Invent. Math. **151**, 167–186 (2003)
9. Goresky, M., MacPherson, R.: Intersection homology theory. Topology **19**, 135–162 (1980)
10. Goresky, M., MacPherson, R.: Stratified Morse Theory. Springer, New York (1988)
11. González, J.A.N., de Salas, J.B.S.: C^∞-Differentiable Spaces. Lecture Notes in Mathematics, vol. 1824. Springer, Berlin (2003)
12. Kozsul, J.L.: Crochet de Schouten-Nijenhuis et cohomologie. "Elie Cartan et les Math. d'Aujour d'Hui", Asterisque hors-serie 251–171 (1985)
13. Huebschmann, J.: Kähler Spaces, Nilpotent Orbits, and Singular Reduction. Memoirs of the AMS, vol. 172(814). American Mathematical Society, Providence (2004). math.DG/0104213
14. Huebschmann, J., Rudolph, G., Schmidt, M.: A gauge model for quantum mechanics on a stratified space. Commun. Math. Phys. **286**, 459–494 (2009). hep-th/0702017
15. Karshon, Y.: An algebraic proof for the symplectic structure of moduli space. Proc. AMS **116**, 591–605 (1992)
16. Lê, H.V., Somberg, P., Vanzura, J.: Smooth structures on conical pseudomanifolds. Acta Math. Vietnam. **38**(1), 33–54 (2013). arxiv:1006.5707
17. Lê H.V., Vanžura, J.: Cohomology theories on locally conformally symplectic manifolds. arXiv:1111.3841 (accepted to Asian J. Math.)
18. Matsumura, H.: Commutative Algebra. Benjamin/Cummings, London (1980)
19. Mather, J.N.: Note on Topological Stability. Mimeographed Lecture Notes. Harvard, Cambridge (1970)
20. Mathieu, O.: Harmonic cohomology classes of symplectic manifolds. Comment. Math. Helv. **70**, 1–9 (1995)
21. Moerdijk, I., Reyes, G.E.: Models for Smooth Infinitesimal Analysis. Springer, New York (1991)
22. Merkulov, S.A.: Formality of canonical symplectic complexes and Frobenius manifolds. Int. Math. Res. Not. **14**, 727–733 (1998)
23. Mostow, M.: The differentiable structure of Milnor classifying spaces, simplicial complexes, and geometric realizations. JDG **14**, 255–293 (1979)
24. Nazaikinskii, V.E., Savin, A.Yu., Sternin, B.Yu., Shultse, B.V.: On the existence of elliptic problems on manifolds with edges. (Russian) Dokl. Akad. Nauk **395**(4), 455–458 (2004)
25. Sjamaar, R., Lerman, E.: Stratified Spaces and Reduction. Ann. Math. **134**, 375–422 (1991)
26. Panyushev, D.: Rationality of singularities and the Gorenstein properties of nilpotent orbits. Funct. Anal. Appl. **25**, 225–226 (1991)
27. Pflaum, M.J.: Analytic and Geometric Study of Stratified Spaces. Lecture Notes in Mathematics, vol. 1768. Springer, Berlin (2001)
28. Tseng, L.S., Yau, S.T.: Cohomology and hodge theory on symplectic manifolds, I, II. J. Diff. Geom. **91**, 383–443 (2012). arXiv:0909.5418, arXiv:1011.1250
29. Yan, D.: Hodge structure on symplectic manifolds. Adv. Math. **120**, 143–154 (1996)

Some Results on Chromaticity of Quasilinear Hypergraphs

Ioan Tomescu

1 Notation and Preliminary Results

A simple hypergraph $H = (V, \mathcal{E})$, with order $n = |V|$ and size $m = |\mathcal{E}|$, consists of a vertex-set $V(H) = V$ and an edge-set $E(H) = \mathcal{E}$, where $E \subseteq V$ and $|E| \geq 2$ for each edge E in $\mathcal{E}$. H is h-uniform, or is an h-hypergraph, if $|E| = h$ for each E in $\mathcal{E}$, and H is linear if no two edges intersect in more than one vertex [1]. H is said to be antilinear if for every two edges E, F of H we have $|E \cap F| \neq 1$. Let $r \geq 1$ and $h \geq 2r + 1$. H is said to be r-quasilinear (or shortly quasilinear) [13] if any two edges intersect in 0 or r vertices. Examples of quasilinear hypergraphs are t-stars [5,8], also called sunflower hypergraphs [7,11,12]. We say that a hypergraph S is a t-star with kernel K where $K \subseteq V(S)$ and $t \geq 1$ if S has exactly t edges and $e \cap e' = K$ for all distinct edges e and e' of S. A system of t pairwise disjoint edges (matching) is a t-star with empty kernel. In [12] a sunflower hypergraph was denoted by $SH(n, p, h)$; it is an h-hypergraph having a kernel of cardinality $h - p$, n vertices, and k edges, where $n = h + (k - 1)p$ and $1 \leq p \leq h - 1$.

A hypergraph for which no edge is a subset of any other is called Sperner. Two vertices $u, v \in V(H)$ belong to the same component if there are vertices $x_0 = u, x_1, \ldots, x_k = v$ and edges $E_1, \ldots, E_k$ of H such that $x_{i-1}, x_i \in E_i$ for each i $(1 \leq i \leq k)$ [1]. H is said to be connected if it has only one component. An h-uniform hypertree is a connected linear h-hypergraph without cycles. We shall define two classes of quasilinear uniform hypergraphs called quasilinear elementary paths and quasilinear elementary cycles and denoted by $P_m^{h,r}$ and $C_m^{h,r}$, respectively, as follows: $P_m^{h,r}$ consists of m edges $E_1, \ldots, E_m$ such that $|E_1| = \cdots = |E_m| = h$,

---

I. Tomescu (✉)
Faculty of Mathematics and Computer Science, University of Bucharest, Bucharest, Romania

Abdus Salam School of Mathematical Sciences, GC University, Lahore, Pakistan
e-mail: ioan@fmi.unibuc.ro

P. Cartier et al. (eds.), *Mathematics in the 21st Century*, Springer Proceedings in Mathematics & Statistics 98, DOI 10.1007/978-3-0348-0859-0_12

$|E_k \cap E_l| = r$ if $\{k,l\} = \{i, i+1\}$ for any $1 \le i \le m-1$ and 0 otherwise. Cycles $C_m^{h,r}$ are defined analogously, by also imposing $|E_m \cap E_1| = r$.

If $\lambda \in \mathbb{N}$, a λ-coloring of a hypergraph H is a function $f : V(H) \to \{1, \ldots, \lambda\}$ such that for each edge E of H, there exist x, y in E for which $f(x) \ne f(y)$. The number of λ-colorings of H is given by a polynomial $P(H, \lambda)$ of degree $|V(H)|$ in λ, called the chromatic polynomial of H. $P(H, \lambda)$ can be obtained applying inclusion-exclusion principle, in the same way as for graphs, getting the following formula:

$$P(H, \lambda) = \sum_{W \subseteq E(H)} (-1)^{|W|} \lambda^{c(W)}, \tag{1}$$

where $c(W)$ denotes the number of components of the spanning subhypergraph induced by the edges from W. By rearranging terms in (1), we obtain that if H has order n, then $P(H, \lambda) = \lambda^n + a_{n-1}\lambda^{n-1} + \cdots + a_1\lambda$, where

$$a_i = \sum_{j \ge 0} (-1)^j N(i, j) \tag{2}$$

and $N(i, j)$ denotes the number of spanning subhypergraphs of H with n vertices, i components, and j edges [10].

From (2) one deduces that

$$P\left(P_m^{h,r}, \lambda\right) = \lambda^n - m\lambda^{n-h+1} + \alpha_1\lambda^{n-2h+r+1} + \alpha_2\lambda^{n-2h+2} - \alpha_3\lambda^{n-3h+2r+1} + Q(\lambda), \tag{3}$$

where $Q(\lambda)$ is a polynomial of degree at most equal to $n - 3h + r + 2$, $\alpha_1 = m - 1$ is the number of subpaths $P_2^{h,r}$ of length 2, $\alpha_2 = \binom{m}{2} - m + 1$ is the number of pairs of pairwise disjoint edges, and $\alpha_3 = m - 2$ is the number of subpaths $P_3^{h,r}$ of length 3 in $P_m^{h,r}$. Also since any spanning subhypergraph of $P_m^{h,r}$ induced by less than m edges is not connected, it follows that in (3) the coefficient of λ is $(-1)^m$.

The chromatic polynomial of $P_m^{h,r}$ may be computed in polynomial time since the following property holds:

Theorem 1 ([14]). *The chromatic polynomial of $P_m^{h,r}$ is given by*

$$P\left(P_m^{h,r}, \lambda\right) = f_m(\lambda) + g_m(\lambda),$$

where

$$f_m(\lambda) = \left(\lambda^{h-2r+1} - 1\right) f_{m-1}(\lambda) + \lambda^{h-2r+1} g_{m-1}(\lambda),$$
$$g_m(\lambda) = \left(\lambda^{h-r} - \lambda^{h-2r+1}\right)\left(f_{m-1}(\lambda) + g_{m-1}(\lambda)\right)$$

and $f_1(\lambda) = \lambda^{h-r+1} - \lambda$, $g_1(\lambda) = \lambda^h - \lambda^{h-r+1}$.

All h-uniform hypertrees have the same chromatic polynomial.

Lemma 1 ([6]). *If T_k^h is any h-uniform hypertree with k edges, then*

$$P\left(T_k^h, \lambda\right) = \lambda\left(\lambda^{h-1} - 1\right)^k. \tag{4}$$

Two hypergraphs H and G are said to be chromatically equivalent or χ-equivalent, written $H \sim G$, if $P(H, \lambda) = P(G, \lambda)$. Let us restrict ourselves to the class of Sperner hypergraphs. A simple hypergraph H is said to be chromatically unique if H is isomorphic to H' for every simple hypergraph H' such that $H' \sim H$; that is, the structure of H is uniquely determined up to isomorphism by its chromatic polynomial. The notion of χ-unique graphs was first introduced and studied by Chao and Whitehead [4] (see also [9]). It is clear that all h-hypergraphs are Sperner. The notion of χ-uniqueness in the class of h-hypergraphs may be defined as follows: an h-hypergraph H is said to be h-chromatically unique if H is isomorphic to H' for every h-hypergraph H' such that $H' \sim H$.

2 Chromaticity of Some Quasilinear Hypergraphs

Nontrivial chromatically unique hypergraphs are extremely rare. One example of a nontrivial chromatically unique hypergraph was proposed by Borowiecki and Lazuka; it is $SH(n, 1, h)$.

Theorem 2 ([3]). *$SH(n, 1, h)$ is chromatically unique.*

The proof of this result was completed in [11]. Note that for $p = h - 1$, $SH(n, h - 1, h)$ is an h-uniform hypertree. The chromaticity of $SH(n, p, h)$ may be stated as follows:

Theorem 3 ([12]). *Let $n = h + (k - 1)p$, where $h \geq 3$, $k \geq 1$ and $1 \leq p \leq h - 1$. Then $SH(n, p, h)$ is h-chromatically unique for every $1 \leq p \leq h - 2$; for $p = h - 1$ $SH(n, h - 1, h)$ is h-chromatically unique for $k = 1$ or $k = 2$ but it has not this property for $k \geq 3$. Moreover, $SH(n, p, h)$ is not chromatically unique for every $p, k \geq 2$.*

$SH(n, p, h)$ is quasilinear with $r = h - p$ and it is a path for $k = 2$.

Since $P_2^{h,r}$ is a sunflower hypergraph $SH(n, p, h)$ with $p = h - r$ having $k = 2$ edges, from Theorem 4 it follows that $P_2^{h,r}$ is h-chromatically unique for every $1 \leq r \leq h - 1$. Also $P_m^{h,1}$ is an h-uniform hypertree; hence, for $m \geq 3$ it is not h-chromatically unique. Corollary 1 asserts that $P_m^{h,r}$ for every $m \geq 1$ and $C_m^{h,r}$ for every $m \geq 3$ are h-chromatically unique hypergraphs in the set of quasilinear hypergraphs provided $r \geq 2$ and $h \geq 3r - 1$. In [10] it was shown that $C_m^{h,r}$ is h-chromatically unique for $r = 1$ and every $m, h \geq 3$, but it is not chromatically unique for $r = 1$ and $m, h \geq 3$ [2]. The chromaticity of nonuniform hypertrees was studied by Walter [16].

The following property was deduced in [10]:

Theorem 4 ([10]). *If two h-hypergraphs H and G are χ-equivalent and H is linear, then G is linear too.*

The following result extends Theorem 6 to quasilinear hypergraphs.

Theorem 5 ([14]). *If h-hypergraphs H and G are χ-equivalent and H is quasilinear, then any two edges of G intersect in* 0, 1 *or r vertices and this result is tight.*

The proof of Theorem 6 stated below uses the following result about the first coefficients of the chromatic polynomial of a quasilinear h-hypergraph with a particular structure relatively to subhypergraphs induced by three edges.

Lemma 2 ([13]). *Let $r \geq 2$, $h \geq 2r + 1$ and H be a quasilinear h-hypergraph of order n and size m having the property that all subhypergraphs induced by three edges have one of the following patterns: (a) $P_3^{h,r}$, (b) $P_2^{h,r}$ and an isolated edge, or (c) three isolated edges. Then*

$$P(H,\lambda) = \lambda^n - m\lambda^{n-h+1} + \beta_1\lambda^{n-2h+r+1} + \beta_2\lambda^{n-2h+2} - \beta_3\lambda^{n-3h+2r+1} + R(\lambda), \tag{5}$$

where $R(\lambda)$ is a polynomial in λ of degree at most equal to $n - 3h + 2r$, β_2 is the number of pairwise disjoint edges of H, and β_1 and β_3 are the numbers of induced subhypergraphs of H isomorphic to $P_2^{h,r}$ and $P_3^{h,r}$, respectively.

It is trivial to see that $P_m^{h,r}$ for $1 \leq m \leq 3$ and $C_3^{h,r}$ are h-chromatically unique. For $m \geq 4$ this result is extended as follows:

Theorem 6 ([13]). *Let H be an antilinear h-hypergraph such that $P(H,\lambda) = P(G,\lambda)$, where G is $P_m^{h,r}(m \geq 1)$ or $C_m^{h,r}(m \geq 3)$. If $r \geq 2$ and $h \geq 3r - 1$, then H is isomorphic to G.*

The proof also uses two potential functions, α and β, for any h-uniform hypergraph K of size m: $\alpha(K) = \alpha_1(K) - m$ and $\beta(K) = \alpha_3(K) - m$, where $\alpha_1(K)$ and $\alpha_3(K)$ are the numbers of induced subhypergraphs of K isomorphic to $P_2^{h,r}$ and to $P_3^{h,r}$, respectively.

Since every quasilinear hypergraph is antilinear for every $r \geq 2$, we get the following:

Corollary 1 ([13]). *Let $r \geq 2, h \geq 3r - 1, m \geq 3$ and H be a quasilinear hypergraph such that $P(H,\lambda) = P(G,\lambda)$, where G is $P_m^{h,r}$ or $C_m^{h,r}$. Then H is isomorphic to G.*

Note that $P_m^{h,r}$ is not chromatically unique for any $m \geq 3, r \geq 1$ and $h \geq 2r + 1$, since by Theorem 7 stated below, any hypergraph containing a pendant path of length at least 2 is not chromatically unique [15].

Consider the hypergraph H, consisting of a subhypergraph H_1 of H; U and W are two edges such that $U \cap V(H_1) = A \neq \emptyset$, $U \cap W = B \neq \emptyset$, $W \cap V(H_1) = \emptyset$, $|U \backslash (A \cup B)| \geq 1$, and $|W \backslash U| \geq 1$. Such a path consisting of edges U and W will be called a pendant path of length 2.

Theorem 7 ([15]). *Every hypergraph containing a pendant path of length at least 2 is not chromatically unique.*

References

1. Berge, C.: Graphs and Hypergraphs. North-Holland, Amsterdam (1973)
2. Bokhary, S.A., Tomescu, I., Bhatti, A.A.: On the chromaticity of multi-bridge hypergraphs. Graph. Combin. **25**(2), 145–152 (2009)
3. Borowiecki, M., Łazuka, E.: Chromatic polynomials of hypergraphs. Discuss. Math. Graph Theory **20**, 293–301 (2000)
4. Chao, C.Y., Whitehead, E.G., Jr.: On chromatic equivalence of graphs. In: Alavi, Y., Lick, D.R. (eds.) Theory and Applications of Graphs. Lecture Notes in Mathematics, vol. 642, pp. 121–131. Springer, New York/Berlin (1978)
5. Dellamonica, D., Koubek, V., Martin, D.M., Rödl,V.: On a conjecture of Thomassen concerning subgraphs of large girth. J. Graph Theory **67**, 316–331 (2011)
6. Dohmen, K.: Chromatische Polynome von Graphen und Hypergraphen. Dissertation, Düsseldorf (1993)
7. Erdös, P., Rado, R.: Intersection theorems for systems of sets. J. Lond. Math. Soc. **35**, 85–90 (1960)
8. Füredi, Z.: On finite set-systems whose intersection is a kernel of a star. Discrete Math. **47**(1), 129–132 (1983)
9. Koh, K.M., Teo, K.L.: The search for chromatically unique graphs. Graph. Combin. **6**, 259–285 (1990)
10. Tomescu, I.: Chromatic coefficients of linear uniform hypergraphs. J. Comb. Theory Ser. B **2**(72), 229–235 (1998)
11. Tomescu, I.: Sunflower hypergraphs are chromatically unique. Discret. Math. **285**, 355–357 (2004)
12. Tomescu, I.: On the chromaticity of sunflower hypergraphs $SH(n, p, h)$. Discret. Math. **307**, 781–786 (2007)
13. Tomescu, I.: Some results on chromaticity of quasi-linear paths and cycles. Electron. J. Comb. **2**(19), P23 (2012)
14. Tomescu, I., Javed, S.: On the chromaticity of quasi-linear hypergraphs. Graph. Combin. **6**(29), 1921–1926 (2013)
15. Tomescu, I.: On chromaticity of hypergraphs with pendant paths. Discuss. Math. Graph Theory **34**(1), 23–29 (2014)
16. Walter, M.: Some results on chromatic polynomials of hypergraphs. Electron. J. Comb. **16**, R94 (2009)

Lecture on the *abc* Conjecture and Some of Its Consequences

Michel Waldschmidt

1 The Radical of a Positive Integer, *abc*–Triples and *abc*–Hits

According to the fundamental theorem of arithmetic, any integer $n \geq 2$ can be written as a product of prime numbers:

$$n = p_1^{a_1} p_2^{a_2} \cdots p_t^{a_t}.$$

The *radical* (also called *kernel* or *core*) $\mathrm{Rad}(n)$ of n is the product of the distinct primes dividing n:

$$\mathrm{Rad}(n) = p_1 p_2 \cdots p_t.$$

An *abc–triple* is a triple (a, b, c) of three positive coprime integers such that $a+b = c$ with $a < b$. The smallest example of an *abc*–triple is $(1, 2, 3)$.

An *abc–hit* is an *abc*–triple (a, b, c) such that $\mathrm{Rad}(abc) < c$. For instance, $(1, 8, 9)$ is an *abc*–hit, since $1 + 8 = 9$, $\gcd(1, 8, 9) = 1$ and

$$\mathrm{Rad}(1 \cdot 8 \cdot 9) = \mathrm{Rad}(2^3 \cdot 3^2) = 2 \cdot 3 = 6 < 9.$$

Among $15 \cdot 10^6$ *abc*–triples with $c < 10^4$, there are 120 *abc*–hits.[1]

---

[1] See the tables of http://rekenmeemetabc.nl/Synthese_resultaten

M. Waldschmidt (✉)
Université Pierre et Marie Curie-Paris 6, Institut de Mathématiques de Jussieu IMJ UMR 7586
Théorie des Nombres Case Courrier 247, 4 Place Jussieu F–75252, Paris Cedex 05, France
e-mail: michel.waldschmidt@imj-prg.fr

P. Cartier et al. (eds.), *Mathematics in the 21st Century*, Springer Proceedings in Mathematics & Statistics 98, DOI 10.1007/978-3-0348-0859-0_13

There are infinitely many *abc*–hits. Indeed, take $k \geq 1$, $a = 1$, $c = 3^{2^k}$, $b = c - 1$. By induction on k, one checks that 2^{k+2} divides $3^{2^k} - 1$. Hence,

$$\mathrm{Rad}\left(\left(3^{2^k} - 1\right) \cdot 3^{2^k}\right) \leq \frac{3^{2^k} - 1}{2^{k+1}} \cdot 3 < 3^{2^k}.$$

Hence,

$$\left(1, 3^{2^k} - 1, 3^{2^k}\right)$$

is an *abc*–hit. From this argument, one deduces (see [66]):

Lemma 1. *There exist infinitely many abc–triples* (a, b, c) *such that*

$$c > \frac{1}{6 \log 3} R \log R,$$

where $R = \mathrm{Rad}(abc)$.

It is not known (but conjectured in [26]) whether there are *abc*–triples (a, b, c) for which $c > \mathrm{Rad}(abc)^2$. The largest known value of λ for which there exists an *abc*–triple (a, b, c) with $c > \mathrm{Rad}(abc)^\lambda$ is $\lambda = 1.62991\ldots$, which is reached by Reyssat's example with

$$a = 2, \quad b = 3^{10} \cdot 109 = 6{,}436{,}341, \quad c = 23^5 = 6{,}436{,}343.$$

Indeed one checks

$$2 + 3^{10} \cdot 109 = 23^5, \quad \mathrm{Rad}\left(2 \cdot 3^{10} \cdot 109 \cdot 23^5\right) = 2 \cdot 3 \cdot 23 \cdot 109 = 15{,}042.$$

When (a, b, c) is an *abc*–triple, define

$$\lambda(a, b, c) = \frac{\log c}{\log \mathrm{Rad}(abc)}.$$

In 2014, there are 237 known values of $\lambda(a, b, c)$ which are ≥ 1.4. Besides Reyssat's example, the largest value of $\lambda(a, b, c)$ is $1.625990\ldots$, obtained by Benne de Weger

$$a = 11^2, \quad b = 3^2 \cdot 5^6 \cdot 7^3 = 48{,}234{,}375, \quad c = 2^{21} \cdot 23 = 48{,}234{,}496:$$

$$11^2 + 3^2 \cdot 5^6 \cdot 7^3 = 2^{21} \cdot 23, \qquad \mathrm{Rad}\left(2^{21} \cdot 3^2 \cdot 5^6 \cdot 7^3 \cdot 11^2 \cdot 23\right)$$
$$= 2 \cdot 3 \cdot 5 \cdot 7 \cdot 11 \cdot 23 = 53{,}130.$$

According to S. Laishram and T.N. Shorey [34], an explicit version, due to A. Baker [2], of the *abc* conjecture, namely, Conjecture 15 below, yields

Conjecture 1. For any *abc*–triple (a, b, c),

$$c < \mathrm{Rad}(abc)^{7/4}.$$

2 *abc* Conjecture

Here is the *abc* conjecture of Œsterlé [53] and Masser [45].

Conjecture 2. Let $\varepsilon > 0$. Then the set of *abc* triples (a, b, c) for which

$$c > \mathrm{Rad}(abc)^{1+\varepsilon}$$

is finite.

It is easily seen that Conjecture 2 is equivalent to the following statement:

- *For each* $\varepsilon > 0$, *there exists* $\kappa(\varepsilon)$ *such that, for any abc triple* (a, b, c),

$$c < \kappa(\varepsilon)\mathrm{Rad}(abc)^{1+\varepsilon}.$$

This may be viewed as a lower bound for $\mathrm{Rad}(abc)$ in terms of c.

An unconditional result in the direction of the *abc* conjecture has been obtained in 1986 by Stewart and Tijdeman [66] using lower bounds for linear combinations of logarithms, in the complex case as well as in the p–adic case:

$$\log c \le \kappa R^{15}$$

with an absolute constant κ. This estimate has been refined by Stewart and Yukunrui, who proved in 1991 [67]: for any $\varepsilon > 0$ and for c sufficiently large in terms of ε,

$$\log c \le \kappa(\varepsilon) R^{(2/3)+\varepsilon}.$$

In 2001, they refined in [68] the exponent $2/3$ to $1/3$ when they established the best known estimate so far:

Theorem 1 (Stewart-Yu Kunrui). *There exists an absolute constant* κ *such that any abc triple* (a, b, c) *satisfies*

$$\log c \le \kappa R^{1/3} (\log R)^3$$

with $R = \mathrm{Rad}(abc)$. *In other terms,*

$$c \le e^{\kappa R^{1/3}(\log R)^3}.$$

J. Œsterlé and A. Nitaj (see [51]) proved that the *abc* conjecture implies the truth of a previous conjecture by L. Szpiro on the conductor of elliptic curves (see [31] p. 227):

Conjecture 3 (Szpiro's Conjecture). Given any $\varepsilon > 0$, there exists a constant $C(\varepsilon) > 0$ such that, for every elliptic curve with minimal discriminant Δ and conductor N,

$$|\Delta| < C(\varepsilon) N^{6+\varepsilon}.$$

According to [32, 36, 78], the next statement is equivalent to the *abc* conjecture.

Conjecture 4 (Generalized Szpiro's Conjecture). Given any $\varepsilon > 0$ and $M > 0$, there exists a constant $C(\varepsilon, M) > 0$ such that, for all integers x and y such that the number $D = 4x^3 - 27y^2$ is not 0 and such that the greatest prime factor of x and y is bounded by M,

$$\max\left\{|x|^3, y^2, |D|\right\} < C(\varepsilon, M)\mathrm{Rad}(D)^{6+\varepsilon}.$$

In view of Conjecture 3, it is natural to introduce another exponent related with the *abc* conjecture. When (a, b, c) is an *abc* triple, define

$$\varrho(a, b, c) = \frac{\log abc}{\log \mathrm{Rad}(abc)}.$$

From the *abc* conjecture, it follows that for any $\varepsilon > 0$, there are only finitely many *abc*–triples (a, b, c) such that $\varrho(a, b, c) > 3 + \varepsilon$.

Here are the two largest known values for $\varrho(a, b, c)$, both found by A. Nitaj [51]

$a + b = c$	$\varrho(a, b, c)$
$13 \cdot 19^6 + 2^{30} \cdot 5 = 3^{13} \cdot 11^2 \cdot 31$	$4.41901\ldots$
$2^5 \cdot 11^2 \cdot 19^9 + 5^{15} \cdot 37^2 \cdot 47 = 3^7 \cdot 7^{11} \cdot 743$	$4.26801\ldots$

In 2013, there are 47 known *abc*-triples (a, b, c) satisfying $\varrho(a, b, c) > 4$.

In 2006, the Mathematics Department of Leiden University in the Netherlands, together with the Dutch Kennislink Science Institute, launched the `ABC@Home` project, a grid computing system which aims to discover additional *abc*–triples. Although no finite set of examples or counterexamples can resolve the *abc* conjecture, it is hoped that patterns in the triples discovered by this project will lead to insights about the conjecture and about number theory more generally. `ABC@Home` is an educational and nonprofit distributed computing project finding *abc*–triples related to the *abc* conjecture.

Surveys on the *abc* conjecture have been written by S. Lang [36, 37]; (see also §7 p. 194–200 of [38]), by A. Nitaj [52] and W.M. Schmidt [61] Epilogue p. 205.

The *congruence abc conjecture* is discussed in [50], §5.5, and 5.6.

Generalizations of the *abc* conjecture to more than three numbers, namely, to $a_1 + \cdots + a_n = 0$, have been investigated by J. Browkin and J. Brzeziński [14] in 1994 and by Hu, Pei-Chu and Yang, Chung-Chun in 2002 [30].

3 Consequences

3.1 *Fermat's Last Theorem*

Assume x, y, z, n are positive integers satisfying $x^n + y^n = z^n$, $\gcd(x, y, z) = 1$ and $x < y$. Then (x^n, y^n, z^n) is an *abc*–triple with

$$\mathrm{Rad}(x^n y^n z^n) \le xyz < z^3.$$

If the explicit *abc* conjecture $c < \mathrm{Rad}(abc)^2$ of [26] is true, then one deduces $z^n < z^6$, hence $n \le 5$.

3.2 *Perfect Powers*

Define a *perfect power* as a positive integer of the form a^b where a and b are positive integers and $b \ge 2$. The sequence of perfect powers starts with

1, 4, 8, 9, 16, 25, 27, 32, 36, 49, 64, 81, 100, 121, 125, 128,

144, 169, 196, 216, 225, 243, 256, 289, 324, 343, 361, 400,

441, 484, 512, 529, 576, 625, 676, 729, 784, 841, 900, 961,

1,000, 1,024, 1,089, 1,156, 1,225, 1,296, 1,331, 1,369,

1,444, 1,521, 1,600, 1,681, 1,728, 1,764, ...

The reference of this sequence in Sloane's Encyclopaedia of Integer Sequences is http://oeis.org/A001597.

From the *abc* conjecture 2, one easily deduces the following conjecture due to Subbayya Sivasankaranarayana Pillai [55] (see also [56, 57])

Conjecture 5 (Pillai). In the sequence of perfect powers, the difference between two consecutive terms tends to infinity.

Pillai's Conjecture 5 can also be stated in an equivalent way as follows:

- *Let k be a positive integer. The equation*

$$x^p - y^q = k,$$

where the unknowns x, y, p, and q take integer values, all ≥ 2, has only finitely many solutions (x, y, p, q).

For $k = 1$, Mihăilescu's solution of Catalan's Conjecture states that the only solution to Catalan's equation [6, 57]

$$x^p - y^q = 1$$

is $3^2 - 2^3 = 1$. It is a remarkable fact that there is no value of $k \geq 2$ for which one knows that Pillai's equation $x^p - y^q = k$ has only finitely many solutions.

The *abc* conjecture implies the following stronger version of Pillai's Conjecture (see the introduction of Chapters X and XI of [35]):

Conjecture 6 (Lang-Waldschmidt). Let $\varepsilon > 0$. There exists a constant $c(\varepsilon) > 0$ with the following property. If $x^p \neq y^q$, then

$$|x^p - y^q| \geq c(\varepsilon) \max\{x^p, y^q\}^{\kappa - \varepsilon} \quad \text{with} \quad \kappa = 1 - \frac{1}{p} - \frac{1}{q}.$$

The motivation of this conjecture in [35] is the quest for a strong (essentially optimal) lower bound for linear combinations of logarithms of algebraic numbers.

P. Vojta, in [78] Chap.V appendix *ABC*, explained connections between various conjectures. Here is a figure from that reference:

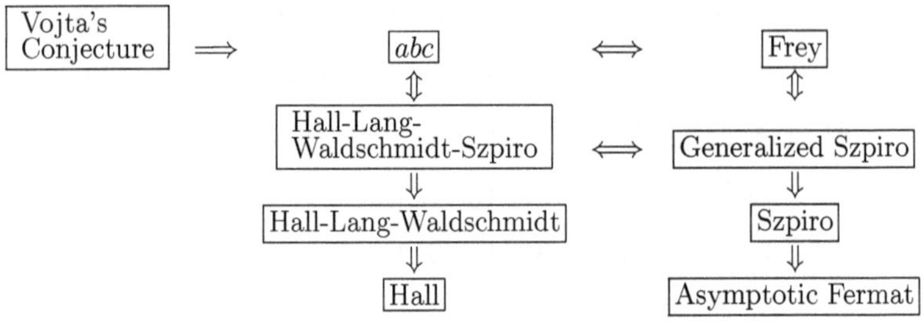

In the special case $p = 3$, $q = 2$, Conjecture 6 reads: *If $x^3 \neq y^2$, then*

$$\left|x^3 - y^2\right| \geq c(\varepsilon) \max\left\{x^3, y^2\right\}^{(1/6) - \varepsilon}.$$

In 1971, Marshall Hall Jr [28] proposed a stronger conjecture without the ε (what is called *Hall's Conjecture* in [78] has the ε):

Conjecture 7 (M. Hall Jr.). There exists an absolute constant $c > 0$ such that, if $x^3 \neq y^2$, then

$$\left|x^3 - y^2\right| \geq c \max\left\{x^3, y^2\right\}^{1/6}.$$

This statement does not follow from the *abc* conjecture 2. In [28], M. Hall Jr discusses possible values for his constant c in Conjecture 7. In the other direction, L.V. Danilov [17] (see also [32]) proved that the inequality

$$0 < \left|x^3 - y^2\right| < 0.971|x|^{1/2}$$

has infinitely many solutions in integers x, y. According to F. Beukers and C.L. Stewart [5], this conjecture maybe too optimistic. Indeed they conjecture:

Conjecture 8 (Beukers–Stewart). Let p, q be coprime integers with $p > q \geq 2$. Then, for any $c > 0$, there exist infinitely many positive integers x, y such that

$$0 < |x^p - y^q| < c \max\{x^p, y^q\}^\kappa \quad \text{with} \quad \kappa = 1 - \frac{1}{p} - \frac{1}{q}\cdot$$

3.3 *Generalized Fermat Equation*

Consider the equation (see for instance [72])

$$x^p + y^q = z^r \tag{1}$$

where the unknowns (x, y, z, p, q, r) take their values in the set of tuples of positive integers for which x, y, z are relatively prime and p, q, r are ≥ 2. Define

$$\chi = \frac{1}{p} + \frac{1}{q} + \frac{1}{r} - 1\cdot$$

If $\chi \geq 0$, then (p, q, r) is a permutation of one of

$$(2,2,k) \quad (k \geq 2), \qquad (2,3,3), \qquad (2,3,4),$$
$$(2,3,5), \qquad (2,4,4), \qquad (2,3,6), \qquad (3,3,3);$$

in each of these cases, all solutions (x, y, z) are known, often there are infinitely many of them (see [4, 16, 33, 34]).

Assume now $\chi < 0$. Then only 10 solutions (x, y, z, p, q, r) with x, y, z relatively prime (up to obvious symmetries) to Eq. (1) are known; by increasing order for z^r, they are:

$$1 + 2^3 = 3^2, \quad 2^5 + 7^2 = 3^4, \quad 7^3 + 13^2 = 2^9, \quad 2^7 + 17^3 = 71^2,$$
$$3^5 + 11^4 = 122^2, \quad 33^8 + 1{,}549{,}034^2 = 15{,}613^3,$$
$$1{,}414^3 + 2{,}213{,}459^2 = 65^7, \quad 9{,}262^3 + 15{,}312{,}283^2 = 113^7,$$
$$17^7 + 76{,}271^3 = 21{,}063{,}928^2, \quad 43^8 + 96{,}222^3 = 30{,}042{,}907^2.$$

Beal's problem, including a 50,000 US$ prize (see [47]), is:

Problem 9 (Beal's Problem). Assume $\chi < 0$. Either find another solution to Eq. (1) or prove that there is no further solution.

A related conjecture, due to R. Tijdeman and D. Zagier [47], is:

Conjecture 10 (Tijdeman-Zagier). Equation (1) has no solution in positive integers (x, y, z, p, q, r) with each of p, q, and r at least 3 and with x, y, z relatively prime.

The next conjecture is proposed by H. Darmon and A. Granville [18]:

Conjecture 11 (Fermat-Catalan Conjecture). The set of solutions (x, y, z, p, q, r) with $\chi < 0$ to Eq. (1) is finite.

It is easy to deduce Conjecture 11 from the *abc* conjecture 2, once one notices that for p, q, r positive integers, the assumption $\chi < 0$ implies

$$\chi \le -\frac{1}{42}.$$

In 1995, H. Darmon and A. Granville [18] proved unconditionally that for fixed (p, q, r) with $\chi < 0$, there are only finitely many (x, y, z) satisfying Eq. (1).

3.4 Wieferich Primes

A *Wieferich prime* is a prime number p such that p^2 divides $2^{p-1} - 1$. Note that the definition in [50], §5.4 is the opposite. The only known Wieferich primes below $4 \cdot 10^{12}$ are 1,093 and 3,511.

J.H. Silverman [64] showed that if the *abc* conjecture 2 is true, given a positive integer $a > 1$, there exist infinitely many primes p such that p^2 does not divide $a^{p-1} - 1$. A consequence is that there are infinitely many primes which are not Wieferich primes, a result which is known only if one assumes the *abc* conjecture. See also [27].

3.5 Erdős–Woods Conjecture

There are infinitely many pairs of positive integers (x, y) with $x < y$ such that x and y have the same radical, and, at the same time, $x + 1$ and $y + 1$ have the same radical. Indeed, for $k \geq 1$, the pair of numbers (x, y) with

$$x = 2^k - 2 = 2\left(2^{k-1} - 1\right) \quad \text{and} \quad y = \left(2^k - 1\right)^2 - 1 = 2^{k+1}\left(2^{k-1} - 1\right)$$

satisfy this condition, since

$$x + 1 = 2^k - 1 \quad \text{and} \quad y + 1 = \left(2^k - 1\right)^2.$$

There is one further sporadic known example, namely, $(x, y) = (75, 1{,}215)$, since

$$75 = 3 \cdot 5^2 \quad \text{and} \quad 1{,}215 = 3^5 \cdot 5 \quad \text{with} \quad \text{Rad}(75) = \text{Rad}(1{,}215) = 3 \cdot 5 = 15,$$

while

$$76 = 2^2{\cdot}19 \quad \text{and} \quad 1{,}216 = 2^6{\cdot}19 \quad \text{with} \quad \text{Rad}(76) = \text{Rad}(1{,}216) = 2{\cdot}19 = 38.$$

It is not known whether there are further examples. It is not even known whether there exist two distinct integers x, y such that

$$\text{Rad}(x) = \text{Rad}(y), \quad \text{Rad}(x{+}1) = \text{Rad}(y{+}1), \quad \text{and} \quad \text{Rad}(x{+}2) = \text{Rad}(y{+}2).$$

The comparatively weaker assertion below [39–42] would have interesting consequences in logic:

Conjecture 12 (Erdős–Woods Conjecture). There exists an absolute constant k such that, if x and y are positive integers satisfying

$$\text{Rad}(x + i) = \text{Rad}(y + i)$$

for $i = 0, 1, \ldots, k - 1$, then $x = y$.

M. Langevin [39, 41, 42] (cf. [32]) proved that this conjecture follows from the *abc* conjecture 2. See also [35] and [3] for connections with Conjectures 6 and 7.

3.6 Warings's Problem

In 1770, a few months before J.L. Lagrange solved a conjecture of Bachet (1621) and Fermat (1640) by proving that every positive integer is the sum of at most four squares of integers, E. Waring wrote (see [81]):

- Omnis integer numerus vel est cubus, vel e duobus, tribus, 4, 5, 6, 7, 8, vel novem cubis compositus, est etiam quadrato-quadratus vel e duobus, tribus, &, usque ad novemdecim compositus, & sic deinceps[2]

Waring's function g is defined as follows: *For any integer* $k \geq 2$, $g(k)$ *is the least positive integer* s *such that any positive integer* N *can be written* $x_1^k + \cdots + x_s^k$.

For each integer $k \geq 2$, define $I(k) = 2^k + \left[(3/2)^k\right] - 2$. It is easy to show that $g(k) \geq I(k)$ (this result is due to J. A. Euler, son of Leonhard Euler). Indeed, write the Euclidean division of 3^k by 2^k, with quotient q and remainder r:

$$3^k = 2^k q + r \quad \text{with} \quad 0 < r < 2^k, \quad q = \left[\left(\frac{3}{2}\right)^k\right]$$

and consider the integer

$$N = 2^k q - 1 = (q-1)2^k + \left(2^k - 1\right) 1^k.$$

Since $N < 3^k$, writing N as a sum of kth powers can involve no term 3^k, and since $N < 2^k q$, it involves at most $(q-1)$ terms 2^k, all others being 1^k; hence it requires a total number of at least $(q-1) + \left(2^k - 1\right) = I(k)$ terms.

The next conjecture [49] is due to C.A. Bretschneider (1853).

Conjecture 13 (Ideal Waring's Theorem). For any $k \geq 2$, $g(k) = I(k)$.

A slight improvement of the upper bound $r < 2^k$ for the remainder $r = 3^k - 2^k q$ would suffice for proving Conjecture 13. Indeed, L.E. Dickson and S.S. Pillai (see, for instance, [29], Chap. XXI or [49] p. 226, Chap. IV) proved independently in 1936 that if $r = 3^k - 2^k q$ satisfies

$$r \leq 2^k - q - 2, \tag{2}$$

then $g(k) = I(k)$. The condition (2) is satisfied for $4 \leq k \leq 471{,}600{,}000$.

According to K. Mahler, the upper bound (2) is valid for all sufficiently large k. Hence the ideal Waring's Theorem

$$g(k) = I(k)$$

holds also for all sufficiently large k. However, Mahler's proof uses a p–adic Diophantine argument related to the Thue–Siegel–Roth Theorem which does not yield effective results.

---

[2] Every integer is a cube or the sum of two, three, ... nine cubes; every integer is also the square of a square, or the sum of up to nineteen such; and so forth. Similar laws may be affirmed for the correspondingly defined numbers of quantities of any like degree.

S. David (see [55]) noticed that the estimate (2) for sufficiently large k follows from the *abc* Conjecture 2. S. Laishram checked that the ideal Waring's Theorem $g(k) = I(k)$ follows from the explicit *abc* conjecture 15.

3.7 A Problem of P. Erdős Solved by C.L. Stewart

Let us denote by $P(m)$ the greatest prime factor of an integer $m \geq 2$. In 1965, P. Erdős conjectured

$$\frac{P(2^n - 1)}{n} \to \infty \quad \text{when} \quad n \to \infty.$$

In 2002, R. Murty and S. Wong [48] proved that this is a consequence of the *abc* conjecture 2.

In 2012, C.L. Stewart [65] proved Erdős's Conjecture (in a wider context of Lucas and Lehmer sequences) in the stronger form:

$$P(2^n - 1) > n \exp\bigl(\log n / 104 \log\log n\bigr).$$

4 Stronger than *abc*: Best Possible Estimate?

Let $\delta > 0$. In 1986, C.L. Stewart and R. Tijdeman [66] proved that there are infinitely many *abc*–triples (a, b, c) for which

$$c > R \exp\left((4 - \delta) \frac{(\log R)^{1/2}}{\log\log R} \right).$$

This is much better than the lower bound $c > R \log R$ obtained in Lemma 1. The coefficient $4 - \delta$ has been improved by M. van Frankenhuijsen [74] into 6.068 in 2000. In the same paper, M. van Frankenhuijsen suggested that there may exist two positive absolute constants κ_1 and κ_2 such that, for any *abc*–triples (a, b, c),

$$c < R \exp\left(\kappa_1 \left(\frac{\log R}{\log\log R} \right)^{1/2} \right),$$

while for infinitely many *abc*–triples (a, b, c),

$$c > R \exp\left(\kappa_2 \left(\frac{\log R}{\log\log R} \right)^{1/2} \right).$$

O. Robert, C.L. Stewart and G. Tenenbaum suggest in [58] the following more precise limit for the *abc* conjecture, which would yield these statements with $\kappa_1 = 4\sqrt{3} + \varepsilon$ for c sufficiently large in terms of ε and $\kappa_2 = 4\sqrt{3} - \varepsilon$ for any $\varepsilon > 0$.

Conjecture 14 (Robert-Stewart-Tenenbaum). There exist positive constants $\kappa_1, \kappa_2, \kappa_3$ such that, for any *abc*–triple (a, b, c) with $R = \mathrm{Rad}(abc)$,

$$c < \kappa_1 R \exp\left(4\sqrt{3}\left(\frac{\log R}{\log\log R}\right)^{1/2}\left(1 + \frac{\log\log\log R}{2\log\log R} + \frac{\kappa_2}{\log\log R}\right)\right)$$

and there exist infinitely many *abc*–triples (a, b, c) for which

$$c > R \exp\left(4\sqrt{3}\left(\frac{\log R}{\log\log R}\right)^{1/2}\left(1 + \frac{\log\log\log R}{2\log\log R} + \frac{\kappa_3}{\log\log R}\right)\right).$$

The only heuristic argument used in [58] is that, whenever a and b are relatively prime positive integers, the radicals of a, b and $a + b$ are statistically independent. The estimates from conjecture 14 are based on the work [59] on the number of positive integers $N(x, y)$ bounded by x whose radical is at most y.

5 Explicit *abc* Conjecture

In 1996, A. Baker [1] suggested the following statement. Let (a, b, c) be an *abc*–triple and let $\varepsilon > 0$. Then

$$c \le \kappa\left(\varepsilon^{-\omega} R\right)^{1+\varepsilon}$$

where κ is an absolute constant, $R = \mathrm{Rad}(abc)$ and $\omega = \omega(abc)$ is the number of distinct prime factors of *abc*.

A. Granville noticed that the minimum of the function on the right-hand side over $\varepsilon > 0$ occurs essentially with $\varepsilon = \omega/\log R$. This incited Baker [2] to propose a slightly sharper form of his previous conjecture, namely,

$$c \le \kappa R \frac{(\log R)^{\omega}}{\omega!}.$$

He made some computational experiments in order to guess an admissible value for his absolute constant κ, and he ended up with the following precise statement:

Conjecture 15 (Explicit *abc* Conjecture). Let (a, b, c) be an *abc*–triple. Then

$$c \le \frac{6}{5} R \frac{(\log R)^\omega}{\omega!},$$

with $R = \mathrm{Rad}(abc)$ and $\omega = \omega(abc)$.

P. Philippon in 1999 [54] (Appendix) pointed out how sharp lower bounds for linear forms in logarithms, involving several metrics, would imply the *abc* conjecture. Effective and explicit versions of the *abc* conjecture have plenty of consequences [8, 10, 13, 34, 60]. Here is a very few set of examples.

The *Nagell–Ljunggren equation* is the equation

$$y^q = \frac{x^n - 1}{x - 1}$$

where the unknowns x, y, n, q take their values in the set of tuples of positive integers satisfying $x > 1$, $y > 1$, $n > 2$ and $q > 1$. This equation means that in basis x, all the digits of the perfect power y^q are 1 (this is a so–called *repunit*).

According to [34], if the explicit *abc* conjecture 15 of Baker is true, then the only solutions are

$$11^2 = \frac{3^5 - 1}{3 - 1}, \quad 20^2 = \frac{7^4 - 1}{7 - 1}, \quad 7^3 = \frac{18^3 - 1}{18 - 1}.$$

Further consequences of the explicit *abc* conjecture 15 are discussed in [34], in particular on the *Goormaghtigh*'s Conjecture, which states that the only numbers with at least three digits and with all digits equal to 1 in two different bases are 31 (in bases 2 and 5) and 8,191 (in bases 2 and 90):

$$\frac{5^3 - 1}{5 - 1} = \frac{2^5 - 1}{2 - 1} = 31 \quad \text{and} \quad \frac{90^3 - 1}{90 - 1} = \frac{2^{13} - 1}{2 - 1} = 8{,}191.$$

In other terms, the Goormaghtigh's Conjecture asserts that if (x, y, m, n) is a tuple of positive integers satisfying $x > y > 1$, $n > 2$, $m > 2$ and

$$\frac{x^m - 1}{x - 1} = \frac{y^n - 1}{y - 1},$$

then (x, y, m, n) is either $(5, 2, 3, 5)$ or $(90, 2, 3, 13)$. Surveys on such questions have been written by T.N. Shorey [62, 63].

6 *abc* for Number Fields

In 1991, N. Elkies [19] deduced Faltings's Theorem on the finiteness of the set of rational points on an algebraic curve of genus ≥ 2 (previously *Mordell's conjecture*) from a generalization he proposed of the *abc* conjecture to number fields. See also [26].

In 1994, E. Bombieri [7] deduced from a generalization of the *abc* conjecture to number fields a refinement of the Thue–Siegel–Roth Theorem on the rational approximation of algebraic numbers

$$\left|\alpha - \frac{p}{q}\right| > \frac{1}{q^{2+\varepsilon}},$$

where he replaces ε by

$$\kappa(\log q)^{-1/2}(\log\log q)^{-1},$$

with κ depending only on the algebraic number α.

A. Granville and H.M. Stark [25] proved that the uniform *abc* conjecture for number fields implies a lower bound for the class number of an imaginary quadratic number field; K. Mahler had shown that this implies that the associated L–function has no Siegel zero. See also [26].

Further work on the *abc* conjecture for number fields (see [8]) are due to M. van Frankenhuijsen [73–75,77], N. Broberg [9], J. Browkin [11,12], A. Granville and H.M. Stark [25], K. Győry, D.W. Masser [46], A. Surroca [70,71], P.C. Hu and C.C. Yang [31] § 5.6 and [32].

7 Further Consequences of the *abc* Conjecture

Further consequences of the *abc* conjecture 2 are quoted in [51], including:

- Erdős's Conjecture on consecutive powerful numbers. The *abc* conjecture 2 implies that the set of triples of consecutive powerful integers (namely, integers of the form a^2b^3) is finite. R. Mollin and G. Walsh conjecture that there is no such triple.
- Dressler's Conjecture: between two positive integers having the same prime factors, there is always a prime.
- Square-free and power-free values of polynomials [15,24].
- Lang's conjectures: lower bounds for heights, number of integral points on elliptic curves [20–22].
- Bounds for the order of the Tate–Shafarevich group [23].
- Vojta's Conjecture for curves [78–80].
- Greenberg's Conjecture on Iwasawa invariants λ and μ in cyclotomic extensions.

- Exponents of class groups of quadratic fields.
- Fundamental units in quadratic and biquadratic fields.

8 *abc* and Meromorphic Function Fields

There is a rich theory related with Nevanlinna value distribution theory. See, for instance, P. Vojta [78–80], Machiel van Frankenhuijsen [75, 76], Hu, Pei–Chu and Yang, Chung-Chun [30–32]. Notice in particular that Vojta's Conjecture on algebraic points of bounded degree on a smooth complete variety over a global field of characteristic zero implies the *abc* conjecture 2.

9 *ABC* Theorem for Polynomials

We end this lecture with a proof of an analog of the *abc* conjecture for polynomials – see, for instance, [26, 38].

Let K be an algebraically closed field. The *radical* of a monic polynomial

$$P(X) = \prod_{i=1}^{n}(X - \alpha_i)^{a_i} \in K[X],$$

with α_i pairwise distinct, is defined as

$$\mathrm{Rad}(P)(X) = \prod_{i=1}^{n}(X - \alpha_i) \in K[X].$$

The following result is due to W.W. Stothers [69] and R. Mason [43, 44]. It can also be deduced from earlier results by A. Hurwitz.

Theorem 2 (*ABC* **Theorem).** *Let A, B, C be three relatively prime polynomials in $K[X]$ with $A + B = C$ and let $R = \mathrm{Rad}(ABC)$. Then*

$$\max\{\deg(A), \deg(B), \deg(C)\} < \deg(R).$$

This result can be compared with the *abc* conjecture 2, where the degree of a polynomial replaces the logarithm of a positive integer.

The proof uses the remark that the radical is related with a gcd: for $P \in K[X]$ a monic polynomial, we have

$$\mathrm{Rad}(P) = \frac{P}{\gcd(P, P')}. \tag{3}$$

Indeed, the common zeroes of

$$P(X) = \prod_{i=1}^{n}(X - \alpha_i)^{a_i} \in K[X]$$

and P' are the α_i with $a_i \geq 2$. They are zeroes of P' with multiplicity $a_i - 1$. Hence (3) follows.

Now suppose $A + B = C$ with A, B, C relatively prime. Notice that

$$\mathrm{Rad}(ABC) = \mathrm{Rad}(A)\mathrm{Rad}(B)\mathrm{Rad}(C).$$

We may suppose, say, $\deg(A) \leq \deg(B) \leq \deg(C)$. Write

$$A + B = C, \qquad A' + B' = C'.$$

Then the three determinants

$$\begin{vmatrix} A & B \\ A' & B' \end{vmatrix} = AB' - A'B, \quad \begin{vmatrix} A & C \\ A' & C' \end{vmatrix} = AC' - A'C, \quad \begin{vmatrix} C & B \\ C' & B' \end{vmatrix} = CB' - C'B$$

take the same value; in particular

$$AB' - A'B = AC' - A'C.$$

Recall $\gcd(A, B, C) = 1$. Since $\gcd(C, C')$ divides $AC' - A'C = AB' - A'B$, it divides also

$$\frac{AB' - A'B}{\gcd(A, A')\gcd(B'B')}$$

which, according to (3), is a polynomial of degree strictly less than

$$\deg\bigl(\mathrm{Rad}(A)\bigr) + \deg\bigl(\mathrm{Rad}(B)\bigr) = \deg\bigl(\mathrm{Rad}(AB)\bigr).$$

Hence

$$\deg\bigl(\gcd(C, C')\bigr) < \deg\bigl(\mathrm{Rad}(AB)\bigr).$$

Using (3) again, we deduce

$$\deg(C) = \deg\bigl(\mathrm{Rad}(C)\bigr) + \deg\bigl(\gcd(C, C')\bigr),$$

hence

$$\deg(C) < \deg\bigl(\mathrm{Rad}(C)\bigr) + \deg\bigl(\mathrm{Rad}(AB)\bigr) = \deg\bigl(\mathrm{Rad}(ABC)\bigr).$$

References

1. Baker, A.: Logarithmic forms and the *abc*-conjecture. In: Number Theory (Eger, 1996), pp. 37–44. de Gruyter, Berlin (1998)
2. Baker, A.: Experiments on the *abc*-conjecture. Publ. Math. Debrecen **65**(3–4), 253–260 (2004)
3. Balasubramanian, R., Shorey, T.N., Waldschmidt, M.: On the maximal length of two sequences of consecutive integers with the same prime divisors. Acta Math. Hungar. **54**(3–4), 225–236 (1989)
4. Beukers, F.: The Diophantine equation $Ax^p + By^q = Cz^r$. Duke Math. J. **91**(1), 61–88 (1998)
5. Beukers, F., Stewart, C.L.: Neighboring powers. J. Number Theory **130**(3), 660–679 (2010). Addendum: J. Number Theory **130**(7), 1571 (2010)
6. Bilu, Y.F.: Catalan's conjecture (after Mihăilescu). Astérisque **294**, vii, 1–26, Séminaire Bourbaki – Volume 2002/2003 – Exposés 909–923 (2004)
7. Bombieri, E.: Roth's Theorem and the abc-Conjecture. ETH, Zürich (1994, unpublished preprint)
8. Bombieri, E., Gubler, W.: Heights in Diophantine Geometry. Cambridge University Press, Cambridge (2006)
9. Broberg, N. Some examples related to the *abc*-conjecture for algebraic number fields. Math. Comput. **69**(232), 1707–1710 (2000)
10. Browkin, J.: A consequence of an effective form of the *abc*-conjecture. Colloq. Math. **82**(1), 79–84 (1999)
11. Browkin, J.: The *abc*-conjecture. In: Number Theory. Trends in Mathematics, pp. 75–105. Birkhäuser, Basel, (2000)
12. Browkin, J.: The *abc*-conjecture for algebraic numbers. Acta Math. Sin. (Engl. Ser.) **22**(1), 211–222 (2006)
13. Browkin, J.: A weak effective *abc*-conjecture. Funct. Approx. Comment. Math. **39**(part 1), 103–111 (2008)
14. Browkin, J., Brzeziński, J.: Some remarks on the *abc*-conjecture. Math. Comput. **62**(206), 931–939 (1994)
15. Browkin, J., Filaseta, M., Greaves, G., Schinzel, A.: Squarefree values of polynomials and the *abc*-conjecture. In: Sieve Methods, Exponential Sums, and Their Applications in Number Theory (Cardiff, 1995). London Mathematical Society Lecture Note Series, vol. 237, pp. 65–85. Cambridge University Press, Cambridge (1997)
16. Cohen, H.: Number Theory. Vol. II. Analytic and Modern Tools. Graduate Texts in Mathematics, vol. 240. Springer, New York (2007)
17. Danilov, L.V.: Letter to the editors: the Diophantine equation $x^3 - y^2 = k$ and a conjecture of M. Hall. . Mat. Zametki **36**(3), 457–458 (1984) (Mat. Zametki **32** (1982), no. 3, 273–275; MR0677595 (84c:10014))
18. Darmon, H., Granville, A.: On the equations $z^m = F(x, y)$ and $Ax^p + By^q = Cz^r$. Bull. Lond. Math. Soc. **27**(6), 513–543 (1995)
19. Elkies, N.D.: *ABC* implies Mordell. Int. Math. Res. Notices **7**, 99–109 (1991)
20. Frey, G.: Links between elliptic curves and solutions of $A - B = C$. J. Indian Math. Soc. (N.S.) **51**, 117–145 (1987, 1988).
21. Frey, G.: Links between solutions of $A - B = C$ and elliptic curves. In: Number theory (Ulm, 1987). Lecture Notes in Mathematics, vol. 1380. Springer, New York, pp. 31–62 (1989)
22. Goldfeld, D.: Modular forms, elliptic curves and the *abc*-conjecture. In: A Panorama of Number Theory or the View from Baker's Garden (Zürich, 1999), pp. 128–147. Cambridge University Press, Cambridge (2002)
23. Goldfeld, D., Szpiro, L.: Bounds for the order of the Tate-Shafarevich group. Compositio Math. **97**(1–2), 71–87 (1995). Special issue in honour of Frans Oort
24. Granville, A.: *abc* allows us to count squarefrees. Int. Math. Res. Notices **1998**(19), 991–1009 (1998)

25. Granville, A., Stark, H.: *ABC* implies no "Siegel zeros" for L-functions of characters with negative discriminant.. Invent. Math. **139**(3), 509–523 (2000)
26. Granville, A., Tucker, T.J.: It's as easy as *abc*. Notices Am. Math. Soc. **49**(10), 1224–1231 (2002). http://www.ams.org/notices/200210/fea-granville.pdf
27. Graves, H., Ram Murty, M.: The *abc* conjecture and non-Wieferich primes in arithmetic progressions. J. Number Theory **133**(6), 1809–1813 (2013)
28. Hall, Jr., M.: The Diophantine equation $x^3 - y^2 = k$. In: Computers in Number Theory. Proceedings of the Science Research Council Atlas Symposium 2, Oxford, 1969, pp. 173–198. Academic, London (1971)
29. Hardy, G.H., Wright, E.M.: An Introduction to the Theory of Numbers, 6th edn. Oxford University Press, Oxford (2008). Revised by D. R. Heath-Brown and J. H. Silverman, With a foreword by Andrew Wiles
30. Hu, P.-C., Yang, C.-C.: A generalized *abc*-conjecture over function fields. J. Number Theory **94**(2), 286–298 (2002)
31. Hu, P.-C., Yang, C.-C.: Distribution Theory of Algebraic Numbers. de Gruyter Expositions in Mathematics, vol. 45. Walter de Gruyter, Berlin (2008)
32. Hu, P.-C., Yang, C.-C.: Roth's theorem and *abc*-conjecture. J. Shandong Univ. Nat. Sci. **44**(8), 1–12 (2009). http://lxbwk.njournal.sdu.edu.cn/EN/volumn/volumn91.shtml.
33. Kraus, A.: On the equation $x^p + y^q = z^r$: a survey. Ramanujan J. **3**(3), 315–333 (1999)
34. Laishram, S., Shorey, T.N.: Baker's explicit *abc*-conjecture and applications. Acta Arith. **155**(4), 419–429 (2012)
35. Lang, S.: Elliptic Curves: Diophantine Analysis. Grundlehren der Mathematischen Wissenschaften [Fundamental Principles of Mathematical Sciences], vol. 231. Springer, Berlin (1978)
36. Lang, S.: Old and new conjectured Diophantine inequalities. Bull. Am. Math. Soc. (N.S.) **23**(1), 37–75 (1990)
37. Lang, S.: Die *abc*-Vermutung. Elem. Math. **48**(3), 89–99 (1993)
38. Lang, S.: Algebra. Graduate Texts in Mathematics, vol. 211, 3rd edn. Springer, New York (2002)
39. Langevin, M.: Cas d'égalité pour le théorème de Mason et applications de la conjecture (abc). C. R. Acad. Sci. Paris Sér. I Math. **317**(5), 441–444 (1993)
40. Langevin, M.: Sur quelques conséquences de la conjecture (abc) en arithmétique et en logique. Rocky Mt. J. Math. **26**(3), 1031–1042 (1996). Symposium on Diophantine Problems (Boulder, 1994)
41. Langevin, M.: Imbrications entre le théorème de Mason, la descente de Belyi et les différentes formes de la conjecture (abc). J. Théor. Nombres Bordeaux **11**(1), 91–109 (1999). Les XXèmes Journées Arithmétiques (Limoges, 1997)
42. Langevin, M.: Liens entre le théorème de Mason et la conjecture (abc). In: Number theory (Ottawa, ON, 1996). CRM Proceedings. Lecture Notes, vol. 19, pp. 187–213. American Mathematical Society, Providence (1999)
43. Mason, R.C.: Diophantine Equations over Function Fields. London Mathematical Society Lecture Note Series, vol. 96. Cambridge University Press, Cambridge (1984)
44. Mason, R.C.: The study of Diophantine equations over function fields. In: New Advances in Transcendence Theory (Durham, 1986), pp. 229–247. Cambridge University Press, Cambridge (1988)
45. Masser, D.W.: Open problems. In: Chen, W.W.L. (ed.) Proceedings of the Symposium on Analytic Number Theory. Imperial College, London (1985)
46. Masser, D.W.: On *abc* and discriminants. Proc. Amer. Math. Soc. **130**(11), 3141–3150 (2002). (electronic)
47. Mauldin, R.: A generalization of Fermat's last theorem: the Beal conjecture and prize problem. Notices Am. Math. Soc. **44**(11), 1436–1437 (1997)
48. Murty, R., Wong, S.: The *abc* conjecture and prime divisors of the Lucas and Lehmer sequences. In: Number Theory for the Millennium, III (Urbana, IL, 2000), pp. 43–54. A K Peters, Natick (2002)

49. Narkiewicz, W.: Classical Problems in Number Theory. Monografie Matematyczne (Mathematical Monographs), vol. 62. Państwowe Wydawnictwo Naukowe (PWN), Warsaw (1986)
50. Nathanson, M.B.: Elementary Methods in Number Theory. Graduate Texts in Mathematics, vol. 195. Springer, New York (2000)
51. Nitaj, A.: The *abc* Conjecture Home Page. http://www.math.unicaen.fr/~nitaj/abc.html
52. Nitaj, A.: La conjecture *abc*. Enseign. Math. (2) **42**(1–2), 3–24 (1996)
53. Oesterlé, J.: Nouvelles approches du "théorème" de Fermat. Astérisque (1988), no. 161–162, Exp. No. 694, 4, 165–186 (1989), Séminaire Bourbaki, Vol. 1987/88
54. Philippon, P.: Quelques remarques sur des questions d'approximation diophantienne. Bull. Aust. Math. Soc. **59**(2), 323–334 (1999). Addendum, idem. **61**(1), 167–169 (2000)
55. Pillai, S.S.: Collected works of S. Sivasankaranarayana Pillai. In: Balasubramanian, R., Thangadurai, R. (eds.) Ramanujan Mathematical Society Collected Works Series, vol. 1. Ramanujan Mathematical Society, Mysore *2010*
56. Ribenboim, P.: 13 Lectures on Fermat's Last Theorem. Springer, New York (1979)
57. Ribenboim, P.: Catalan's Conjecture Are 8 and 9 the Only Consecutive Powers? Academic, Boston (1994)
58. Robert, O., Stewart, C.L., Tenenbaum,G.: A refinement of the *abc* conjecture. Bull. London Math. Soc. (2014). doi: 10.1112/blms/bdu069. First published online: September 2, 2014
59. Robert, O., Tenenbaum, G.: Sur la répartition du noyau d'un entier. Indag. Math. 24, 802–914 (2013)
60. Saradha, N.: Application of the explicit *abc*-conjecture to two Diophantine equations. Acta Arith. **151**(4), 401–419 (2012)
61. Schmidt, W.M.: Diophantine Approximations and Diophantine Equations. Lecture Notes in Mathematics, vol. 1467. Springer, Berlin (1991)
62. Shorey, T.N.: Exponential Diophantine equations involving products of consecutive integers and related equations. In: Number Theory. Trends in Mathematics, pp. 463–495. Birkhäuser, Basel (2000)
63. Shorey, T.N.: An equation of Goormaghtigh and Diophantine approximations. In: Currents Trends in Number Theory (Allahabad, 2000), pp. 185–197. Hindustan Book Agency, New Delhi (2002)
64. Silverman, J.H.: Wieferich's criterion and the *abc*-conjecture. J. Number Theory **30**(2), 226–237 (1988)
65. Stewart, C.L.: On divisors of Lucas and Lehmer numbers. Acta Mathematica **211**(2), 291–314 (2013)
66. Stewart, C.L., Tijdeman, R. On the Oesterlé-Masser conjecture. Monatsh. Math. **102**(3), 251–257 (1986)
67. Stewart, C.L., Yu, K.: On the *abc* conjecture. Math. Ann. **291**(2), 225–230 (1991)
68. Stewart, C.L., Yu, K.: On the *abc* conjecture. II. Duke Math. J. **108**(1), 169–181 (2001)
69. Stothers, W.W.: Polynomial identities and Hauptmoduln. Quart. J. Math. Oxford Ser. (2) **32**(127), 349–370 (1981)
70. Surroca, A.: Siegel's theorem and the *abc* conjecture. Riv. Mat. Univ. Parma (7) **3***, 323–332 (2004)
71. Surroca, A.: Sur l'effectivité du théorème de Siegel et la conjecture *abc*. J. Number Theory **124**(2), 267–290 (2007)
72. Tijdeman, R.: Exponential Diophantine equations 1986–1996. In: Number theory (Eger, 1996), pp. 523–539. de Gruyter, Berlin (1998)
73. van Frankenhuijsen, M.: The *ABC* conjecture implies Roth's theorem and Mordell's conjecture. Mat. Contemp. **16**, 45–72 (1999). 15th School of Algebra (Portuguese) (Canela, 1998)
74. van Frankenhuijsen, M.: A lower bound in the *abc* conjecture. J. Number Theory **82**(1), 91–95 (2000)
75. van Frankenhuijsen, M.: The *ABC* conjecture implies Vojta's height inequality for curves. J. Number Theory **95**(2), 289–302 (2002)
76. van Frankenhuijsen, M.: *ABC* implies the radicalized Vojta height inequality for curves. J. Number Theory **127**(2), 292–300 (2007)

77. van Frankenhuijsen, M.: About the *ABC* conjecture and an alternative. In Number Theory, Analysis and Geometry, pp. 169–180. Springer, New York (2012)
78. Vojta, P.: Diophantine Approximations and Value Distribution Theory. Lecture Notes in Mathematics, vol. 1239. Springer, Berlin (1987)
79. Vojta, P.: A more general *abc* conjecture. Int. Math. Res. Notices **21**, 1103–1116 (1998)
80. Vojta, P.: On the *ABC* conjecture and Diophantine approximation by rational points. Am. J. Math. **122**(4), 843–872 (2000). Correction: Am. J. Math. **123**(2), 383–384 (2001)
81. Wooley, T.D.: Waring's problem, the declining exchange rate between small powers, and the story of 13,792. University of Bristol. 19/11/2007 http://www.maths.bris.ac.uk/~matdw/bristol%20b16.pdf.

Further Resources

- Additional information about the *abc* conjecture is available at http://www.astro.virginia.edu/~eww6n/math/abcConjecture.html.
- `ABC@Home`, a project led by Hendrik W. Lenstra Jr., B. de Smit and W. J. Palenstijn http://www.abcathome.com/
- Ivars Peterson. — The Amazing ABC Conjecture. http://www.sciencenews.org/sn_arc97/12_6_97/mathland.htm
- Pierre Colmez. — $a + b = c$? Images des Mathématiques, CNRS, 2012. http://images.math.cnrs.fr/a-b-c.html
- Bart de Smit/ABC triples. http://www.math.leidenuniv.nl/~desmit/abc/
- Reken mee met *abc* http://rekenmeemetabc.nl/Synthese_resultaten

  Reken mee met *abc* is een project dat gericht is op scholieren en andere belangstellenden. Op deze website vind je allerlei interessante artikelen, wedstrijden en informatie voor een praktische opdracht of profielwerkstuk voor het vak wiskunde. Daarnaast kun je je computer laten meerekenen aan een groot rekenproject gebaseerd op een algoritme om abc-drietallen te vinden.

  Reken mee met abc is a project aimed at students and other interested parties. On this website you can find all sorts of interesting articles, contests and information for a practical assignment or workpiece profile for mathematics. In addition, you can take your computer to a large project based on an algorithm to *abc*–triples.

- Greg Martin and Winnie Miao: abc Triples; Arxiv:1409.2974v1 [math.NT] 10 sep 2014.

Approximation on Curves

Rein L. Zeinstra

Dedicated to Professor Jaap Korevaar on the Occasion of his Ninetieth Birthday.

1 Classical Müntz-Szasz Theorem

1.1 Weierstrass Approximation Theorem

Probably, the oldest and best-known approximation theorem is the famous result of Weierstrass from 1885, contained in what is nowadays usually known as the Weierstrass-Stone theorem. Its original statement is that the set of all polynomials $P(t) = \sum_0^n a_k\, t^k$ is uniformly dense in $C([a,b])$, the space of continuous functions on a real interval $[a,b]$: For every $f \in C([a,b])$ and $\epsilon > 0$, there is a polynomial P such that $|f(t) - P(t)| < \epsilon$ for all $t \in [a,b]$. We express this by saying that the monomials t^k, $k \geq 0$, span $C([a,b])$.

Does one need all monomials in the Weierstrass theorem? If $0 \notin [a,b]$, we can apparently omit any finite number of exponents. And the simple transformation $t = \sqrt[m]{u}$, $m \geq 2$, shows that on the interval $[0,b]$ the monomials t^{km}, $k \geq 0$, alone span $C([0,b])$.

1.2 Müntz-Szasz Theorem

Müntz [18] established the following surprising result.

The author is grateful to Abdus Salam School of Mathematical Sciences in Lahore for its warm hospitality and excellent atmosphere.

R.L. Zeinstra (✉)
Abdus Salam School of Mathematical Sciences, GC University, Lahore, Pakistan
e-mail: rl.zeinstra@gmail.com

P. Cartier et al. (eds.), *Mathematics in the 21st Century*, Springer Proceedings in Mathematics & Statistics 98, DOI 10.1007/978-3-0348-0859-0_14

Let $0 < p_1 < p_2 < \ldots$ be positive numbers (not necessarily integers!) increasing to ∞ and let $b > 0$. Then the (generalized) monomials t^{p_n} span $C_0([0,b])$ if and only if $\sum_{k=1}^{\infty} \frac{1}{p_k} = \infty$.

Here $C_0([0,b])$ denotes the subspace of all continuous functions $f(t)$ with $f(0) = 0$. (To obtain the whole space $C([a,b])$, we only have to add the constant monomial $t^0 = 1$ to the sequence t^{p_k}.) For example, the rather sparse sequence p_k of the prime numbers spans $C_0([0,b])$ because $p_k \asymp k \log k$ by the prime number theorem. Still it is possible to omit infinitely many primes, e.g., every second one, and still keep a spanning system; however, the slightly thinner sequence $t^{k \log^2 k}$ gives a non-spanning set.

Szasz [21] even extended the theorem of Müntz to complex monomials t^{λ_k} with $\mathrm{Re}\,(\lambda_k) > 0$. For λ_k lying in a sector $|\mathrm{Arg}\, z| < \theta < \frac{\pi}{2}$, his corresponding spanning condition is $\sum_{k=1}^{\infty} \frac{1}{|\lambda_k|} = \infty$. For convenience, we shall often restrict to positive exponents.

Original Proof of MS. It is rather interesting that Müntz and Szasz came to their result by first considering the corresponding approximation problem in the L^2- instead of the L^∞-norm. In fact, they found an explicit formula for the L^2-distance of a monomial t^z to the linear span S_n of the first n monomials t^{p_j}, $1 \leq j \leq n$. Namely, when $b = 1$ and z a complex number with $\mathrm{Re}(z) > -1/2$, then

$$\mathrm{dist}_{L^2}(t^z, S_n) = \inf_{a_j \in \mathbb{C},\, 1 \leq j \leq n} \left(\int_0^1 \left| t^p - \sum_1^n a_j t^{p_j} \right|^2 dt \right)^{1/2}$$

$$= \frac{1}{2\,\mathrm{Re}(z) + 1} \left| \prod_1^n \frac{(p_j - z)}{(z + p_j + 1)} \right|$$

The formula is obtained by minimizing the quadratic form arising under the infimum in $n-$dimensional space and involves the evaluation of certain determinants. When $n \to \infty$, the finite products on the right diverge for $n \to \infty$ to the function $\equiv 0$ if $\sum \frac{1}{p_k}$ diverges, and it will follow that approximation in L^2 is possible (it is possible to pass from L^2-approximation to uniform approximation); however, if $\sum \frac{1}{p_k}$ converges, they converge absolutely for $\mathrm{Re}(z) > -1/2$ to the corresponding infinite product (a half-plane "Blaschke" product!) which has only the p_k as its zeros so that even every complex monomial t^z with z different from the p_k has positive L^2-distance and certainly positive L^∞-distance to the span. In particular, omitting just one power t^{p_j} strictly diminishes the closed span! This suggests that the class of approximable functions must be very small in this case (cf. Sect. 5).

1.3 Functional Analysis Proof Using Laplace Transforms

The first proof without determinants is due to Carleman [4] who combined complex analysis with new developed methods from functional analysis. In fact, by the

Hahn-Banach theorem, the system t^{λ_k} spans $C_0([0,1])$ (or $C([a,1])$, $0 < a < 1$, we have again taken $b = 1$) iff the only finite (complex) Radon measure μ supported on $(0,1]$ (on $[a,1]$) satisfying $\int e^{-\lambda_k t}\, d\mu(t) = 0$, $k \geq 1$, is the zero measure. Indeed, these measures form the corresponding dual space according to the Riesz representation theorem. (For the corresponding L^p-approximation problem, $1 \leq p < \infty$, measures should be replaced by functions in L^q, $q := p/(p-1)$.)

For easier formulation, let us make the substitution $t = e^{-u}$. (This moves the special point 0 to ∞.) Replacing u again by t, the original problem becomes one about general Dirichlet polynomials $\sum_1^n a_k\, e^{-p_k t}$ on the unbounded interval $[0,\infty)$ (or on $[0,\beta]$, where $\beta = -\log a$). We define the Laplace transform of the finite Borel measure μ on $[0,\infty)$ by

$$\mathcal{L}\mu(z) := \int_{[0,\infty]} e^{-zt}\, d\mu(t).$$

This is a bounded analytic function in the right half-plane $x = \mathrm{Re}(z) > 0$; if the support of μ lies in a bounded subinterval $[\alpha,\beta]$, it is even an entire function. Assuming now $\mathcal{L}\mu \not\equiv 0$, its complex zeros in the half-plane (here and later counted with multiplicities) must satisfy the Blaschke-Carleman condition for a half-plane: $\sum \frac{\mathrm{Re}(\lambda_k)}{1+|\lambda_k|^2} < \infty$; in particular, the *restricted* sum $\sum' \frac{1}{|\lambda_k|}$ over the zeros of modulus ≥ 1 on the positive axis or in a proper subsector is finite! This shows that $\mathcal{L}\mu$ cannot vanish identically unless $\mu = 0$ (Weierstrass theorem!: f would in particular vanish at all positive integers); this immediately implies the sufficiency of Szasz's spanning condition.

Conversely, under Szasz's non-spanning condition, the existence of a complex measure $\mu \neq 0$ in fact of a smooth bounded integrable function on $[0,\infty)$ whose Laplace transform vanishes at the μ_k can be obtained from the Cauchy theorem or from the half-plane Paley-Wiener theorem (see, e.g., Rudin [19]). One starts from the infinite (convergent!) Blaschke product $B(z) := \prod_1^\infty (1 - z/\lambda_k)/(1 + z/\overline{\lambda_k})$ (which essentially appears in the L^2-distance formula!) with zeros λ_k.

2 Approximation on Curves

2.1 Walsh's Theorem and Curves of Bounded Slope

Walsh [22] showed that the Weierstrass theorem extends to a Jordan arc (homeomorphic image of the interval $[0,1]$): Given a Jordan arc Γ in the complex ζ−plane, the polynomials $\sum_0^n a_k\, \zeta^k$ are uniformly dense in $C(\Gamma)$. This result is a forerunner of the famous theorem of Mergelyan (cf. Rudin [19]) involving polynomial approximation on a general compact plane set with connected complement.

The obvious problem is whether the Walsh theorem admits interesting analogues of the MS theorem. More precisely, let Γ have 0 as an end point; we ask under what further conditions the monomials ζ^{p_k} (defined by some continuous choice of the logarithm on Γ) span $C(\Gamma)$, given the divergence condition $\sum \frac{1}{p_k} = \infty$. Here $p_k > 0$ is a sequence as before. Unfortunately, the MS method of proof with determinants does not work here.

It is somewhat easier to formulate results for the corresponding unbounded Jordan arc $\gamma =" - \log \Gamma"$ in terms of exponentials $e^{-p_k z}$. It turns out that we need the rather strong condition on γ that it has a parametrization $\gamma(t) = t + ih(t)$, $a \leq t < \infty$, where $h(t)$ is Lipschitz continuous. In other words, γ has its slopes

$$\frac{h(t_1) - h(t_2)}{t_1 - t_2}$$

bounded in absolute value by a finite constant M. If $c := a + ih(a)$, the begin point of γ, then the translated curve $\gamma_c := \gamma - c$ starts at 0 and lies entirely in the closed angle $|\text{Arg } z| \leq \alpha$, where $\alpha := \arctan M$. This guarantees that the exponentials e^{-pz} are in $C_0(\gamma)$ and in all L^p-spaces.

The main reason for considering only curves of bounded slope is that the posed problem can have a positive answer only if γ does not contain two distinct points ζ_1, ζ_2 such that $\zeta_1 - \zeta_2 = ic$ is purely imaginary. Otherwise, the choice $p_k := 2\pi/c$ leads to a non-spanning system – the $e^{-p_k z}$ take the same value at the ζ_j ! – but the p_k even have *positive density* $= \frac{1}{2}$ wrt the positive integers. Phenomena like these which are well known in the theory of trigonometric approximation (where all exponents are *purely imaginary*) suggest that any verticality in γ should be avoided.

For a finite (complex) Radon measure μ on γ, the Laplace transform

$$\mathcal{L}\mu(z) := \int_\gamma e^{-z\zeta}\, d\mu(\zeta)$$

is defined and continuous in the closed "dual" sector $|\text{Arg } z| \leq \alpha' := \frac{\pi}{2} - \alpha$, analytic in its interior. It follows from the theorem of Walsh that it is $\equiv 0$ only if $\mu = 0$. In the following, it is to be assumed that $\mu \neq 0$ and that multiplicities of the zeros of $\mathcal{L}\mu$ are taken into account.

2.2 *Main Results on Curves*

We only state results for Laplace transforms here. They differ in additional conditions on the steepness of γ and/or on the zeros of $\mathcal{L}\mu$. The corresponding Müntz or Müntz-Szasz type approximation results (like Theorem 3 in Sect. 4) in $C_0(\gamma)$ should be clear. They hold actually as well in the spaces $L^p(\gamma)$, $1 \leq p < \infty$ (defined wrt arclength):

1. $\alpha < \frac{\pi}{4}$: The positive zeros $p_k \geq 1$ of $\mathcal{L}\mu$ satisfy the convergence condition $\sum \frac{1}{p_k} < \infty$ (Malliavin-Siddiqi [17]; also Leont'ev [13] for the case where γ is real analytic).

2. $\alpha \leq \frac{\pi}{4}$: The complex zeros λ_k of $\mathcal{L}\mu$ that satisfy $|\lambda_k| \geq 1$, $|\text{Arg } \lambda_k| \leq \alpha' - \frac{\pi}{4}$ satisfy the convergence condition $\sum \frac{1}{|\lambda_k|} < \infty$ (Korevaar [10]).
3. General $\alpha < \frac{\pi}{2}$ but for "regular" complex zeros: Suppose $\mathcal{L}\mu$ vanishes on a sequence λ_k in some sector $|\text{Arg } z| \leq \alpha' - \delta$, $0 < \delta \leq \alpha'$ (but $\mathcal{L}\mu$ may have infinitely many other zeros in that sector). If $|\lambda_k| = kL(k)$ where $L(t)$ is a slowly oscillating function, then $\sum \frac{1}{|\lambda_k|} < \infty$ (Zeinstra [23], cf. also [24]).

   By a slowly oscillating function, we mean a positive measurable function $L(t)$ on an interval $[b, \infty)$ that satisfies $\lim_{t\to\infty} \frac{L(ct)}{L(t)} = 1$ for every $c > 0$. Typical examples are the functions $(\log t)^q$, $(\log t)^q (\log \log t)^r, \ldots, \exp((\log t)^s)$, $(q, r, s \text{ real}, |s| < 1)$.

   The following result considerably extends (1), (2) above:
4. General $\alpha < \frac{\pi}{2}$ and for γ piecewise C^1: Let λ_k denote all complex zeros of modulus ≥ 1 in the angle regions $|\text{Arg } z| \leq \alpha' - \delta$, $0 < \delta \leq \alpha'$. Then $\frac{1}{|\lambda_k|} < \infty$ (Korevaar-Zeinstra [12], cf. also [23]). Moreover, if $\alpha < \frac{\pi}{4}$, the smoothness condition on γ may be dropped [23].

An "amalgam" of these two cases is the following result. Apart from the cases just mentioned, this includes the further (overlapping) case where the function $h(t)$ in the definition of γ is piecewise monotonic.

Theorem 1 (Zeinstra [24]). *Suppose that γ can be subdivided in finitely or countably many subarcs γ_j such that on each γ_j the directions of the tangents (which exist a.e. wrt t or* arc *length) vary strictly less than $\frac{\pi}{2}$. Then $\sum \frac{1}{|\lambda_k|} < \infty$, where λ_k are as in (4) above.*

3 Müntz and Quasi-analyticity: Its Use in the Proofs

The proofs of the above results use the concept of quasi-analyticity as an essential ingredient. Recall that given a sequence (M_n), $n \geq 0$, of positive numbers and a real interval I, the class $C(M_n) = C_I(M_n)$ consists of all C^∞-functions f on I that satisfy $|f^{(n)}(t)| \leq A^{n+1} M_n$ $(t \in I,\ n \geq 0)$ for some positive constant A depending on f. It is usually assumed that $M_n^2 \leq M_{n-1} M_{n+1}$ (logarithmic convexity). The class $C(M_n)$ is said to be quasi-analytic if it contains no function (other than the zero function) all of whose derivatives vanish at some point of I. For example, $C_I(n!)$ is quasi-analytic since it consists of (real-)analytic functions which are restrictions to I of some analytic function. According to the Denjoy-Carleman theorem (cf. Rudin [19]), the class $C(M_n)$ (where (M_n) is log convex) is q.a. iff $\sum \frac{M_n}{M_{n+1}} = \infty$, a characterization very similar to the Müntz condition for approximation! In fact, if p_k are as in the original Müntz theorem, then the numbers $M_n := p_1 p_2 \ldots p_{n+1}$ are log convex and determine a q.a. class iff the Müntz (divergence) condition holds.

Classes $C(M_n)$ can be defined similarly on arbitrary curves γ and on other plane sets without isolated points, e.g., closed sectors, when derivatives are defined via the familiar complex differential quotients. For example, restrictions of analytic functions to any such subset are certainly infinitely differentiable and (locally) in $C(n!)$. On a locally rectifiable arc – like our curves of bounded slope – the divergence condition of the DC theorem still implies quasi-analyticity: Parametrizing by arclength, the proof is almost the same as the "real-variable" proof of the sufficiency part of the DC theorem (cf. [6], cf. also [23]).

Short Outline of Proofs Related to Sect. 2.1. All proofs of the mentioned results argue indirectly, assuming divergence of the sum of reciprocals, and rest on the technique of "dividing out" the considered zeros (or sufficiently many of them) of $\mathcal{L}\mu$ to arrive at a situation where the initial measure μ is replaced by a C^∞–function ϕ on γ with a zero of infinite order at the begin point of γ and whose new Laplace transform $\mathcal{L}\phi$ $(= \mathcal{L}(\phi\, d\zeta))$ divides the original one. To accomplish this, one needs a sufficiently big infinite product $F(z)$ with the zeros one wants to divide out and then defines ϕ as the convolution (along γ!) of μ with what is essentially the complex Fourier transform of $1/F$.

In the results (1)–(3) above, the even infinite product $F(z) := \prod(1 - z^2/\mu_k^2)$ can be used; however, in (4), rather complicated products of even rational functions had to be constructed to obtain sufficiently large angles where $|F|$ is big. The construction then shows that ϕ belongs to the quasi-analytic class $C(M_n)$, where M_n are defined essentially as above. Thus, one concludes successively that ϕ, $\mathcal{L}\phi$, $\mathcal{L}\mu$ are $\equiv 0$, and from Walsh's theorem, one finally infers $\mu = 0$.

4 Regular Growth of Laplace Transforms

What more can be said about the Laplace transforms $\mathcal{L}\mu$? As a rule, one would expect that an analytic function with "few" zeros – by "few" we are thinking of the convergence condition – is not very often "small." This is made precise in the theorem below that in the half-plane case is part of the classical Ahlfors-Heins-Azarin theorem [1,2] and is in fact derived from it. To state it in the simplest possible way, we shall assume without restriction that γ is normalized; that is, γ starts at the origin ($a = h(a) = 0$). We further assume that 0 is in the support of μ. The Laplace transform $\mathcal{L}\mu$ is then bounded in the dual angle $|\text{Arg } z| \leq \alpha' := \frac{\pi}{2} - \alpha$; moreover, it is of "type" 0 on the rays $z = re^{i\theta}$:

$$\limsup_{r\to\infty} \frac{\log|f(re^{i\theta})|}{r} = 0, \qquad |\theta| < \alpha'.$$

The result we give below is much stronger: f cannot fall off exponentially in any proper subsector with the possible exception of a small exceptional set related to its zeros.

Theorem 2 (cf. [12,23,24]). *Under the same conditions as in Theorem 1 and under the additional assumptions that γ be normalized and that the origin belong to the support of μ, we have, given δ, $0 < \delta \le \alpha'$, that*

$$\frac{\log|f(re^{i\theta})|}{r} \to 0$$

when $r \to \infty$, uniformly in θ for $|\theta| \le \alpha' - \delta$, either without restriction or, otherwise, outside the union of a countable sequence of disks $D(a_j, r_j)$ such that $\sum \frac{r_j}{|a_j|} < \infty$, $a_j \to \infty$. (The centers a_j can be taken to be zeros of f.)

From the theorem, it easily follows that the limsup above is a limit for a.e. θ and that a "typical" circle $|z| = r$ will not intersect E. Another consequence is a (further) approximation result of Müntz type for curves which are not uniformly Lipschitz over the interval $[0, \infty)$, like the curves $\gamma(t) := t^s$ or $\gamma(t) := t^s \sin t$, $t \ge 0$, $s > 1$.

Theorem 3 (Zeinstra [24]). *Suppose γ is as in Theorem 1 except that it is only locally Lipschitz. Let $0 < p_1 < p_2 < \dots$ with the (additional) separation condition $p_{k+1} - p_k \ge c > 0$. If $\sum \frac{1}{p_k} = \infty$, then the exponentials $e^{-p_k \zeta}$ span $C_0(\gamma)$ and $L^p(\gamma)$, $1 \le p < \infty$.*

5 The Span in the Non-spanning Case

5.1 *Non-spanning Systems on a Closed Interval*

It is far from trivial that the system $e^{-p_k t}$ with $\sum 1/p_k < \infty$ is still non-spanning on all *bounded* intervals $[\alpha, \beta]$, $\alpha < \beta$. This corresponds to constructing a function g (or a measure μ) with support in the given interval, and this requires the Paley-Wiener theorem for entire functions. A very satisfactory approach to this problem can be found in Luxemburg-Korevaar [15]. Here the case of complex exponents λ_k in a sector has also been considered.

What is the exact closed uniform span or the L^p-span ($1 \le p < \infty$) in the non-spanning case? Clarkson-Erdös [5], Schwartz [20], and Korevaar [9] have given a complete answer to these questions: If we assume the separation condition $p_{k+1} - p_k \ge c > 0$, then a continuous (or L^p) function $f(x)$ on $[\alpha, \beta]$ is in the uniform span (in the L^p-span) iff it is the restriction of (coincides a.e. with) an analytic (!) function $\tilde{f}$ in the half-plane $\mathrm{Re}(z) > \alpha$ that has a (unique) representation in that half-plane as the sum of a Dirichlet series $\sum c_k e^{-p_k z}$ that converges uniformly (in L^p) to $f(x)$ on each interval $(\alpha + \delta, \infty)$, $\delta > 0$. If the separation condition on the p_k does not hold, there is a similar characterization, but the involved Dirichlet series must be summed by certain grouping of its terms. Moreover, everything extends to the case of a double-sided **real** sequence p_k, $k \in \mathbb{Z}$, where the Müntz convergence condition holds for the positive and negative terms separately. The corresponding extension domain will now be the vertical strip $\alpha < \mathrm{Re}(z) < \beta$.

A very thorough account of these and related kinds of problems is given in [20]. In [15], these results have been extended to the case of complex sequences in an angle or double angle. More recently, Borwein and Erdelyi [3] have, by entirely different methods, shown that the non-spanning property of $e^{-p_k t}$ (and a corresponding characterization for the approximable functions) holds in $C(K)$, where K is an arbitrary compact subset of the reals of positive Lebesgue measure!

5.2 *Non-spanning Systems on Arcs*

For an analytic arc, the above non-spanning results extend: The Müntz condition is still necessary and sufficient for approximation! The necessity has been shown by Malliavin-Siddiqi [16] and Korevaar [10]. It should be no surprise that the complete Denjoy-Carleman theorem remains true here. In fact here as in the theorem below, the essential point is the construction of a C^∞-function $\phi \not\equiv 0$ all of whose derivatives vanish at the end point of γ (hence is in some non-quasi-analytic class!). The analytic extendability here and in the following of the approximable functions in the non-spanning case (naturally in smaller sectors) has been proved in Dixon-Korevaar [8]. For general Lipschitz arcs γ, we mention a final result, stated and (partially) proved in Korevaar-Dixon [11]. (A complete but different proof, again based on the construction of a "non-quasi-analytic" function, has been given by Lundin [14].)

Theorem 4 (Korevaar-Dixon [11]). *Suppose (λ_k) is a sequence of complex numbers. (No restriction on the argument!) If there is a positive increasing function $L(t)$ such that $|\lambda_k| \geq k\,L(k)$ and such that $\sum \frac{1}{k\,L(k)} < \infty$, then the system $e^{\pm\lambda_k \xi}$ does not span $C(\gamma)$.*

We finish with two questions which as far as I know are still unanswered: Could the additional monotonicity condition in Theorem 4 be dropped, at least on Lipschitz arcs? Is perhaps the sufficiency condition in the DC theorem in the case of Lipschitz arcs also necessary? Couture ([7]) proves this under an additional monotonicity condition on the M_n.

References

1. Ahlfors, L., Heins, M.: Questions of regularity connected with the Phragmen-Lindelöf principle. Ann. Math. **50**, 341–346 (1949)
2. Azarin, V.S.: Generalization of a theorem of Hayman on subharmonic functions in an n-dimensional cone (Russian). Mat.Sb. **66**, 248–264 (1965). English translation in AMS Transl. **80**, 119–138 (1969)
3. Borwein, P.B., Erdelyi, T.: Generalizations of Müntz's theorem via a Remez type inequality for Müntz spaces. J. Am. Math. Soc. **10**, 327–349 (1997)

4. Carleman, T.: Über die Approximation analytischer Funktionen durch lineare Aggregate von vorgegebenen Potenzen. Ark. f. Mat. Astr. Fys. **17**(No. 9), 1–30 (1922)
5. Clarkson, J.A., Erdös, P.: Approximation by polynomials. Duke Math. J. **10**, 5–11 (1943)
6. Cohen, P.J.: A simple proof of the Denjoy-Carleman theorem. Am. Math. Monthly **75**, 26–31 (1978)
7. Couture, R.: Un theoreme de Denjoy-Carleman sur une courbe du plan complexe. Proc. AMS **85**, 401–406 (1982)
8. Dixon, M., Korevaar, J.: Nonspanning sets of powers on curves: analyticity theorem. Duke Math. J. **45**, 543–559 (1978)
9. Korevaar, J.: A characterization of the sub-manifold of $C[a, b]$ spanned by the sequence x^{n_k}. Indag. Math. **9**, 360–368 (1947)
10. Korevaar, J.: Approximation on curves by linear combinations of exponentials. In: Lorentz, G.G. (ed.) Approximation Theory, pp. 387–393. Academic Press, New York (1973)
11. Korevaar, J., Dixon, M.: Nonspanning sets of exponentials on curves. Acta Math. Acad. Sci. Hung. **33**, 89–100 (1979)
12. Korevaar, J., Zeinstra, R.: Theoreme d'approximation du type Müntz-Szasz pour des courbes lipschitziennes. C. R. Acad. Sci. Paris Ser. A **301**, 695–698 (1985)
13. Leont'ev, A.F.: On the completeness of a system of exponentials on a curve (Russian). Sib. Mat. Zh. **15**, 1103–1114 (1974)
14. Lundin, M.: A new proof of a Müntz-type theorem of Korevaar and Dixon. Department of Mathematics, University of Göteborg, report no. 1979-7.
15. Luxemburg, W.A.J., Korevaar, J.: Entire functions and Müntz-Szasz type approximation. Trans. AMS **157**, 23–37 (1971)
16. Malliavin, P., Siddiqi, J.A.: Approximation polynomiale sur un arc analytique dans le plan complexe. C.R. Acad. Sci. Paris Ser. A **273**, 105–108 (1971)
17. Malliavin, P., Siddiqi, J.A.: Classes de fonctions monogenes et approximation par des sommes d'exponentielles sur un arc rectifiable de $\mathbb{C}$. C.R. Acad. Sci. Paris Ser. A **282**, 1091–1094 (1976)
18. Müntz, C.H., Über den Approximationssatz von Weierstrass, pp. 303–312. H.A. Schwarz Festschrift, Berlin (1914)
19. Rudin, W.: Real and complex analysis, 3rd edn. McGraw-Hill, New York (1986)
20. Schwartz, L.: Etude des sommes d'exponentielles reelles. Hermann, Paris (1943). (2nd ed. 1959)
21. Szasz, O., Über die Approximation stetiger Funktionen durch lineare Aggregate von Potenzen. Math. Ann. **77**, 482–496 (1916)
22. Walsh, J.L.: Über die Entwicklung einer Funktion einer komplexen Vernderlichen nach Polynomen. Math. Ann. **96**, 437–450 (1927)
23. Zeinstra, R.L.: Müntz-Szasz approximation on curves and area problems for zero sets, Thesis, University of Amsterdam (1985)
24. Zeinstra, R.L.: Zeros and regular growth of Laplace transforms along curves. J. reine angew. Math. **424**, 1–15 (1992)

MIX
Papier aus verantwortungsvollen Quellen
Paper from responsible sources
FSC® C105338

FSC
www.fsc.org

If you have any concerns about our products,
you can contact us on
ProductSafety@springernature.com

In case Publisher is established outside the EU,
the EU authorized representative is:
Springer Nature Customer Service Center GmbH
Europaplatz 3, 69115 Heidelberg, Germany

Printed by Libri Plureos GmbH
in Hamburg, Germany